Angiogenesis
Molecular Biology, Clinical Aspects

NATO ASI Series

Advanced Science Institutes Series

A series presenting the results of activities sponsored by the NATO Science Committee, which aims at the dissemination of advanced scientific and technological knowledge, with a view to strengthening links between scientific communities.

The series is published by an international board of publishers in conjunction with the NATO Scientific Affairs Division

A	**Life Sciences**	Plenum Publishing Corporation
B	**Physics**	New York and London
C	**Mathematical and Physical Sciences**	Kluwer Academic Publishers
D	**Behavioral and Social Sciences**	Dordrecht, Boston, and London
E	**Applied Sciences**	
F	**Computer and Systems Sciences**	Springer-Verlag
G	**Ecological Sciences**	Berlin, Heidelberg, New York, London,
H	**Cell Biology**	Paris, Tokyo, Hong Kong, and Barcelona
I	**Global Environmental Change**	

Recent Volumes in this Series

Series A: Life Sciences

Angiogenesis
Molecular Biology, Clinical Aspects

Edited by
Michael E. Maragoudakis
University of Patras Medical School
Patras, Greece

Pietro M. Gullino
University of Torino Medical School
Torino, Italy

and
Peter I. Lelkes
University of Wisconsin Medical School
Milwaukee, Wisconsin

Plenum Press
New York and London
Published in cooperation with NATO Scientific Affairs Division

Proceedings of a NATO Advanced Study Institute on
Angiogenesis: Molecular Biology, Clinical Aspects,
held June 16–27, 1993,
in Rhodes, Greece

NATO-PCO-DATA BASE

The electronic index to the NATO ASI Series provides full bibliographical references (with keywords and/or abstracts) to more than 30,000 contributions from international scientists published in all sections of the NATO ASI Series. Access to the NATO-PCO-DATA BASE is possible in two ways:

—via online FILE 128 (NATO-PCO-DATA BASE) hosted by ESRIN, Via Galileo Galilei, I-00044 Frascati, Italy

—via CD-ROM "NATO Science and Technology Disk" with user-friendly retrieval software in English, French, and German (©WTV GmbH and DATAWARE Technologies, Inc. 1989). The CD-ROM also contains the AGARD Aerospace Database.

The CD-ROM can be ordered through any member of the Board of Publishers or through NATO-PCO, Overijse, Belgium.

Library of Congress Cataloging-in-Publication Data

Angiogenesis : molecular biology, clinical aspects / edited by Michael
    E. Maragoudakis, Pietro M. Gullino, and Peter I. Lelkes.
        p.   cm. -- (NATO ASI series. Series A: Life sciences, vol.
    263.)
        "Proceedings of a NATO Advanced Study Institute on angiogenesis:
    molecular biology, clinical aspects, held June 16-27, 1993, in
    Rhodes, Greece" -- T.p. verso.
        Includes bibliographical references and index.
        ISBN 0-306-44713-4
        1. Neovascularization--Congresses.   I. Maragoudakis, Michael E.
    II. Gullino, Pietro M.   III. Lelkes, Peter I.   IV. Series: NATO
    advanced sciences institutes series. Series A, Life sciences ; v.
    263.
    QP106.6.A55   1994
    612.1'3--dc20
                                                              94-10036
                                                                 CIP

ISBN 0-306-44713-4

©1994 Plenum Press, New York
A Division of Plenum Publishing Corporation
233 Spring Street, New York, N.Y. 10013

Printed in the United States of America

PREFACE

Angiogenesis is a multistep process, which involves activation, proliferation and directed migration of endothelial cells to form new capillaries from existing vessels. Under physiological conditions, in the adult organisms angiogenesis is extremely slow, yet it can be activated for a limited time only in situations such as ovulation or wound healing. In a number of disease states, however, there is a derangement of angiogenesis, which can contribute to the pathology of these conditions. Hence, understanding the molecular biology of endothelial cell activation and differentiation and the mechanisms involved in the regulation of angiogenesis, could explain the derangement in disease states and also provide the basis for developing promoters or suppressors of angiogenesis for clinical applications.

This book contains the proceedings of the NATO Advanced Study Institute on "Angiogenesis: Molecular Biology, Clinical Aspects" held in Rhodes, Greece, from June 16-27, 1993. This meeting was a comprehensive review of the various aspects of angiogenesis such as embryonic development, endothelial cell heterogeneity and tissue specificity, molecular biology of endothelial cell, mechanisms for the regulation of angiogenesis, disease states in which angiogenesis is involved and potential application of promoters or suppressors of angiogenesis. The presentations and discussions of the meeting provided an opportunity for investigators from many different areas of basic science and medicine to exchange information, evaluate the present status and provide future research directions in the field of angiogenesis.

I wish to thank Drs. Pietro Gullino and Peter Lelkes, co-Directors of this meeting and the International Organizing Committee that included Drs. Robert Auerbach, Francoise Dieterlen, Juliana Denekamp and Peter Polverini for their help in organizing the meeting. I thank also all the participants for their enthusiastic participation and their complimentary comments on the success of this conference. In addition, I thank the Scientific Affairs Division of NATO who provided the funds for publication of this book and the major portion of the funds for the organization of the meeting. The contribution of the following organizations: Boehringer (Greece), British Council (Greece), Bristol-Meyers-Squibb (Greece), Ciba-Geigy (Greece), Council of Tabaco Res. (USA), Farmitalia (Italy), Galenica (Greece), Genentech (U.S.A.), Genetech (USA), Gibco Life Technology (USA), Help (Greece), Hoechst (Greece), Jensen (Greece), Lilly Ltd (UK), Pfizer (USA), Promega (USA), Spinx Pharm. Corp. (USA), Sandoz (Greece), Smith Kline and French Beecham (Greece), Upjohn (U.S.A.), and Zeneca Pharmac. (UK), which was used to support the participation of many young scientists, is gratefully acknowledged. For travel arrangements and the daily operations of the conference I am thankful to Mrs. Lydia Argyropoulou. I am particularly grateful to Mrs. Anna Marmara for her dedicated and enthusiastic work throughout the organization of the meeting and the editing of this monograph.

Michael E. Maragoudakis (Greece)

CONTENTS

REGULATION OF ANGIOGENESIS

ANGIOGENESIS AND DISEASE STATES

METHODOLOGY

EMBRYOLOGY OF THE ENDOTHELIAL NETWORK: IS THERE AN HEMANGIOBLASTIC ANLAGE ?

Françoise Dieterlen-Lièvre, Dominique Luton and
Luc Pardanaud

Institut d'Embryologie cellulaire et moléculaire
du CNRS et du Collège de France
49bis, av. de la Belle Gabrielle
94736 Nogent s/Marne, cédex - France

INTRODUCTION

Endothelial cells (EC) play a leading role in the modelling of blood vessels. In the embryo they form the blueprint of the vascular tree, the differentiation of which occurs by regionalization of EC properties and apposition of appropriate wall cells. In the adult animal, the turnover of EC, an extremely slow process, accelerates only in a few physiological situations and during tumor progression: formerly quiescent EC becoming reactivated, start multiplying, rupture their basement membrane and migrate. This process, termed angiogenesis (see Folkman, 1974) has been extensively studied at the cellular and biochemical levels. The interplay of growth factors involved either in maintenance of the quiescent state or in reactivation has been investigated. *De novo* appearance of EC during ontogeny is less well understood. This process, designated vasculogenesis (Pardanaud et al. 1987; Risau and Lemon, 1988) has been partly circumscribed but the cellular, biochemical and molecular events underlying it are still largely elusive.

At the beginning of the century, several investigators severed the embryonic area or part of it from the area vasculosa, and demonstrated that blood vessels form *in situ* in the whole surface of the blastodisc (Miller and McWorther, 1914; Reagan, 1915). A strictly local origin of EC in regions of the blastodisc has been confirmed since in avian "yolk sac chimeras", associating an extraembryonic area to a foreign embryo (Beaupain et al. 1979; Cuadros et al. 1992). Appropriate markers such as that provided by the quail/chick system, (Le Douarin, 1969) or monoclonal antibodies MB1 (Péault et al. 1983) and QH1 (Pardanaud et al. 1987) were used to trace cell origins in the chimeras. These antibodies recognize EC and hemopoietic cells (HC) in the quail and not in the chick, thus being markers both for the hemangioblastic lineage and for one of the avian species combined in the quail/chick experimental paradigm. The *in situ* progressive appearance of EC during day E1 could also be observed in quail blastodiscs treated with QH1 *in toto*. QH1+ cells could thus be seen to appear as isolated units at the level of the first pair of somites, to progress in both cephalic and caudal directions in parallel with somitogenesis and to associate into an endothelial network (Pardanaud et al. 1987; Coffin and Poole, 1988; Péault et al. 1988).

We established a few years ago that, while EC form in the whole surface of the blastodisc, the capacities of the two mesodermal layers of the lateral plate are different. Only the splanchnopleural layer (lining the internal organs) produces EC, while the somatopleural layer (lining the body wall) becomes colonized by extrinsic EC precursors (Pardanaud et al. 1989, Pardanaud and Dieterlen-Lièvre, 1993a). The emergence of hemopoietic stem cells (HSC) on the other hand follows different rules. These cells also appear in the splanchnopleural mesoderm, but only in some sites and at several distinct periods, the area opaca at E1-2 and the surroundings of the dorsal aorta at E3-4. We have shown in yolk sac chimeras, in transplantation and culture experiments that the HSC produced in this latter location are responsible for the colonization of definitive hemopoietic organs (see Dieterlen-Lièvre et al. 1990 for a review). Tight association between EC and HC are evident in these two sites. These aspects lead us to revive the hypothesis that a common progenitor, the hemangioblast, may give rise to the two lineages (Murray, 1932).

We will review here the approaches that we have undertaken in order to settle some pending questions. The first concerns the capacity of other subdivisions of the mesoderm, namely the segmental plate and the somites, to give rise to EC and HSC. The second approach concerns the expression of two protooncogenes, that code for transcription factors. These protooncogenes, c-*ets*1 and c-*myb*, are relevant to our studies because they are respectively expressed in EC and HSC during the amplification phase of these two lineages (Vandenbunder et al. 1989).

VASCULOGENESIS CAPACITIES OF THE SOMITES AND SEGMENTAL PLATE

Several investigators have reported that EC were produced from transplanted somites (see Noden, 1989). However somites become individualized at the same time as the primitive bilateral aortae, which differentiate in tight association with their ventral aspect. We deemed it useful to determine whether the vasculogenic capacity of the somites was not due to attached aortic cells. Somites were dissected out from quail embryos and submitted to a series of treatments, the presence of QH1+ cells being controlled at each step. After mechanical separation, aortic profiles indeed remained attached to the somites. After pancreatin treatment of this preparation, 5 to 6 QH1+ cells still clung to the somite. The tissues were then treated successively with QH1 and with complement. The somite then appeared to have lost its rigid structure but it was rid of QH1+ cells. It was grafted into the limb bud of a chick host. The three step treatment diminished the vasculogenic capacity of the somite without however eliminating it. While the non-treated somite gave rise in this location to many QH1+ endothelial profiles, the treated somite gave rise to less numerous QH1+ profiles.

The segmental plate is the paraxial strip of the mesoderm that will become segmented into somites. Stern et al. (1988) have analyzed the fate of cells in this structure by marking an individual cell through injection of the fluorescent stain Rhodamine-Lysine-Dextran. When the injected cell was in the vicinity of the somites, it gave rise two days later to a clone that comprised classical somite derivatives, i.e. sclerotomal or dermomyotomal cells. In contrast when the cell was injected in the caudalmost end of the segmental plate, the clones comprised EC and HC in the lumen of the aorta. Intriguingly the EC were always located in the ventral aspect of the aorta, which seemed to indicate that the segmental plate contributed partially to this vessel. It would be useful to map the origin of the whole aortic endothelium. As a first step in this direction, we decided to establish the fate of the segmental plate by grafting this structure from the quail into a chick host in orthotopic position. The anteriormost portion of the plate corresponding to the length of two somites was exchanged in this way. This type of graft gave rise to numerous endothelial cells in the corresponding portion of the embryo. These EC were restricted to the side of the embryo that received the graft; they constituted most of the network in the body section that developed at the level of the graft, i.e., lined vessels irrigating the neural tube, the dermomyotome and the kidney and surrounding the sclerotome. When transplanted plate was grafted at the limb bud level, most of the endothelial network in the limb originated from the transplant. In the aorta at the grafted level, some foreign EC were intercalated among host EC, but they were rarely in a ventral or ventrolateral position and HC were exceptional. Thus the results obtained with this approach do not fit exactly with the expectations raised by the report of Stern's group.

DIFFERENTIAL EXPRESSION OF C-ETS1 IN EARLY CHICK ONTOGENY

This oncogene has been found expressed preferentially in EC of the embryo (Vandenbunder et al. 1989). The signal is present in all the cells of the very early blood islands prior to their compaction then becomes restricted to the peripheral cells, which are the future EC while disappearing from the central future HC. First expressed all over the developing endothelial network, the signal faints away from the endothelia of differentiating blood vessels. For instance at E6 the signal is extinct in the aorta, while still on in the segmental arteries. The c-myb probe, on the other hand, hybridizes to immature HC. However it is absent from the primitive area vasculosa blood 'islands, appearing on the E3 intra-aortic HSC aggregates and on the E6 para-aortic foci. We will detail here our recent analysis bearing on the expression of c-etsl in the early mesoderm. (Pardanaud and Dieterlen-Lièvre, 1993b). The signal first lights up as mesoderm becomes

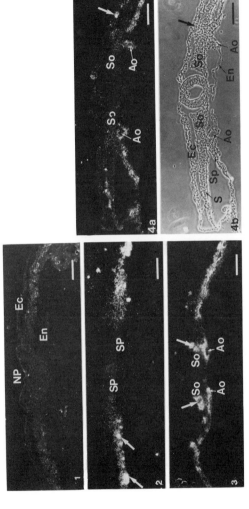

c-*ets*1 expression in relationship to maturation of the mesoderm.

Transverse sections depicting successive steps of this maturation are from embryos at two different stages. At the older stage, sections correspond to progressing maturation in the caudal to rostral direction. Bar= 100μm

Figure 1. Ten-somite stage chick embryo: segmental plate level. c-*ets*1 messengers are present in the mesoderm while the ectoderm (Ec) and the endoderm (En) do not express the gene. NP: neural plate.

Figure 2. Nineteen-somite stage chick embryo: segmental plate (SP) level. c-*ets*1 is restricted to the lateral mesoderm where future blood islands display a higher signal (arrows).

Figure 3. Nineteen-somite stage chick embryo: caudal pair of somites level. c-*ets*1 expression is present in lateral mesoderm. The somites (So) do not transcribe c-*ets*1 while intermediate plate (thick arrows) gives a strong signal. Ventrally to the somites, the aortic (Ao) endothelia are very positive.

Figure 4. Nineteen-somite stage chick embryo: median somitic level. (4a) At this level the mesoderm is split: the somatopleural layer (S) displays a few positive patches (thick arrow) whereas the messengers are very abundant in the splanchnopleural layer (Sp). The aortic endothelia (Ao) are strongly positive, while silver grains in the intermediate plate are few in this case. Ectoderm (Ec), endoderm (En), neural tube and somits (So) do not transcribe c-*ets*1. (4b) Phase contrast.

established by ingression through the primitive streak. It is first distributed homogeneously throughout the whole germ layer (Figure 1). When vasculogenesis begins, the expression becomes restricted to the ventral cells of the mesoderm, in close contact with the endoderm, then intensifies in the blood islands (Figures 2). When mesoderm splits into two layers encasing the coelom, c-*ets*1 remains ubiquitous in the splanchnopleural layer, while it is weak and restricted to small spots in the somatopleural layer. c-*ets*1 is also expressed strongly in the intermediate plate (forerunner of the kidney) (Figure 3) and in the ventral aspect of somites when they undergo the epithelio-mesenchymal transition (Figure 4).

CONCLUSIONS

The data described above are part of an endeavour to understand how the endothelial network and HSC arise during ontogeny. One of the points that we lay importance on is to define a hemangioblastic anlage, i.e. to determine where the capacities to produce these two lineages become located, as the original mesodermal sheet becomes subdivided into structures with a well defined fate. We had already determined that the two sheets of lateral plate mesoderm have definitely different capacities in this regard (Pardanaud et al. 1989). We now show that, while the segmental plate has an important potential to give rise to EC, this potential decreases when the plate become segmented into somites, presumably by segregation of EC. Mapping the origin of the whole aorta remains to be accomplished, the best tool to achieve this goal appearing to be the exchange of defined components of the mesoderm between quail and chick embryo. Whether and how EC and HSC segregate from a common progenitor may also be dealt with using similar approaches. The combination of such experimental devices with the detection of the complementary patterns of expression of c-*ets*1 and c-*myb*, used as early markers for the dichotomy of the two lineages should help powerfully.

REFERENCES

Beaupain, D., Martin, C., Dieterlen-Lièvre, F. (1979). Are developmental hemoglobin changes related to the origin of stem cells and site of erythropoiesis? *Blood* 53: 212-225.

Coffin J. D. & Poole T. J. (1988). Embryonic vascular development: immunohistochemical identification of the origin and subsequent morphogenesis of the major vessel primordia in quail embryos. *Development* 102: 735-748.

Cuadros, M.A., Coltey, P., Nieto, M.C., Martin, C. (1992). Demonstration of a phagocytic cell system belonging to the hemopoietic lineage and originating from the yolk sac in the early avian embryo. *Development* 115: 157-168.

Dieterlen-Lièvre, F., Pardanaud, L., Bolnet C. and Cormier, F. (1990). Development of the hemopoietic and vascular system studied in the avian embryo. In : "The avian model in developmental biology", éditions du CNRS: pp.319.

Folkman, J. (1974). Tumour angiogenesis. *Adv. Cancer Res.* 19: 331-358.

Le Douarin N.M. (1969). Particularités du noyau interphasique chez la Caille japonaise (*Coturnix coturnix japonica*). Utilisation de ces particularités comme "marquage biologique" dans les recherches sur les interactions tissulaires et les migrations cellulaires au cours de l'ontogenèse. *Bull. Biol. Fr. Belg.* 103: 435-452.

Miller A.M. & Mc Whorter J.E. (1914). Experiments on the development of blood vessels in the area pellucida and embryonic body of the chick. *Anat. Rec.* 8: 203-227.

Murray P.D.F. (1932). The development *in vitro* of blood of the early chick embryo. *Strangeways Res. Labor. Cambridge*: 497-521.

Noden DM (1989) Embryonic origins and assembly of embryonic blood vessels. *Ann. Rev. Pulmon. Dis.* 140: 1097-1103

Pardanaud, L. and Dieterlen-Lièvre, F. (1993a). Emergence of endothelial and hemopoietic cells in the avian embryo. *Anat. & Embryol.* 187: 107-114.

Pardanaud, L. and Dieterlen-Lièvre, F. (1993b). Expression of c-*ets*1 in early chick embryo mesoderm: relationship to the hemangioblastic lineage. *Cell Adhesion & Commun* .(In press).

Pardanaud, L., Altmann, C., Kitos, P., Dieterlen-Lièvre, F. and Buck, C. (1987). Vasculogenesis in the early quail blastodisc as studied with a monoclonal antibody recognizing endothelial cells. *Development* 100: 339-349.

Pardanaud, L., Yassine, F. and Dieterlen-Lièvre, F. (1989). Relationship between vasculogenesis, angiogenesis and haemopoiesis during avian ontogeny. *Development* 105: 473-485.

Péault B., Coltey M. & Le Douarin N.M. (1988). Ontogenic emergence of a quail leukocyte/endothelium cell surface antigen. *Cell Diff.* 23: 165-174.

Péault B., Thiery J.P. & Le Douarin N.M. (1983). A surface marker for the hemopoietic and endothelial cell lineage in the quail species defined by a monoclonal antibody. *Proc. Natl. Acad. Sci. USA.* 80: 2976-2980.

Reagan F.P. (1915). Vascularization phenomena in fragments of embryonic bodies completely isolated from yolk-sac blastoderm. *Anat. Rec.* 9: 329-341.

Risau, W. and Lemmon, V. (1988). Changes in the vascular extracellular matrix during embryonic vasculogenesis and angiogenesis. *Dev. Biol.* 125: 441-450.

Stern, C.D., Fraser, S.E., Keynes R.G. and Primmett, D.R.N. (1988). A cell lineage analysis of segmentation in the chick embryo. *Development* 104 Suppl.: 231-244.

Vandenbunder, B., Pardanaud, L., Jaffredo, T., Mirabel, M.A. and Stéhelin, D. (1989). Complementary patterns of expression of c-ets1, c-myb and c-myc in the blood-forming system of the chick embryo. *Development* 106: 265-274.

CELLULAR AND MOLECULAR BIOLOGY OF ENDOTHELIAL CELL DIFFERENTIATION DURING EMBRYONIC DEVELOPMENT

Thomas J. Poole

Department of Anatomy & Cell Biology
SUNY Health Science Center at Syracuse
766 Irving Avenue
Syracuse, NY 13210 USA

INTRODUCTION

How does the pattern of the rudiments of the major blood vessels establish itself in the developing embryo? To understand the cellular biology of these events we must know how the angioblasts, the precursors of endothelial cells, segregate from the mesoderm, migrate, and cohere to one another to form the cords and tubes which are the earliest embryonic blood vessels. We have been using a monoclonal antibody (QH-1) and microsurgery to determine where angioblasts originate and how they assemble into vessel rudiments (Coffin and Poole, 1988; Poole and Coffin, 1989; 1991). The extent and type of directed angioblast migration define three distinct modes of vessel morphogenesis (Poole and Coffin, 1991; Poole, 1993). Vessel rudiments may organize in place, a process termed **vasculogenesis**, either from angioblasts originating at the rudiment's location (vasculogenesis type I) or from angioblasts which migrate as individual cells or small groups to that site from different locations (vasculogenesis type II). The dorsal aortae form by the first type of vasculogenesis (Coffin and Poole, 1988; DeRuiter et al., 1993; Pardanaud et al., 1987; Poole and Coffin, 1988; 1989; 1991). The endocardium, ventral aortae and posterior cardinal veins form by the second type (Coffin and Poole, 1991; DeRuiter et al., 1993; Drake and Jacobson, 1988; Poole and Coffin, 1991). New vessels may also form by sprouting from preexisting vessels, a process called **angiogenesis**. The intersomitic and vertebral arteries are the first vessels to form by angiogenesis, sprouting off the rudiments of the dorsal aortae (Coffin and Poole, 1988; Poole and Coffin, 1988; 1989; 1991). Figure 1 illustrates the different roles of endothelial cells in vasculogenesis and angiogenesis.

LUMEN STUDIES (1900's) vs. ANTIBODY STUDIES (1980's and 90's)

There was a great deal of work done on vascular development early in this century which until the last decade served as the basis for our understanding of blood vessel origin (reviews: Evans, 1912; McClure, 1921; Sabin, 1917). The injection of opaque fluids in the early 1900's revealed much of what we now know about embryonic vascular anatomy. The limit of this technique was that only patent vessels could be visualized. Since the vascular pattern results from the remodeling of complex capillary plexuses arising by the anastomosis of solid endothelial cords, the inability to resolve endothelial precursors lead to a heated argument about the sites and mechanisms of endothelial cell origin (McClure, 1921; Sabin, 1920). The production of monoclonal antibodies which labelled angioblasts in the quail embryo (Peault et al., 1983; Pardanaud et al., 1987) allowed the recent progress in our understanding of endothelial cell differentiation.

Angiogenesis: Molecular Biology, Clinical Aspects
Edited by M.E. Maragoudakis *et al*, Plenum Press, New York 1994

GRAFTING EXPERIMENTS AND MESODERM ORIGINS

The construction of quail/chick chimeras in the 1980s has begun to delineate the details of endothelial cell lineage and its close relationship to the hematopoietic lineages.These studies have been reviewed in detail elsewhere (Noden, 1989; Pardanaud et al., 1989; Poole and Coffin, 1991; Poole, 1993). An interesting aspect which emerges from recent work is the variation in angioblast differentiation from quail mesoderm grafted to different sites in chick embryos. Blocks of tissues the size of a single somite grafted beneath the otic placode resulted in angioblast differentiation from all intraembryonic mesoderm except the prechordal plate and notochord (Noden, 1988;1989). Similar blocks of tissues grafted to older chick embryos in the limb bud or coelom demonstrated a striking difference between splanchnopleural mesoderm (the portion of mesoderm adjacent to embryonic endoderm) and somatopleural mesoderm (the mesoderm adjacent to ectoderm). Splanchnopleural mesoderm gives rise to abundant angioblasts in these grafting experiments;

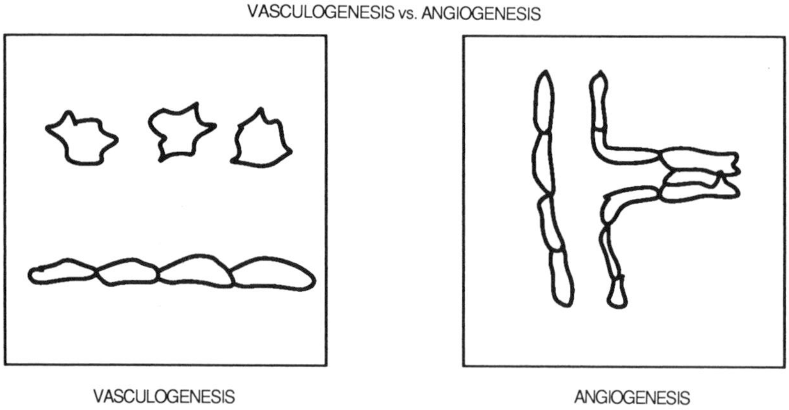

VASCULOGENESIS vs. ANGIOGENESIS

VASCULOGENESIS

ANGIOGENESIS

Figure 1. Vasculogenesis forms the earliest embryonic vessels as single angioblasts arise from the splanchnic mesoderm and migrate and cohere to one another to form cords of endothelial cells at the sites of vessel origin. Angiogenesis occurs only after the vascular pattern is established, but dominates later stages of vessel origin as sprouting off of preexisting vessels continues throughout fetal and adult life.

whereas, somatopleural mesoderm produces no angioblasts or only a few (Pardanaud and Dieterlen-Lievre, 1993). We have transplanted a single quail somite or a piece of lateral mesoderm the size of a somite into the head or trunk of chick hosts at the same stage of development (10 somite stage) and found that a somite graft to the head produces many angioblasts, a somite graft to the trunk produces very few, and lateral mesoderm grafts (containing both splanchnopleural and somatopleural mesoderm) produce many angioblasts in both locations (Poole, 1991 and in preparation). Figure 2 shows an example of a somite graft result in each location. These differences between somite and lateral mesoderm are also seen *in vitro*. Quail mesodermal cells, dissociated with trypsin, cultured for 20 hours result in 5% QH-1 labelled cells of somite origin and 25% QH-1 labelled cells from the lateral mesoderm. The addition of basic fibroblast growth factor (bFGF) to these cultures at 25 ng/ml produced a ten-fold increase in angioblasts from somite mesoderm (50% QH-1 positive), but did not significantly affect lateral mesoderm cultures (25% QH-1 positive).

MOLECULAR APPROACHES TO ENDOTHELIAL CELL DIFFERENTIATION

The complexity of developmental events involved in the morphogenesis of embryonic vascular pattern is summarized in Figure 3. It can be appreciated from even such a simplified summary that there must be many molecular events involved in the formation of embryonic blood vessels. Cell adhesion, cell migration, extracellular matrix synthesis, cell proliferation and cell death are all involved. The determination of the endothelial cell lineage and the recent discovery of molecular markers for this lineage is all that will be dealt with here. The grafting experiments discussed in the previous section suggest that by the early somite stage, when angioblasts are first recognized by monoclonal antibodies, there are already some differences in the abilities of mesoderm from various intraembryonic locations to form endothelial cells under experimental conditions. Mesoderm originates in avian embryos by gastrulation from the embryonic epiblast. Fragments of the epiblast or dissociated epiblast cells will form angioblasts in culture. There is some conflict in the literature as to whether the addition of basic fibroblast growth factor (bFGF) is necessary to induce angioblast differentiation. Flamme and Risau (1992) found that bFGF addition was necessary to produce blood island-

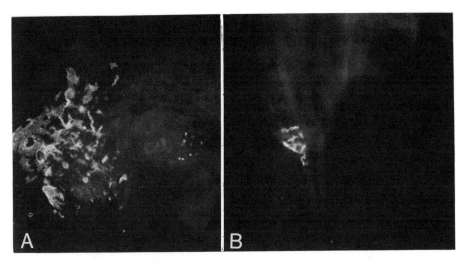

Figure 2. A single quail somite from a 10 somite stage embryo grafted into the head of a similar stage chick embryo produces many angioblasts (A). A quail somite grafted in the trunk in place of chick somite number 5 forms only a few angioblasts (B).

like structures and endothelial tubes from dissociated epiblast cells plated into 96-well plates. These authors only scored for blood islands at early time points (3 days) and found nothing but endothelial cells (QH-1 labelling and DiI-ac-LDL uptake) remaining in the cultures after 4 weeks of culture.Christ et al. (1991) treated blastodiscs with cytochalasin B to prevent gastrulation and found labelled angioblasts arising from fragments of blastodisc grafted into limb buds. We have found many QH-1 labelled cells arising from dissociated epiblast cells during the first 3 days of culture without the addition of bFGF to the culture medium. (Poole and Martini, unpublished observations). A role for bFGF in the initial induction of mesoderm from the epiblast similar to bFGF induction of mesoderm from the cells of the animal cap of blastula-stage frog embryos (Shiurba et al., 1991; Slack, 1993) has been reported (Mitrani et al., 1990a). Both the message and the protein are present in pre-gastrula stage chick embryos and suramin and heparin which can inhibit FGF action both inhibited mesoderm formation (Mitrani et al., 1990a). Mesodermal axial structures such as the notochord and somites are induced by activin, a peptide growth factor of the transforming growth factor beta family (Mitrani et al., 1990b). Activin is a dimeric protein composed of inhibin subunits which are secreted by early chicken endoderm in culture (Kokan-Moore et al., 1991).

Previous early markers for the endothelial cell lineage in mammalian embryos (Coffin et al., 1991) labelled only a subset of developing angioblasts. Recently, receptor tyrosine kinases (RTKs) have been discovered that are specific to endothelial cells and label angioblasts in the early mouse embryo. *Flk-1* is a high affinity receptor for vascular endothelial growth factor (VEGF) isolated using PCR from embryonic (Millauer et al., 1993) and ES cell (Yamaguchi et al., 1993) cDNA libraries utilizing degenerate oligonucleotides with conserved sequences from the kinase domain of RTKs. Sequence analysis revealed this RTK to be a member of the *c-kit/pdgfra/flt* family of RTKs. Another RTK from a different subfamily of receptors which is also endothelial-specific is *tek.* (Dumont et al., 1992; 1993). A direct comparison of *flt-1* and *tek* expression showed similar patterns, but that *flt-1* expresssion preceded *tek* expression (Yamaguchi et al., 1993). Since RTKs are involved in signal transduction pathways, the study of their expression may give us some clues to the cascade of gene expression involved in endothelial cell determination and differentiation. A more direct approach to an understanding of the molecular biology of endothelial cell differentiation is possible through the use of PCR-based subtractive hybridization protocols. We are using one such protocol which has been called a gene expression screen (Wang and Brown, 1991) to identify mRNAs that differ in abundance between somites cultured in the presence and absence of bFGF. The rationale behind this experimental approach comes from our results discussed in the previous section which showed an induction of angioblast differentiation from somite mesoderm by bFGF. This phenomena allows us to screen cDNAs from two populations of quail mesodermal cells which differ primarily in the presence of angioblasts. We are hoping in this way to identify genes that are differentially expressed early in the endothelial cell lineage. Such genes may be involved in endothelial determination or might serve as useful new markers for angioblasts in avian embryos.

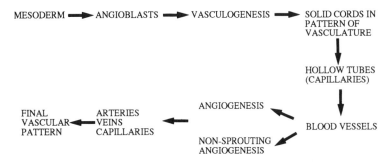

Figure 3. A simplified flow chart of the events in vascular development. The embryonic vasculature arises by the assembly of angioblasts into cords by the process of vasculogenesis. Further expansion occurs by sprouting off of the walls of primitive vessels through the process of angiogenesis.

ACKNOWLEDGEMENTS

We thank Marisa Martini and Vanaja Bandaru for technical assistance. The work in the author's laboratory was supported by The Brownstein Foundation and the National Science Foundation.

REFERENCES

Christ, B., Grim, M., Wilting, J., von Kirschhofer, K., and Wachtler, F., 1991,

Differentiation of endothelial cells in avian embryos does not depend on gastrulation, *Acta Histochem.* 91: 193-199.

Coffin, J.D., Harrison, J., Schwartz, S., and Heimark, R., 1991, Angioblast differentiation and morphogenesis of the vascular endothelium in the mouse embryo, *Develop. Biol.* 148: 51-62.

Coffin, J.D. and Poole, T.J., 1988, Embryonic vascular development: Immunohistochemical identification of the origin and subsequent morphogenesis of the major vessel primordia, *Development* 102: 735-748.

Coffin, J.D. and Poole, T.J., 1991, Endothelial cell origin and migration in embryonic heart and cranial blood vessel development, *Anat. Rec.* 231: 383-395.

DeRuiter, M.C., Poelmann, R.E., Mentink, M.M.T., Vaniperen, L., and Gittenberger-De Groot, A.C., 1993, Early formation of the vascular system in quail embryos, *Anat. Rec.* 235: 261-274.

Drake, C.J. and Jacobson, A.G., 1988, A survey by scanning electron microscopy of the extracellular matrix and endothelial components of the primordial chick heart, *Anat. Rec.* 222: 391-400.

Dumont, D.J., Yamaguchi, T.P., Conlon, R.A., Rossant, J., and Breitman, M.L., 1992, *tek*, a novel tyrosine kinase gene located on mouse chromosome 4, is expressed in endothelial cells and their presumptive precursors, *Oncogene* 7: 1471-1480.

Dumont, D.J., Gradwohl, G.J., Fong, G.-H., Auerbach, R., and Breitman, M.L., 1993, The endothelial-specific receptor tyrosine kinase, *tek*, is a member of a new subfamily of receptors, *Oncogene* 8: 1293-1301.

Evans, H.M., 1912, The development of the vascular system, In "Manual of Human Embryology, Vol. II," *in*: E. F. Keibel and F.P. Mall, eds., J.B. Lippincott Company, Philadelphia, pp. 570-709.

Flamme, I. and Risau, W., 1992, Induction of vasculogenesis and hematopoiesis in vitro, *Development* 116: 435-439.

Kokan-Moore, N.P., Bolender, D.L., and Lough, J., 1991, Secretion of inhibin beta by endoderm cultured from early embryonic chicken, *Dev. Biol.* 146:242-245.

McClure, C.F.W., 1921, The endothelial problem, *Anat. Rec.* 22: 219-237.

Millauer, B., Wizigmann-Voos, S., Schnurch, H., Martinez, R., Moller, N.P.H., Risau, W., and Ullrich, A., 1993, High affinity VEGF binding and developmental expression suggest flk-1 as a major regulator of vasculogenesis and angiogenesis, *Cell* 72: 835-846.

Mitrani, E., Gruenbaum, Y., Shohat, H., and Ziv, T., 1990a, Fibroblast growth factor during mesoderm induction in the early chick embryo, *Development* 109: 387-393.

Mitrani, E., Ziv, T., Thomsen, G., Shimoni, Y., Melton, D.A., and Bril, A., 1990b, Activin can induce the formation of axial structures and is expressed in the hypoblast of the chick, *Cell* 63: 495-501.

Noden, D.M., 1988, Interactions and fates of avian cranio-facial mesenchyme, *Development* 103 suppl.: 121-140.

Noden, D.M., 1989, Embryonic origins and assembly of blood vessel, *Am. Rev. Respir. Dis.* 140: 1097-1103.

Pardanaud, L., Altmann, C., Kitos, P., Dieterlen-Lievre, F., and Buck, C.A., 1987, Vasculogenesis in the early quail blastodisc as studied with a monoclonal antibody recognizing endothelial cells, *Development* 100:339-349.

Pardanaud, L., Yassine, F., and Dieterlen-Lievre, F., 1989, Relationship between vasculogenesis, angiogenesis and hematopoiesis during avian ontongeny, *Development* 105: 473-485.

Pardanaud, L. and Dieterlen-Lievre, F., 1993, Emergence of endothelial and hemopoietic cells in the avian embryo, *Anat. Embryol.* 187: 107-114.

Poole, T.J., 1991, Fibroblast growth factor influences the differentiation and migration of endothelial cells in avian embryos, *J. Cell Biol.* 115: 366a.

Poole, T.J., 1993, Cell migration in embryonic blood vessel assembly, *in* Homing Mechanisms and Cellular Targeting. (B.R. Zetter, ed.); publ. Marcel Decker Inc., New York, New York.

Poole, T.J. and Coffin, J.D., 1988, Developmental angiogenesis: Quail embryonic vasculature, *Scanning Microsc.* 2: 443-448.

Poole, T.J. and Coffin JD. 1989. Vasculogenesis and angiogenesis: Two distinct morphogenetic mechanisms establish embryonic vascular pattern, *J. Exp. Zool.* 251: 224-231.

Poole, T.J. and Coffin, J.D., 1991, Morphogenetic mechanisms in avian vascular development, *in*: The Development of the Vascular System. Issues Biomed. (R.N. Feinberg, G.K. Sherer, R. Auerbach, eds.); publ. S. Karger AG, Basel, Switzerland. vol. 14, pp 25-36.

Sabin, F.R. ,1917, Origin and development of the primitive vessels of the chick and of the pig, *Contrib. Embryol. Carnegie Inst. Wash.* 6: 61-124.

Sabin, F.R., 1920, Studies on the origin of blood vessels and of red blood corpuscles as seen in the living blastoderm of chicks during the second day of incubation, *Contrib. Embryol. Carnegie Inst. Wash.* 36: 213-259.

Schnurch, H.G. and Risau W., 1991, Differentiating and mature neurons express the acidic fibroblast growth factor gene during chick neural development, *Development* 111: 1143-1154.

Shiurba, R.A., Jing, N., Sakakura, T., and Godsave, S.F., 1991, Nuclear translocation of fibroblast growth factor during Xenopus mesoderm induction, *Development* 113: 487- 493.

Slack, J.M.W., 1993, Embryonic induction, *Mechanisms of Development* 41: 91-107.

Wang, Z. and Brown, D.D., 1991, A gene expression screen, *Proc. Natl. Acad. Sci. USA*: 11505-11509.

Yamaguchi, T.P., Dumont, D.J., Conlon, R.A., Breitman, M.L., and Rossant, J., 1993, *flk*-1, an *flt*-related receptor tyrosine kinase is an early marker for endothelial cell precursors, *Development* 118: 489-498.

ENDOTHELIAL CELL HETEROGENEITY AND ORGAN - SPECIFICITY

Peter I. Lelkes[*], Vangelis G. Manolopoulos[*,#], Dawn Chick[*],
and Brian R. Unsworth[#],

([*]) Univ. Wisconsin Med School, Milwaukee Clinical Campus,
University, Dept. Biology, Milwaukee, WI U.S.A.
and ([#])Marquette

INTRODUCTION

Endothelial cell (EC) biology and physiology play a prominent role in studying angiogenesis, i.e. the establishment of new blood vessels from existing ones. According to current perceptions, major features of the angiogenic cascade involve mainly EC-related phenomena such as dissolution of the subendothelial basement membrane, migration of the EC, and formation of a new vessel lumen by establishing tight interendothelial cell contacts. Angiogenesis *in vivo* occurs primarily at the level of the microvasculature (capillaries, arterioles, venules) and yet most *in vitro* models have convincingly employed EC isolated from large vessels. Such seemingly discordant approaches raise the question whether EC derived from different vascular beds can be used interchangeably to study common "vascular" phenomena. Over the past few years a large body of experimental findings has been accumulated to the effect that "an EC is not an endothelial cell is not an endothelial cell"[1]. Rather, EC phenotypic and functional diversity is differentially regulated by a plethora of microenvironmental and/or hemodynamic cues.

Based on our own work we will first discuss different aspects of EC heterogeneity and organ-specificity *in vivo*. In the second part of this paper, we will introduce the concept that EC heterogeneity may be related to the heterogeneity of signal transduction cascades, in particular of the adenylyl cyclase signaling system.

ENDOTHELIAL CELL HETEROGENEITY

Endothelial cells (EC) are pluripotent cells which perform pivotal physiological functions e.g., in hemostasis, inflammation, immunology[2]. All EC are presumably derived from common mesenchymal precursors[3], and yet they develop profound differences in terms of morphological phenotypes in various locations within the vascular tree[4]. Moreover, phenotypical differences have been noted in EC within similar vessels in different areas in the same organ[5] and even between different segments of the same vessel[6].

These differences comprise distinct expression of "EC-specific" markers[7], and also profound functional differences, e.g. at the level of receptors[8], ion channels[9], vasomotor responses[10], balance of procoagulant and anticoagulant activities[11], and responses to mechanical stimulation[12]. Functional EC heterogeneity arises primarily from the variation of EC phenotype within the vascular tree[13] and, to a lesser extent, also from species differences[14].

Unlike most other cells, vascular EC are directly exposed to a variety of flow-related mechanical forces in vivo, mainly flow-induced shear stress, cyclic deformation of the substratum and oscillatory pressure.[15] Flow-induced shear force was found to be an important regulator of

EC gene expression[16], as manifested in the modulation of a host of basic EC functions, including vasomodulatory[17,18] and thrombomodulatory[19,20] responses. Cyclic strain was also found to regulate fundamental EC responses, such as signal transduction, gene expression, etc.[21]. Since the waveform and the magnitude of the cyclic strain varies widely according to the local hemodynamics, this differential exposure to strain might contribute to EC heterogeneity. For example, it was recently reported that venous but not arterial EC align perpendicular to the force vector in an uniaxial strain system[22]. Other examples of heterogeneous EC responses to cyclic strain will be discussed below.

The distinct effects of the various physical mechanical forces on EC physiology suggest that EC are equipped with an exquisite mechanism for sensing and transducing mechanical forces. This latter notion is obviously of importance in understanding the possible role of mechanical forces in angiogenic processes in arterioles and/or venules.

ENDOTHELIAL CELL ORGAN-SPECIFICITY

The above described phenotypic diversity has primarily been studied in EC derived from large vessels. More recently an even more remarkable degree of diversity was postulated for EC lining the microvasculature: Spanel-Borowski and coworkers isolated five morphologically, immuno-histochemically and functionally different microvessel-derived EC phenotypes from the bovine corpus luteum[23]. Similarly, we have isolated 5 morphologically distinct microvascular EC phenotypes from the bovine adrenal medulla (Figure 1).

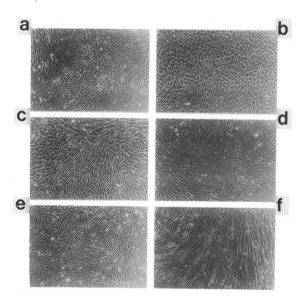

Figure 1: Phenotypic heterogeneity of bovine adrenal medullary microvascular endothelial cells. Panel a: primary isolate, Panels b-f: individual phenotypes cloned by limited dilution. Original magnification 100 x.

The primary culture, obtained by plating capillary fragments after differential filtration, showed a mixture of distinct phenotypes. Upon cloning by limited dilution, populations of homogeneous appearance were obtained: some of the cell types exhibited "typical cobble-stone" morphology, others were clearly more spindle shaped. This finding corroborates recent reports on the morphology of other microvascular EC[24]. The various adrenal medullary EC cell types all take up a "universal" EC marker, diI-tagged acetylated LDL, albeit to different degrees. In line with recent reports[7], we failed to detect consistent staining with von Willebrand factor (vWF), another "sure" marker for endothelial cells. In the absence of specific markers, it is at present not known whether the distinct EC phenotypes in endocrine organs represent EC from different microvascular beds, or whether they might represent EC heterogeneity within a single vessel type.

Pivotal factors which determine the organ-specificity of the EC are locally derived from the microenvironment, most probably originating from neighboring (parenchymal) cells. The nature of these organ-specific factors is not well understood: they might be secreted products (growth factors, hormones, etc.), or factors residing in the common basement membrane or in extracellular matrix which separates EC from the underlying parenchymal cells. In addition, some of the regulatory factors are believed to act juxtapositionally by direct heterotypic contacts with other cell types[25,26].

Irrespective of the nature of these differentiating factors, the overall result of such "organ-specific" modulations is the expression of some unique EC phenotype, which is manifested morphologically and functionally through the expression of organ-specific gene products. Four examples will illustrate this principle:

1. Parenchymal cell-derived humoral factors: Interactions between EC and parenchymal cells have been described in several organs *in situ* and/or with cells isolated from these organs[27-30]. For example, in the adrenal medulla, parenchymal chromaffin cells synthesize, store and secrete catecholamines (CA). Recently, adrenal medullary EC were shown to exhibit a high affinity CA uptake mechanism[31]. Yet in contrast to chromaffin cells, in which the re-utilization of CA via re-uptake and re-storage appears to be part of the cell-specific housekeeping, adrenal medullary EC are equipped with the capacity to efficiently metabolize excess CA[31]. Thus in this particular endocrine gland, there exists a functional, organ-specific symbiosis between parenchymal cells and EC. It will be instructive to study similar relationships in other endocrine tissues, such as the adrenal cortex, pancreas, thyroid, parathyroid, etc.

The molecular mechanisms which lead to the expression of organ-specific traits in diverse EC are largely unknown. It is intriguing to speculate that secretory products and/or other microenvironmental cues derived from parenchymal cells are amongst the factors which differentially activate certain "organ-specific" genes in EC. In support of this notion, recent transplantation experiments of endocrine parenchymal cells (pancreatic ß-cells, adrenal medullary chromaffin cells, PC12 pheochromocytoma cells) have shown that cues derived from these cells are capable of modulating the phenotype of the host-derived EC *in vivo*[32,33].

2. Heterotypic Cell Contacts: Direct physical contacts occur between EC and neighboring cells, both during development as well as in mature blood vessels. In peripheral blood vessels such contacts have been demonstrated between EC and smooth muscle cells (myo-endothelial cell junctions), or between EC and pericytes[4]. Such heterotypic cell contacts between vascular wall cells, currently being explored *in vitro*[34], might be of particular importance for mutual restraint of uncontrolled cell proliferation[25].

On the other hand, establishing heterotypic contacts with neighboring cells might be a prerequisite for, and/or a contributing factor to, the induction of organotypic EC differentiation (e.g. EC-astrocyte contacts in the brain). Recently we have presented evidence that such heterotypic contacts also occur between putative EC precursor cells of mesenchymal origin and parenchymal cells during the post-natal development of the rat adrenal-medulla[29].

In order to evaluate the role of such heterotypic interactions in detail we isolated and co-cultured adrenal medullary EC and parenchymal cells. As previously reported these co-cultures were found to induce the phenotypic differentiation of undifferentiated parenchymal chromaffin cell-derived PC12 pheochromocytoma tumor cells towards the neuroendocrine phenotype[28]. Concomitantly, 2-D gel electrophoresis revealed induction of a limited number of specific gene products in the endothelial cells[29]. In analogy to the junctional contacts seen in the developing gland, we also observed direct heterotypic interactions between adrenal medullary EC and PC12 cells in long-term co-culture. Shown in Figure 2 is an electron micrograph of such an 11 day old co-culture. The lack of a basement membrane between the two cells is noteworthy and reminiscent of the situation *in situ*: the development of a subendothelial basement membrane separating parenchymal cells from the vascular lining is a late event in post-natal organ development.

This particular (static) micrograph is remarkable, in that it captures three different modes of dynamic intercellular communications: 1. direct juxtapositional (cell-cell) contact 2. exocytosis and 3. endocytosis. Obviously, each of these dynamic pathways of communication may carry different signals: a junctional contact may mediate electrical signals across cell boundaries and/or permit the transfer of small molecules, whereas exocytotic/endocytotic processes might facilitate the secretion and uptake of large peptides or proteins, which either lack signal peptides (e.g. FGFs) or for which there are no specific receptors on the cell surface.

Organ-specific adhesive and/or differentiative interactions between EC and parenchymal or tumor cells have been previously demonstrated to occur *in vitro*[35]. Others have also demonstrated functional consequences of such heterotypic cell-cell signaling: For example, EC plasminogen activator and plasminogen activator inhibitor type I expression are modulated in co-cultures between retinal pigment epithelial cells and EC from both human retinal microvessels and from fetal bovine aortic endothelium[27]. Similarly, the mRNA levels for both TGF-ß and preproendothelin are upregulated in cardiac microvascular EC co-cultured with ventricular myocytes[30].

Figure 2: Intercellular heterotypic communication between a PC12 cell (top) and a bovine adrenal medullary endothelial cell in co-culture. Original magnification 16,000 x.

3. Organ-specific glycoconjugates (lectins): Endothelial cell organ-specificity and heterogeneity is also supported by the existence of highly differentiated glycosylated cell surface antigens which provide for the selective, organ-specific homing of leukocytes. Examples are the homing molecules such as the addressins and selectins which belong to different families of cell adhesion molecules[36,37]. Similarly, organ-specificity of metastasis requires recognition by circulating tumor cells of distinct glycoconjugates on the surface of EC in the target organs[38]. Indeed, the specificity of these interactions is also maintained *in vitro*: binding experiments have demonstrated preferential adherence of certain tumor cells to EC derived from organs to which the primary cancer commonly metastasizes[39].

Recently, several laboratories have begun to identify some unique EC specific antigens by generating organ-specific anti-endothelial cell antibodies[40,41]. Much of the ongoing work is being devoted to clarifying the nature of the organ-specific EC antigens and to develop strategies to exploit individual antigens, e.g. for highly selective organ-specific pharmaceutical interventions. Remarkably, certain traits of organ-specificity are not only maintained, but can also be induced *in vitro*, by growing "generic" EC on "organ-specific" extracellular matrix proteins[42].

Plant lectins bind with high specificity to particular sugar residues and have been used to identify and to map unique markers for organ-specific EC and/or blood vessels.[43,44] We recently tested a panel of commercially available fluorescence-conjugated lectins in frozen sections of various visceral and endocrine (rat) organs, and found that both *solanum tuberosum* and *lycopersicon esculentum* stained distinct blood vessels in different organs: microvessels in the rat adrenal and large vessels in the rat liver (Figure 3). Moreover, *ulex europaeus I* lectin, a useful marker for all human vascular EC, was found to specifically stain large vessels in the rat liver and glomeruli in the rat kidney, while no staining was observed in the rat adrenal. Thus, another level of organ-specificity and heterogeneity is introduced: some sugar residues are expressed on

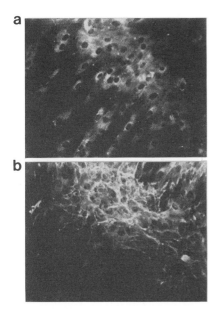

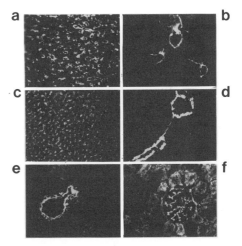

Figure 3: Organ-specific staining of frozen sections with fluorescence-conjugated lectins. Panels a,c, rat adrenal medulla; Panels b, d, e: rat liver, panel f: rat kidney; Top row: *solanum tuberosum* lectin, Middle row: *lycopersicon esculentum* lectin, Bottom row: *Ulex europaeus* lectin. Original magnification 200 x.

Figure 4: Specific binding of fluorescence-conjugated lectins to isolated adrenal medullary endothelial cells. Panel a: *lycopersicon esculentum* stain cell-associated antigen, Panel b: *solanum tuberosum* stains extracellular matrix component. Original magnification 200 x.

particular types of blood vessels but only in certain organs. The factors that induce this high degree of specialization are totally unknown.

Specific staining of distinct blood vessels *in situ* warrants further exploration of the nature and the localization of the antigens that bind those lectins. In testing the above mentioned panel of lectins on isolated adrenal medullary EC, *lycopersicon esculentum* was found to bind to a-cell-associated glycoconjugate. Hence, this particular lectin might be a useful marker for cellular organ-specific gene products. By contrast, *solanum tuberosum* stained a glycoconjugate which was deposited into the extracellular matrix (Figure 4). These findings support the notion that endothelial-cell derived molecules contribute to the establishment of organ-specific cues which might induce the organ-specific differentiation of parenchymal cells[29]. Therefore, such lectins might serve as markers for organ-specific extracellular matrix molecules.

4. Extracellular-matrix resident cues: Over the past decade, numerous investigators have shown that extracellular matrix proteins forming the subendothelial basement membrane contain epigenetic cues, which on their own or in conjunction with soluble factors secreted from neighboring cells can induce EC phenotypic differentiation[45,46].

One of the most remarkable modulations of EC phenotype *in vitro* is the (ECM protein-mediated) transition from a two-dimensional (monolayer) culture to the formation of three-dimensional tubes (*in vitro* angiogenesis)[47]. Such a phenotypic transdifferentiation was first observed to occur spontaneously in "aged, post-confluent" cultured EC[48]. More recently, spontaneous *in vitro* angiogenesis was also observed when EC were confined to or induced to migrate into 3-D collagen or fibrin gels.[49,50]

The involvement of ECM-derived cues in this process is suggested by the fact that profound quantitative and qualitative changes occur in the composition of the subendothelial ECM during the "maturation" of EC cultures *in vitro*. Remarkably, some of these changes in ECM proteins quite accurately reflect similar changes in ECM composition during angiogenesis *in vivo*, e.g. in the chick chorioallantoic membrane[51]. However, as a caveat, not all EC are equal in their ability to form "tubes *in vitro*", even though they might be derived from the same vessel[52]. For example, when plated onto a murine EHS tumor-derived basement membrane (Matrigel), large vessel-derived EC form chord-like structures within 12-16 hours[53]; by contrast, microvascular EC assemble into complex networks in as little as 3-4 hours[24,54], suggesting a remarkable degree of heterogeneity in the mechanisms of perception and transduction of "angiogenic" signals in the various EC.

At present there is little information as to the molecular and cellular mechanisms that modulate a particular EC phenotype and lead to the induction of *in vitro* angiogenesis[53,55,56]. The angiogenic EC phenotype seems to be associated with the expression of a particular set of oncogenes and/or differential activation of some not (yet) further characterized genes and gene products (see paper by Grant et al., this volume). However, little is known about differences in the signal transduction mechanisms related to EC heterogeneity, in particular to the transformation from a quiescent to the angiogenic phenotype. Initial studies suggest some involvement of protein kinase C in angiogenesis, both *in vitro*[57] and, presumably, also *in vivo*[58] (see also paper by Maragoudakis et al. this volume).

HETEROGENEITY OF ENDOTHELIAL CELL SIGNAL TRANSDUCTION

Most of the known signal transduction pathways have been shown to operate in EC and to mediate their responses to extracellular stimuli[59]. We focused our attention on the adenylyl cyclase (AC) signaling system: in response to various extracellular stimuli, this plasma membrane-bound enzyme is activated via G-proteins and catalyzes the production of cAMP from ATP[60]. Cyclic AMP then activates cAMP-dependent protein kinases which in turn phosphorylate several substrates in the cytoplasm, the nucleus and the cell membrane[61,62]. In addition, cAMP is degraded by cAMP-specific phosphodiesterases (PDE).

Cyclic AMP regulates multiple ubiquitous cellular functions like growth, proliferation, differentiation, relaxation, secretion, etc. In addition, cAMP is involved in the regulation of several EC-specific functions. For example, elevation of intracellular cAMP level results to an increase in the expression of thrombomodulin and a decrease in the expression of tissue factor activity, on the EC surface[63,64]. Further, cAMP increases the synthesis of 13-hydroxyoctadecadienoic acid[65], and also affects the balance of tPA/PAI-1 secretion[66]. These EC-specific effects suggest that elevation of cAMP might result in a coordinate enhancement of the antithrombotic property of the endothelium.

As mentioned earlier in this paper, EC derived from various locations within the vascular tree exhibit several differences in both morphology and function. As cAMP is an active participant in the regulation of many of these processes, it is tempting to speculate that differences in the cAMP signaling might be related to EC heterogeneity. In support of that notion, some earlier reports have noted that the effect of cAMP on EC proliferation depends critically on their origin: elevation in cAMP inhibits the growth of bovine aortic EC[67], has no effect on human umbilical vein EC, while it accelerates the rate of proliferation of human dermal microvascular EC[68].

We have formulated a working hypothesis relating EC phenotypic diversity to the heterogeneity of the signal transduction machinery in particular to that of AC signaling. To test this hypothesis and to identify at which step(s) of the AC cascade this heterogeneity might occur, we stimulated EC of various origin with compounds that interfere at specific sites of the cascade and measured the intracellular levels of cAMP before and after stimulation. The methods used in our laboratory for measuring the cAMP levels in EC have recently been described in detail[69].

As mentioned above, the intracellular levels of cAMP are regulated by the concerted action of AC and PDE. In some cases, a stimulus might activate AC resulting in an increase in cAMP which then initiates a cellular response. However, this increase could be "masked" by parallel activation of PDE resulting in a quick degradation of cAMP. For that reason, our experiments measuring cAMP accumulation were done both in the presence and absence of the potent, ubiquitous PDE inhibitor 3-isobutyl-1-methyl-xanthine (IBMX). In the absence of IBMX, the basal levels of cAMP were similar in pig pulmonary artery endothelial cells (PPAEC), bovine aortic endothelial cells (BAEC), and human umbilical vein endothelial cells (HUVEC), but significantly higher in human adipose microvascular endothelial cells (HAMVEC) (figure 5). In the presence of IBMX, basal levels of cAMP were elevated and varied considerably among these four cell types (Figure 5). These results suggest a differential degree of involvement of the PDE in the regulation of basal cAMP levels in EC isolated from various locations within the vascular tree.

We recently reported that the levels of cAMP in PPAEC, BAEC, HUVEC, and HAMVEC stimulated with forskolin, in the presence of IBMX, exhibit considerable differences ranging from 4- fold increase over basal in PPAEC to almost 50- fold increase in HAMVEC[69]. When we compared the effect of forskolin on these cells in the presence and absence of IBMX, differences in the ability of PDE to degrade cAMP have emerged (Figure 6). Taken together, these results constitute evidence for heterogeneity in the regulation of cAMP levels in phenotypically diverse EC at the PDE level.

We also studied the effect of a physiological cyclic strain regimen (24% maximal strain, 60 cycles/min) in the same EC types. Mechanical stimulation of EC resulted in a time dependent increase in cAMP levels which reached a maximum after 5 min of stimulation in BAEC, PPAEC, and HAMVEC while in HUVEC, no significant increase in cAMP levels was detected even after 15 min of stimulation[69]. However, when the combined action of cyclic strain and forskolin was assessed, a potentiation of the effect of forskolin by cyclic strain was found in HUVEC but not in the other EC types[69]. These results raise the possibility of differential regulation of the cAMP signaling in phenotypically diverse EC by mechanical forces.

Possible sites of heterogeneity of the AC signaling system in endothelial cells

Based on our results as well as emerging evidence in the literature, we have compiled a working model which comprises the multitude of possible sites of heterogeneity in the AC signaling machinery in endothelial cells (Figure 7). These sites include:

1. G-protein coupled-receptors: The occurence of several types of receptors positively coupled to AC through G_s-proteins has been shown in endothelial cells[70]. It is possible that differences in the number of receptor sites present in each EC type or differences in the efficiency of their coupling to AC might contribute to heterogeneity of the AC signaling system in various types of endothelial cells. We will use the examples of ß-adrenergic and H_2-histamine receptors to illustrate this point:

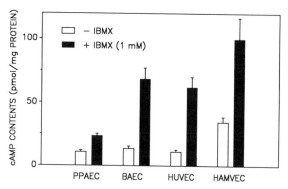

Figure 5: Basal cAMP levels in phenotypically diverse endothelial cells in the presence and absence of 1mM IBMX.

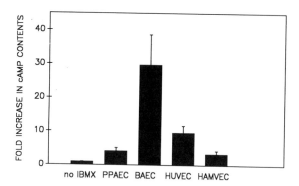

Figure 6: Effect of 1mM IBMX on 100μM forskolin-stimulated cAMP levels in phenotypically diverse endothelial cells.

a) ß-adrenergic receptors: ß-adrenergic receptors have been described on EC from many species including humans[70]. While most of those studies were done in EC from large vessels, a recent report confirms their presence also on cerebral microvascular EC[71]. However, to date, no studies comparing the ability of ß-adrenergic stimulation to raise cAMP in different EC types have been reported. Recently, we investigated the effect of the ß-adrenergic agonist isoproterenol on the intracellular cAMP levels in the above mentioned EC types and found differences in both the sensitivity and the responsiveness between the various cell types[72]. Thus, it appears that ß-adrenergic receptors might be a site of heterogeneity of the AC signaling system in endothelial cells.

b) H₂-histamine receptors: It has been reported that histamine increases cAMP levels in BAEC[73] and HUVEC[74] via activation of H_2 receptors. In BAEC, the H_2-receptor type seems to be the predominant (if not exclusive) type responsive to histamine stimulation, yielding a 4 fold increase in cAMP over basal levels[73]. On the other hand, in HUVEC, there is an abundance of H_1 receptors and the increase in cAMP due to maximal histamine stimulation is modest (approx.

1.5 fold over basal levels)[74]. We recently confirmed these findings and went on to show that in two other EC types (PPAEC and HAMVEC), histamine has no effect on cAMP levels[72]. Therefore, we suggest that H_2 receptors are another candidate for heterogeneity of the AC signaling system in various types of EC at the receptor level.

2. G-Proteins: The activity state of AC is dually regulated by stimulatory (G_s) and inhibitory (G_i)-proteins[60]. The α subunits of G_s-and G_i-proteins contain sites which can be ADP-ribosylated by cholera toxin (α_s) or by pertussis toxin (α_i), respectively. Therefore, these toxins provide us with excellent tools for the study of the G-protein involvement in the regulation of the activity of adenylyl cyclase.

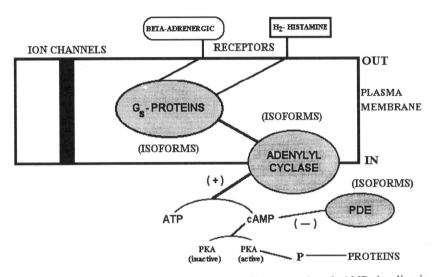

Figure 7: Working model for the possible sites of heterogeneity of cAMP signaling in endothelial cells.

Several substrates for bacterial toxin-mediated ADP-ribosylation have been found in endothelial cells. For example, in human and bovine pulmonary arterial EC, cholera toxin induces the ADP- ribosylation of 39-, 45-, and 52-kD proteins which were confirmed immunologically to be G_s-related proteins[75], although in an earlier study[76], no substrate ADP-ribosylated by cholera toxin could be detected in bovine pulmonary arterial EC. To the best of our knowledge, no studies on the ADP-ribosylation substrates of cholera toxin in arterial and microvascular EC have been published. We have recently obtained results showing considerable quantitative differences among BAEC, PPAEC, HUVEC, and HAMVEC in cAMP accumulation following cholera toxin stimulation[72]. These results suggest that differencial regulation of the AC signaling system in various EC types might also be occuring at the G-protein level.

3. Adenylyl Cyclase Isoforms: Recently, a new aspect in the complexity of the AC signaling system has been introduced at the level of adenylyl cyclase. Over the past 5 years, cDNAs encoding eight AC isoforms have been identified in mammalian cells[77]. Since the physiological actions of agonists that bind to AC-stimulating receptors are quite disparate, it is tempting to speculate that some of these receptors might activate different AC isoform(s). However, at present no studies are available on the AC isoforms in endothelial cells. We have obtained preliminary evidence, using RT-PCR analysis, that the AC isoform content of EC derived from various rat tissues is not identical (Manolopoulos VG, Liu J and Lelkes PI, unpublished observations).

4. Mechanical stimuli: The mechanism(s) by which vascular EC perceive and transduce signals generated by hemodynamic forces may involve a variety of intracellular second messenger systems, including calcium, PKC and also cAMP[78]. We recently showed differential regulation of the AC signaling system in phenotypically diverse EC by mechanical forces[69]. Our results extend and explain previous observations by Sumpio and colleagues, who reported an increase in AC activity and intracellular cAMP levels induced by cyclic strain in BAEC[79], but found no effect of cyclic strain in EC from human saphenous vein[80]. It is currently not known how the AC of vascular cells perceives the mechanical signal induced by cyclic strain. It could be a result of the activation of one or some of the above mentioned components of the AC cascade, such as specific plasma membrane receptors (mechanoreceptors?) or physical perturbations of the plasma membrane resulting in direct activation of AC or G_s-proteins.

5. Cyclic nucleotide phosphodiesterases: This family of enzymes provides the only known mechanism of degrading cAMP and, as the majority of cells are unable to extrude cAMP, it supplies the cells with the only inactivation route for this second messenger. Five PDE isoenzymes have been found in mammalian cells[81]. Our results obtained with the potent, non-isoform specific PDE inhibitor IBMX show clear differences in the ability of PDEs to degrade both basal and stimulated cAMP in phenotypically diverse EC[69,72], and support the notion that heterogeneity in the AC signaling in phenotypically diverse EC might also occur at the PDE level. It could depend on the nature of stimulus, the isoform of AC involved, or it might be a result of different types of PDE present in each cell type. Studies with isoenzyme-specific PDE inhibitors are needed in order to elucidate this point.

CONCLUSIONS

Recent studies have solidified our notion of EC heterogeneity and organ-specificity. A variety of microenvironmental cues have been implicated in establishing and maintaining EC phenotypic diversity. At present, work is under way to establish by molecular, immunological and cell biological approaches endogenous EC-specific parameters which regulate the expression of a particular EC phenotype. Some of the issues to be tackled in the upcoming years include:

1. Isolation of EC-specific agonists: There are many ubiquitous differentiation factors *in vivo*, which have shown promise as "angiogenic factors" *in vitro*. What are the factors (alone or in combination), which specifically trigger endothelial cell heterogeneity and organ-specificity *in vivo*? In particular, which factor(s) will induce angiogenesis by switching on the EC angiogenic phenotype?

2. The issue of *in vitro* angiogenesis raises a host of questions, for example: What is the "correct" EC phenotype to be studied? How can we maintain and study a particular EC phenotype in culture? Is there an EC phenotypic transdifferentiation, for example can a "generic" HUVEC become the equivalent of an organ-specific microvessel-derived EC? How sophisticated should we get in combining microenvironmental cues, such as ECM, cytokines, and mechanical forces, in order to attain a proper model for studying EC biology *in vitro*?

3. How do the various EC phenotypes differ in terms of cellular and molecular biology? Which are the unique genes and gene products which characterize a particular (angiogenic, activated, pathological, etc.) EC phenotype? What is the control mechanism that switches such genes on or off? How do these various EC differ in the ways in which they perceive and transduce external stimuli?

Clearly, answers to these questions will have far reaching implica- tions for our understanding of fundamental mechanisms of EC heterogeneity and organ-specificity, also from the vantage point of angiogenesis. In particular, we should learn not to expect identical or even similar responses when testing some of the "angiogenic" agents in diverse vascular beds *in vivo* or when comparing "generic" cultured EC vs. microvascular EC isolated from different tissues.

ACKNOWLEDGEMENTS

The original work described in this paper was supported by grants in aid (to PIL) from the American Heart Association (National Center and Wisconsin Affiliate).

REFERENCES

1. **Lelkes, P.I.** 1991. New aspects of endothelial cell biology. *J. Cell. Biochem.* 45:242-244.
2. **Simionescu, N. and M. Simionescu (eds).** 1992. Endothelial cell dysfunctions. Plenum Press, New York, NY.
3. **Noden, D.M.** 1989. Embryonic origins and assembly of blood vessels.*Am. Rev. Respir. Dis.* 140:1097-1103.
4. **Simionescu, N. and M. Simionescu.** 1988. The cardiovascular system.In *Cell and tissue biology: a textbook of histology.* L. Weiss, editor. Urban & Schwarzenberg, Baltimore, MD. pp 353-400.
5. **Bumbasirevic, V., G.D. Pappas, and R.P. Becker.** 1990. Endocytosis of serum albumin-gold conjugates by microvascular endothelial cells in rat adrenal gland: regional differences between cortex and medulla. *J. Submicrosc. Cytol. Pathol.* 22:135-145.
6. **Repin, V.S., V.V. Dolgov, O.E. Zaikina, I.D. Novikov, A.S. Antonov,M.A. Nokolaeva, and V.N. Smirnov.** 1984. Heterogeneity of endothelium in human aorta: a quantitative analysis by scanning electron microscopy. *Atherosclerosis* 50:35-52.
7. **McCarthy, S.A., I. Kuzu, K.C. Gatter, and R. Bicknell.** 1991. Heterogeneity of the endothelial cell and its role in organ preference of tumor metastasis. *Trends Pharmacol. Sci.* 12:462-467.
8. **Allsup, D.J. and M.R. Boarder.** 1990. Comparison of P_2 purinergic receptors of aortic endothelial cells with those of adrenal medulla: evidence for heterogeneity of receptor subtype and of inositol phosphate response. *Mol. Pharmacol.* 38:84-91.
9. **Bossu, J.L., a. Elhamdani, and A. Feltz.** 1992. Voltage-dependent calcium entry in confluent bovine capillary endothelial cells. *FEBS Let.* 299(3):239-242.
10. **Homma, S., Y. Miyauchi, Y. Sugishita, K. Goto, M. Sato, and N. Ohshima.** 1992. Vasoconstrictor effects of endothelin-1 on myocardium microcirculation studied by the Langendorff perfusion method: differential sensitivities among microvessels. *Microvasc. Res.* 43:205-217.
11. **Speiser, W., E. Anders, K.T. Preissner, O. Wagner, and G. Muller-Berghaus.** 1987. Differences in coagulant and fibrinolytic activities of cultured human endothelial cells derived from omental tissue microvessels and umbilical veins. *Blood* 69:964-967.
12. **Zetter, B.R.** 1988. Endothelial heterogeneity: influence of vessel size, organ localization, and species specificity on the properties of cultured endothelial cells. In *Endothelial cells: volume II.* U.S. Ryan, editor. CRC Press, Boca Raton, FL. pp 63-79.
13. **Dupuy, E., A. Bikfalvi, F. Rendu, S.L. Toledano, and G. Tobelem.** 1989. Thrombin mitogenic responses and protein phosphorylation are different in cultured human endothelial cells derived from large and microvessels. *Exp. Cell. Res.* 185:363-372.
14. **Toda, N., T. Matsumoto, and K. Yoshida.** 1992. Comparison of hypoxia-induced contraction in human, monkey and dog coronary arteries. *Am. J. Physiol.* 262:H678-H683.
15. **Nerem, R.M. and P.R. Girard.** 1990. Hemodynamic influences on vascular endothelial biology. *Toxic. Path.* 18:572-582.
16. **Nollert, M.U., N.J. Panaro, and L.V. McIntire.** 1992. Regulation of genetic expression in shear stress-stimulated endothelial cells. *Ann. NY Acad. Sci.* 665:94-104.
17. **Lamontagne, D., U. Pohl, and R. Busse.** 1992. Mechanical deformation of vessel wall and shear stress determine the basal release of endothelium-derived relaxing factor in the intact rabbit coronary vascular bed. *Circ. Res.* 70:123-130.
18. **Malek, A. and S. Izumo.** 1992. Physiological fluid shear stress causes downregulation of endothelin-1 mRNA in bovine aortic endothelium. *Am. J. Physiol.* 263:C389-C396.
19. **Diamond, S.L., S.G. Eskin, and L.V. McIntire.** 1989. Fluid flow stimulates tissue plasminogen activator secretion by cultured human endothelial cells. *Science* 243:1483-1485.

20. **Grabowski, E.F., E.A. Jaffe, and B.B. Weksler.** 1985. Prostacyclin production by cultured endothelial cell monolayers exposed to step increases in shear stress. *J. Lab. Clin. Med.* 105:36-43.

21. **Sumpio, B.E.(ed.)** 1993. Hemodynamic forces and vascular cell biology. R.G. Landes Company, Austin, TX.

22. **Iba, T., S. Maitz, A. Vogt, O. Rosales, M. Widmann, and B.E. Sumpio.** 1990. Alignment of human endothelial cells (EC) with cyclic stretch in vitro. *FASEB J.* A415. Abstract.

23. **Spanel-Borowski, K. and J. van der Bosch.** 1990. Different phenotypes of cultured microvessel endothelial cells obtained from bovine corpus luteum. *Cell. Tissue. Res.* 261:35-47.

24. **Carley, W.W., M.J. Niedbala, and M.E. Gerritsen.** 1992. Isolation, cultivation and partial characterization of microvascular endothelium derived from human lung. *Am. J. Respir. Cell. Mol. Biol.* 7:620-630.

25. **Orlidge, A. and P.A. D'Amore.** 1987. Inhibition of capillary endothelial cell growth by pericytes and smooth muscle cells. *J. Cell Biol.* 105:1455-1462.

26. **Tagami, M., K. Yamagata, H. Fujino, A. Kubota, Y. Nara, and Y. Yamori.** 1992. Morphological differentiation of endothelial cells co-cultured with astrocytes on type-I or type-IV collagen. *Cell. Tis. Res.* 268:225-232.

27. **Moisseiev, J., J.A. Jerdan, K. Dyer, A. Maglione, and B.M. Glaser.** 1990. Retinal pigment epithelium cells can influence endothelial cell plasminogen activators. *Invest. Ophthalmol. Vis. Sci.* 31:1070-1078.

28. **Mizrachi, Y., J. Narranjo, B.-Z. Levi, H.B. Pollard, and P.I. Lelkes.** 1990. PC12 cells differentiate into chromaffin cell like phenotype in co-culture with adrenal medullary endothelial cells. *Proc. Natl. Acad. Sci. USA* 87:6161-6165.

29. **Lelkes, P.I. and B.R. Unsworth.** 1992. Role of heterotypic interactions between endothelial cells and parenchymal cells in organospecific differentiation: a possible trigger of vasculogenesis. In *Angiogenesis in health and disease.* M.E. Maragoudakis, P. Gullino, and P.I. Lelkes, eds. Plenum Press, New York, NY. pp 27-43.

30. **Nishida, M., J.P. Springhorn, R.A. Kelly, and T.W. Smith.** 1993. Cell-cell signaling between adult rat ventricular myocytes and cardiac microvascular endothelial cells in heterotypic primary culture. *J. Clin. Invest.* 91:1934-1941.

31. **Youdim, M.B.H., D.K. Banerjee, K. Kelner, L. Offutt, and H.B. Pollard.** 1989. Steroid regulation of monamine oxidase activity in the adrenal medulla. *FASEB J.* 3:1753-1759.

32. **Hart, T.K. and R.M. Pino.** 1986. Pseudoislet vascularization: induction of diaphragm-fenestrated endothelia from the hepatic sinusoids. *Lab. Invest.* 54:304-313.

33. **Jaeger, C.B.** 1991. Fenestration of cerebral microvessels induced by PC12 cells grafted to the brain of rats. *Ann. NY Acad. Sci.* 361-364.

34. **Saunder, K.B. and P.A. D'Amore.** 1992. An in vitro model for cell-cell interactions. *In Vitro Cell. Dev. Biol.* 28A:521-528.

35. **Alby, L. and R. Auerbach.** 1984. Differential adhesion of tumor cells to capillary endothelial cells in vitro. *Proc. Natl. Acad. Sci. USA* 81:5739-5743.

36. **Berg, E.L., L.A. Goldstein, M.A. Jutila, M. Nakache, L.J. Picker, P.R. Streeter, N.W. Wu, D. Zhou, and E.C. Butcher.** 1989. Homing receptors and vascular addressins: cell adhesion molecules that direct lymphocyte traffic. *Immunol. Rev.* 108:5-18.

37. **Lasky, L.A.** 1991. Lectin cell adhesion molecules (LEC-CAMs): a new family of cell adhesion proteins involved with inflammation. *J. Cell. Biochem.* 45:139-146.

38. **Pauli, B.U., H.G. Augustin-Voss, M.E. el-Sabbah, R.C. Johnson, and D.A. Hammer.** 1990. Organ-preference of metastasis: the role of endothelial cell adhesion molecules. *Cancer Metastasis Rev.* 9:175-189.

39. **Auerbach, R.** 1991. Interactions between cancer cells and the endothelium. In *Microcirculation in cancer metastasis.* F.W. Orr, M. Buchanan, and L. Weiss, eds. CRC Press, Boca Raton, FL. pp 169-181.

40. **Auerbach, R., L. Alby, L.W. Morrissey, M. Tu, and J. Joseph.** 1985. Expression of organ-specific antigens on capillary endothelial cells. *Microvasc. Res.* 29:401-411.

41. **Risau, W., R. Hallmann, U. Albrecht, and S. Henke-Fahle.** 1986. Brain induces the expression of an early cell surface marker for blood-brain barrier-specific endothelium. *EMBO J.* 1:3179-3183.

42. **Pauli, B.U. and C.L. Lee.** 1988. Organ preference of metastasis: the role of organ-specifically modulated endothelial cells. *Lab. Invest.* 58:379-387.

43. **Augustin-Voss, H.G., R.C. Johnson, and B.U. Pauli.** 1991. Modulation of endothelial cell surface glycoconjugate expression by organ-derived biomatrices. *Exp. Cell Res.* 192:346-351.

44. **Plendl, J., L. Hartwell, and R. Auerbach.** 1993. Organ-specific change in *Dolichos biflorus* lectin binding by myocardial endothelial cells during in vitro cultivation. *In Vitro Cell. Dev. Biol.* 29A:25-31.

45. **Madri, J.A., L. Bell, M. Marx, J.R. Merwin, C. Basson, and C. Prinz.** 1991. Effects of soluble factors and extracellular matrix components on vascular cell behavior in vitro and in vivo: models of de-endothelialization and repair. *J. Cell. Biochem.* 45:123-130.

46. **Schubert, D.** 1992. Collaborative interactions between growth factors and the extracellular matrix. *Trends cell biol.* 2:63-66.

47. **Montesano, R.** 1992. Regulation of angiogenesis in vitro. *Eur. J. Clin. Invest.* 22:504-515.

48. **Folkman, J. and C. Haudenschild.** 1980. Angiogenesis in vitro. *Nature* 288:551-556.

49. **Montesano, R., L. Orci, and P. Vassalli.** 1983. In vitro rapid organization of endothelial cells into capillary-like networks is promoted by collagen matrices. *J Cell Biol.* 97:1648-1652.

50. **Fournier, N. and C.J. Doillon.** 1992. In vitro angiogenesis in fibrin matrices containing fibronectin or hyaluronic acid. *Cell Biol. Int. Rep.* 16:1251-1263.

51. **Papadimitriou, E., B.R. Unsworth, M.E. Maragoudakis, and P.I. Lelkes.** 1993. Time-course and quantitation of extracellular matrix maturation in the chick chorioallantoic membrane and in cultured endothelial cells. *Endothelium,* in press.

52. **Canfield, A.E., F.E. Wren, S.L. Schor, M.E. Grant, and A.M. Schor.** 1992. Aortic endothelial cell heterogeneity in vitro. Lack of association between morphological phenotype and collagen biosynthesis. *J. Cell Sci.* 102:807-814.

53. **Grant, D.S., P.I. Lelkes, K. Fukuda, and H.K. Kleinman.** 1991. Intracellular mechanisms involved in basement membrane induced blood vessel differentiation in vitro. *In Vitro Cell. Dev. Biol.* 27A:327-336.

54. **Nishida, M., W.M. Carley, M.E. Gerritsen, O. Ellingsen, R.A. Kelly, and T.W. Smith.** 1993. Isolation and characterization of human and rat cardiac microvascular endothelial cells. *Am. J. Physiol.* 264:H639-H652.

55. **Carley, W.W., A.J. Milici, and J.A. Madri.** 1988. Extracellular matrix specificity for the differentiation of capillary endothelial cells. *Exp. Cell Res.* 178:426-434.

56. **Pepper, M.S., and R. Montesano.** 1990. Proteolytic balance and capillary morphogenesis. *Cell Differ. Dev.* 32:319-328

57. **Kinsella, J.L., D.S. Grant, B.S. Weeks, and H.K. Kleinman.** 1992. Protein kinase C regulates endothelial cell tube formation on basement membrane matrix, matrigel. *Exp. Cell Res.* 199:56-62.

58. **Wright, P.S., D. Cross-Doersen, J.A. Miller, W.D. Jones, and A.J. Bitonti.** 1992. Inhibition of angiogenesis in vitro and in ovo with an inhibitor of cellular protein kinases, MDL 27032. *J. Cell. Physiol.* 152:448-457.

59. **Catravas, J.D., C.N. Gills, and U.S. Ryan (eds).** 1989. Vascular endothelium: receptors and transduction mechanisms. Plenum Press, New York, NY.

60. **Levitzki, A.** 1988. From epinephrine to cyclic AMP. *Science* 241:800-806.

61. **Shenolikar, S.** 1988. Protein phosphorylation: hormones, drugs, and bioregulation. *FASEB J.* 2:2753-2764.

62. **Lee, K.A.W.** 1991. Transcriptional regulation by cAMP. *Curr. Opinion Cell Biol.* 3:953-959.

63. Ishii, H., K. Kizaki, H. Uchiyama, S. Horie, and M. Kazama. 1990. Cyclic AMP increases thrombomodulin expression on membrane surface of cultured human umbilical vein endothelial cells. *Thromb. Res.* 59:841-850.

64. Lybert, G. 1984. Intercellular signal mechanisms in induction of thromboplastin synthesis. *Haemostasis* 14:393-399.

65. Haas, T.A., M.C. Bertomeu, E. Bastida, and M.R. Buchanan. 1990. Cyclic AMP regulation of triacylglycerol turnover, 13-hode synthesis and endothelial cell thrombogenicity. *Biochem. Biophys. Acta* 1051:174-178.

66. Santell, L. and E.G. Levin. 1988. Cyclic AMP potentiates phorbol ester stimulation of t-PA release and inhibits secretion of PA inhibitor-1 from human endothelial cells. *J. Biol. Chem.* 263:16802-16808.

67. Leitman, D.C., R.P. Fiscus, and F. Murad. 1986. Forskolin, phosphodiesterase inhibitors and cyclic AMP analogs inhibit proliferation of cultured bovine aortic endothelial cells. *J. Cell. Physiol.* 127:237-243.

68. Davison, P.M. and M.A. Karasek. 1981. Human dermal microvascular endothelial cells in vitro: effect of cyclic AMP on cellular morphology and proliferation rate. *J. Cell. Physiol.* 106:253-258.

69. Manolopoulos, V.G. and P.I. Lelkes. 1993. Cyclic strain and forskolin differentially induce cAMP production in phenotypically diverse endothelial cells. *Biochem. Biophys. Res. Commun.* 191:1379-1385.

70. McEwan, J.R., H. Parsaee, D.C. LeFroy, and J. McDermot. 1990. Receptors linked to adenylate cyclase in endothelial cells. In *The Endothelium: An Introduction to Current Research.* J.B. Warren, editor. Wiley-Liss, Inc., 45-51.

71. Bacic, F., R.M. McCarron, S. Uematsu, and M. Spatz. 1992. Adrenergic receptors coupled to adenylate cyclase in human cerebromicrovascular endothelium. *Metab. Brain Dis.* 7(3):125-137.

72. Manolopoulos, V.G., M.M. Samet, and P.I. Lelkes. 1993. Regulation of the adenylyl cyclase signalling system in various types of cultured endothelial cells. Submitted.

73. Hekimian, G., S. Cote, J.V. Sande, and J.M. Boeynaems. 1992. H_2 receptor-mediated responses of aortic endothelial cells to histamine. *Am. J. Physiol.* 262:H220-H224.

74. Takeda, T., Y. Yamashita, S. Shimazaki, and Y. Mitsui. 1992. Histamine decreases the permeability of an endothelial cell monolayer by stimulating cyclic AMP production through the H_2-receptor. *J. Cell. Sci.* 101:745-750.

75. Garcia, J.G.N., and V. Natarajan. 1992. Signal transduction in pulmonary endothelium (implications for lung vascular dysfunction). *Chest* 102:592-607.

76. Voyno-Yasenetskaya, T.A., V.A. Tkachuk, E.G. Cheknyova, M.P. Panchenko, G.Y. Grigiorian, R.J. Vavrek, J.M. Stewart, and U.S. Ryan. 1989. Guanine nucleotide-dependent, pertussis toxin-insensitive regulation of phosphoinositide turnover by bradykinin in bovine pulmonary artery endothelial cells. *FASEB J.* 3:44-51.

77. Iyengar, R. 1993. Molecular and functional diversity of mammalian G_s-stimulated adenylyl cyclases. *FASEB J.* 7:768-775.

78. Davies, P.F. and S.C. Tripathi. 1993. Mechanical stress mechanisms and the cell. *Circ. Res.* 72:239-245.

79. Letsou, G.V., O. Rosales, S. Maitz, A. Vogt, and B.E. Sumpio. 1990. Stimulation of adenylate cyclase activity in cultured endothelial cells subjected to cyclic stretch. *J. Cardiovasc. Surg.* 31:634-639.

80. Iba, T., I. Mills, and B.E. Sumpio. 1992. Intracellular cyclic AMP levels in endothelial cells subjected to cyclic strain in vitro. *J. Surg. Res.* 52:625-630.

81. Giembycz, M.A. 1992. Could isoenzyme-selective phosphodiesterase inhibitors render bronchodilator therapy redundant in the treatment of bronchial asthma? *Biochem. Pharmacol.* 43:2041-2051.

INHIBITORS OF NEOVASCULARIZATION: CRITICAL MEDIATORS IN THE COORDINATE REGULATION OF ANGIOGENESIS

Peter J. Polverini

University of Michigan School of Dentistry
Section of Oral Pathology
Laboratory of Molecular Pathology, Rm 5217
Ann Arbor , Michigan 48109-1078

INTRODUCTION

The processes of tissue regeneration and repair, the cyclical proliferation of the nutrient-rich endometrial lining in preparation for implantation of the fertilized egg; and the complex developmental program that characterizes embryogenesis are biological processes that are strictly dependent on the rapid yet temporary ingrowth of new capillary blood vessels. In contrast, disorders such as neoplasia, proliferative vascular lesions, rheumatoid arthritis and psoriasis, and glaucoma are all characterized by disregulated angiogenesis. The mechanisms underling inappropriate neovascularization have been the subject of considerable investigation. Although there is ample evidence implicating the "overproduction" of normal and/or aberrant forms of angiogenic mediators in the pathogenesis of several well characterized disorders, only recently has attention been given to the role of naturally occurring inhibitors of angiogenesis and the consequences that result from a deficiency in the production of one or more of these "angiostatic" mediators (Moses and Langer, 1991a; Bouck, 1990 and 1993; DiPietro and Polverini, 1993, in press). This report will describe recent studies that support the assertion that angiogenesis is a process that is dependent upon the coordinate production of growth stimulatory and inhibitory molecules and that any disruption in this finely tuned regulatory circuit can result in the development of a number of diseases now classified as "angiogenesis-dependent".

INHIBITORS OF ANGIOGENESIS: MEDIATORS THAT REGULATE THE TEMPORAL AND SPATIAL DISTRIBUTION OF CAPILLARY BLOOD VESSELS

In the absence of disease neovascularization occurs relatively infrequently in normal adult tissues. The rate of capillary endothelial cell turn over in adult organisms is typically measured in months or years (Engerman et al., 1967; Tannock and Hayashi, 1972). When stimulated however normally quiescent endothelial cells that line venules, will systematically degrade their basement membrane and proximal extracellular matrix, migrate directionally, divide, and organize into new functioning capillaries all within a matter of days. This dramatic amplification of the microvasculature is nevertheless temporary for as rapidly as they are formed they virtually disappear with similar swiftness, returning the tissue vasculature to the status quo. It is this unique feature of orderly growth and subsequent regression of capillaries that primarily distinguishes physiological from pathological angiogenesis. While the mechanisms responsible for the dramatic upregulation of new capillary growth has been the subject of extensive investigation for the past twenty years (Folkman and Klagsbrun, 1987; Klagsbrun and Folkman, 1990; Klagsbrun and D'Amore, 1991), only recently has attention focused on the mechanisms and mediators responsible for the timely down-regulation of angiogenesis (Bouck 1990; Klagsbrun and D'Amore 1991; Moses and Langer 1991a).

One of the earliest observations implicating inhibitory molecules in the control of neovascularization was made by Brem and Folkman (1975) and Langer et al, (1976).They showed that fragments of cartilage when place in close proximity to tumors placed on either the chick chorioallantoic membrane or implanted into the rabbit cornea, were able to potently block tumor neovascularization and growth. Cartilage and cartilage-like extracts derived from several other animal sources were subsequently found to have similar angioinhibitory activity (Eisenstein et al., 1975; Pauli et al., 1981; Lee and Langer, 1983). These cartilage-derived inhibitors, CDIs, have recently been characterized and reported to belong to the metalloproteinase family of enzyme inhibitors (Moses and Langer, 1991b). These early studies helped establish the emerging field of "antiangiogenesis" research. Table I is a list of some of the more well characterized naturally occurring inhibitors of angiogenesis. Many of these inhibitors however are anything but endothelial-specific. Most of them have a broad range of biological activities and since few of these agents have been thoroughly characterized their mechanism of action and full spectrum of biological functions remain to be determined.

Table I. Endogenous Inhibitors of *in Vivo* Angiogenesis

Angiostatic steroids & sulfated polysaccharides (Crum et al., 1985; Folkman et al., 1989; Inoue et al., 1988)
Eosinophilic major basic protein (Blood and Zetter, 1990)
High molecular weight hyaluronan (West and Kumar, 1989)
Interferons (Sidky and Borden, 1987; Orchard et al., 1989; White et al., 1991)
Interleukin-1 (Cozzolino et al., 1990)
Laminin peptides (Grant et al., 1989)
Placental RNase (angiogenin) inhibitor (Shapiro and Vallee, 1987)
Platelet factor 4 (Maione et al., 1990)
Prostaglandin synthesis inhibitors (Peterson 1986, Haynes et al., 1989; Lynch et al., 1978; Ziche et al., 1982; Robin et al., 1985)
Protamine (Taylor and Folkman 1982)
Somatostatin (Woltering et al., 1991)
Thrombospondin 1 (Rastinejad et al., 1989; Good et al., 1991)
Tissue inhibitors of metalloproteinases (Moses and Langer, 1991b)
Vitamin A & retinoids (Ingber and Folkman 1988; Arensman and Stolar, 1979; Oikawa et al., 1989)
Vitreous fluid (Brem et al., 1977; Lutty et al., 1983; Taylor and Weiss, 1985)

THROMBOSPONDIN 1: A PROTOTYPE INHIBITOR OF ANGIOGENESIS

Thrombospondin (TSP) is a large trimeric 450 kD molecule of mosaic construction and function. It is present in great abundance in the platelet α granules and is secreted by a wide variety of epithelial and mesenchymal cells (Lawler, 1986; Sage and Bornstein, 1991; Frazier, 1987; Frazier, 1991, Bornstein, 1992). It mediates cell to cell and cell to substrate interactions where it has been shown to have both adhesive and anti-adhesive properties. It also has been reported to promote neurite outgrowth and stimulate as well as inhibit cell proliferation and migration (Good et al., 1990; Sage and Bornstein, 1991). It is now known that there are three TSPs, TSP1, TSP2, and TSP3, each encoded by three homologous genes in both human and mouse and that differ in their amino acid composition and pattern of expression. Intuitively this would suggest that each of these closely related molecules has discrete functions.

One of the first reports implicating TSP as an inhibitor of angiogenesis came from the laboratory of Noel Bouck where an anti-angiogenic hamster protein whose secretion was controlled by a tumor suppressor gene was found to have an amino acid sequence similar to human platelet TSP1 (Rastinejad et al., 1989). Authentic TSP1 was then purified from platelets and shown to block neovascularization *in vivo* (Good et al., 1990). A role for TSP1 in the inhibition of angiogenesis is supported by several other observations. It is presence adjacent to mature quiescent vessels and is absence from actively growing sprouts both *in vivo* (O'Shea and Dixit, 1988) and *in vitro* (Iruela-Arispe et al., 1991). Hemangiomas which consist of rapidly proliferating endothelial cells fail to make detectable TSP1 (Sage & Bornstein, 1982). Antibodies to TSP1 added to endothelial cell cultures decrease the formation of endothelial sprouts *in vitro* (Iruela-Arispe et al., 1991) and endothelial cells in which TSP1 production has been blocked following the introduction of antisense TSP1 exhibit an accelerated rate of growth, enhanced chemotactic activity and an increase in the number of capillary-like cords [DiPietro et al, submitted for publication (a)]. Using small peptides derived from the parent TSP1 molecule and truncated versions of the TSP1 Tolsma et al., (1993, in press) have localized the inhibitory activity to two domains of the 70-kD central stock region, the procollagen homology region and the properdin-like type 1 repeats. How these domains actually inhibit neovascularization remains to be elucidated.

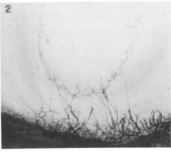

Figure 1. A positive rat corneal neovascular response 7 days after implanting a Hydron pellet containing 50 ng of bFGF.
Figure 2. A markedly suppressed neovascular response when 125 ng of TSP1 was combined with 50 ng of bFGF

There also appears to be a role for TSP1 in the control of wound neovascularization. Previous investigations have shown that macrophages, which are potent angiogenesis effector cells, also produce TSP1 (Jaffe et al., 1986). DiPietro and Polverini (1992, in press) have found an approximately 6-fold increase in steady state levels of TSP1 mRNA in the murine monocyte line WEHI-3 when the cells were treated for 24 hrs with the potent activating agent lipopolysaccharide (LPS) with peak secretion of TSP1 protein occurring by 8 hrs. The finding that activated macrophages produce the angiogenesis inhibitor TSP1 would seem paradoxical since induction of angiogenesis is a well established property of activated macrophages (Polverini, 1989). There are several possible explanation for this apparent functional dichotomy. The angiogenic potential of macrophages *in vivo* is most likely the result of the balanced production of both positive and negative regulators of angiogenesis. Alternatively macrophage-derived TSP1 may not exert significant effects on endothelial cells, particularly if TSP1 is rapidly degraded or sequestered into the extracellular matrix. In this instance the influence of activated macrophages may be swayed toward diffusable mediators of angiogenesis rather than inhibitors such as TSP1. Several other functions of macrophage derived TSP1 may influence wound repair and angiogenesis. The adhesive capacity of TSP1 may facilitate migration of activated macrophages. Macrophages have surface receptors for TSP1 (Silverstein and Nachman, 1987), and thus might lay down TSP1 upon the existing extracellular matrix as a scaffold upon which to migrate. A second possibility is that macrophages produce TSP1 as a functional protease inhibitor, a function that has recently been ascribed to TSP1 *in vitro* and which may be important to the activated macrophage as it produces soluble mediators (Hogg et al., 1992). Finally TSP1 has been shown to enhance neutrophil chemotaxis, and to facilitate the phagocytosis of senescent neutrophils by macrophages (Mansfield et al., 1990; Savill et al., 1991). The idea that activated macrophages might produce TSP1 to effect neutrophil migration and clearance is an intriguing one, as it suggests a functional link between macrophages and neutrophils and the inflammatory process. The potential significance of these observations to *in vivo* wound neovascularization is just beginning to emerge. More recently we have also observed a tight correlation between the level of TSP production by monocytes during wound repair and the timely down regulation of wound neovascularization (DiPietro and Polverini, unpublished observations). Further investigations of the significance of TSP1 production by monocytes and macrophages during the evolution and regression phases of angiogenesis may determine whether its production is a common characteristic of activated macrophages and other angiogenesis accessory cells. Such investigations will no doubt lead to a clearer understanding of the functional significance of TSP1, as well as other as yet unidentified inhibitory molecules in the angiogenic response.

DISPARATE DISEASES PROCESSES SHARE COMMON MECHANISMS OF DISREGULATED ANGIOGENESIS

The acquisition of angiogeneic activity is a hallmark of cells that have undergone neoplastic transformation. It can occur anytime during the initiation, promotion or progression phases of tumor development. It is well established that the "transformed phenotypes" which collectively are the signature of neoplastic cells are a consequence of a series of genetic lesion, some of which involve activation of dominantly acting oncogenes and with others resulting from the loss or inactivation of recessive genes termed tumor suppressor genes or "anti-oncogenes" (Moroco et al., 1990; Bouck, 1990 and Bouck, 1993). A number of well known angiogenic mediators appear to be overexpressed in transformed cells and at least three tumor derived oncogenes *int-* 2, *hst* /K-*fgf* /KS3, and FGF-5, encode proteins that are structurally and functionally homologous to the well known angiogenic mediator bFGF (Thomas, 1988 and Goldfarb, 1990). Still other oncogenes rather than inducing angiogenesis directly appear to recruit and/or induce other cells to produce mediators of angiogenesis.

Where do inhibitors or rather the lack of inhibitors fit into this scheme. This is where tumor suppressor gene enter the picture. In an elegant series of studies Bouck and colleagues (1986, 1989, 1990, and 1993 in press) demonstrated with two different hamster tumor systems that loss of a suppressor gene was directly linked to the acquisition of angiogenic activity. In one system cultured BHK fibroblasts that were transformed in a single step to anchorage independence and tumorigenicity, loss of a tumor suppressor gene was accompanied by the acquisition of angiogenic activity. The reintroduction of an active tumor suppressor gene was found to be associated with production of elevated levels of thrombospondin. These studies suggest that when the suppressor is inactivated, the levels of TSP decrease and the amount of secreted protein becomes inadequate to overcome constitutively produced growth factors (Bouck, 1993). Similarly during the *in vivo* progression of carcinogen-initiated hamster keratinocytes to squamous carcinomas, we have shown by cell fusion that a tumor suppressor gene is lost as cells undergo neoplastic conversion and gain the ability to induce angiogenesis (Moroco et al, 1990) and fail to produce an inhibitor of angiogenesis (Polverini, unpublished data). There are also several human tumor systems where expression of angiogenic activity coincidental with loss of an inhibitory factor appears to be under the control of a tumor suppressor gene (Bader et al., 1991, Huang et al., 1988, and Tolsma et al., unpublished data).

A combined defect in the overproduction of positive regulators of angiogenesis and a deficiency in inhibitors of this process is a feature not unique to tumor angiogenesis. We have recently reported that psoriasis, a common inherited inflammatory skin disease characterized by hyperkinetic epidermal keratinocytes and excessive angiogenesis of dermal vascular capillaries (Folkman, 1972) is associated with elevated levels of several proangiogenic mediators and a deficiency in the production of TSP1 (Nickoloff et al, submitted for publication). Using phenotypically "mutant" keratinocytes obtained from spontaneously arising psoriatic lesions we observed that psoriatic keratinocytes as well as keratinocytes from the symptomless skin of psoriatic patients exhibited a dramatic upregulation in the proangiogenic cytokine IL-8 and a two to eight fold decrease in TSP1 production (Koch et al., 1992, Strieter et al., 1992).

FUTURE CONSIDERATIONS AND THERAPEUTIC IMPLICATIONS

It is now well established that angiogenesis is a tightly regulated process that is under the control of both growth stimulatory and inhibitory molecules. Although the complement of positive and negative regulators of angiogenesis may vary among different physiologic and pathologic settings, the recognition of this dual mechanism of control is necessary if we are to gain a more thorough understanding of this complex process and its significance in disease. The recent elucidation of the cooperative interaction among positive and negative regulatory molecules during normal physiological angiogenesis and the apparent disruption of this program in disorders such as solid tumor formation (Rastinejad et al., 1989) rheumatoid arthritis (Koch et al., 1986) and psoriasis (Folkman, 1972; Nickoloff et al, submitted for publication) suggest that future studies of pathological angiogenesis must focus on the interaction of positive and negative regulators in this process.

The implications of these recent findings for the treatment of disorders of neovascularization are obvious. The use of inhibitors of neovascularization for the treatment of solid tumors and chronic inflammatory diseases have long be envisioned as a possible mode of therapy (Folkman, 1985; Folkman and Klagsbrun, 1987). Several examples already exist where this approach has met with success (Orchard et al., 1989; White et al., 1991). As the tools of genetic engineering move from the laboratory to the bedside and as the molecular and biochemical basis for the functional diversity of many positive and negative regulators of angiogenesis are defined, it may be possible to up- or down-regulate angiogenic responses with exquisite precision. For example it may be possible customize therapy by instructing angiogenesis accessory cell populations such as macrophages that normally enter a focus to inflammation or infiltrate tumors to either produce or even overproduce endogenous regulators of angiogenesis or alternatively block production of these mediators using the antisense strategy. Regardless of the diverse settings in which angiogenesis is found and the great redundancy of mediator systems that participate in this process, the sorting out of the mechanisms which control the balanced production of positive and negative regulators of angiogenesis will no doubt prove to be a fruitful area of investigation.

REFERENCES

Bader, S.A., Fasching, C., Brodeur, G.M., and Stanbridge, E.J., Dissocoation of suppression of tumorigenicity and differentiation in vitro effected by transfer of a single human chromosome into neuroblastoma cells. Cell Growth Differ. 2:245.

Blood, C.H., and Zetter, B.R., 1990, Tumor interactions with the vasculature: Angiogenesis and tumor metastasis. Biochim. Biophys. Acta 1032:89

Bornstein, P., 1992, Thrombospondins: structure and regulation of expression. FASEB Jr. 6:3290.

Bouck, N., 1990, Tumor angiogenesis: the role of oncogenes and tumor suppressor genes. Cancer Cells 2, 179.

Bouck, N., Angiogenesis: a mechanism by which oncogenes and tumor suppressor genes regulate tumorigenesis, in " Oncogenes and Tumor Suppressor Genes" C.C. Benz and E.T. Liu, eds, Kluwer Academic Publishers, Boston

Bouck, N. P., Stoler, A. and Polverini, P. J., 1986, Coordinate control of anchorage independence, actin cytoskeleton and angiogenesis by human chromosome 1 in hamster-human hybrids. Cancer Res. 46:5101.

Brem, H., and Folkman, J., 1975, Inhibition of tumor angiogenesis mediated by cartilage. J. Exp. Med. 141:427.

Brem, S., Preis, I., Langer, R., Folkman, J., and Patz, A., 1977, Inhibition of neovascularization by an extract derived from vitreous. Am. J. Ophthalmol. 84:323.

Cozzolino, F., Torcia, M., Aldinucci, D., Ziche, M., Almerigogna, F., Bani, D., and Stern, D.M., 1990, Interleukin 1 is an autocrine regulator of human endothelial cell growth. Proc. Natl. Acad. Sci. USA., 87:6487.

Crum, R., Sazbo, S., and Folkman, J., 1985, A new class of steroids inhibits angiogenesis in the presence of heparin or a heparin fragment. Science 230:1375.

DiPietro, L. A., and Polverini, P.J., 1993, Role of the macrophage in the positive and negative regulation of wound neovascularization, Behring Inst. Mitt., (in press)

DiPietro, L. A., Nebgen, D. R., and Polverini, P.J., [submitted for publication (a)]. Down-regulation of endothelial cell thrombospondin 1 enhances in vitro angiogenesis

DiPietro, L. A., and Polverini, P. J., 1993, Angiogenic macrophages produce the angiogenic modulator thrombospondin 1. Am. J. Pathol. in press

Eisenstein, R., Kuettner, K.E., Neopolitan, C., Sobel, L.W., and Sorgente, N., 1975, The resistance of certain tissues to invasion III. Cartilage extracts inhibit the growth of fibroblasts and endothelial cells in culture. Am. J. Pathol. 81:337.

Engerman R.L., Pfaffenbach, D., and Davis, M.D., 1967, Cell turnover of capillaries. Lab. Invest. 17:738.

Folkman, J., 1972, Angiogenesis in psoriasis: Therapeutic implications. J. Invest. Dermatol. 59:40.

Folkman, J., and Klagsbrun, M., 1987, Angiogenic factors. Science 235:442.

Folkman, J., Weisz, P. B., Joullie, M. M., Li, W. W., and Ewing, W. R., 1989, Control of angiogenesis wiyh systemic heparin substitutes. Scienec 243:1490.

Frazier, W. A., 1987, Thrombospondin: a modular adhesive glycoprotein of platelets and nucleated cells. J. Cell Biol. 105:625.

Frazier, W. A., 1991, Thrombospondin. Curr. Opinions in Cell Biol. 3:792.

Goldfarb, M., 1990, The fibroblast growth factor family. Cell Growth Differ. 1:439.

Good, D. J., Polverini, P. J., Rastinejad, F., Le Beau, M. M., Lemons, R. S., Frazier, W. A., and Bouck, N. P., 1990, A tumor suppressor-dependent inhibitor of angiogenesis is immunologically and functionally indistinguishable from a fragment of thrombospondin. Proc. Natl. Acad. Sci. USA 87:6624.

Grant, D. K., Tashiro, K-I., Segui-Real, B., Yamada, Y., Martin, G. R., and Kleinman, H. K., 1989, Two different laminin domains mediate the differentiation of human endothelial cells into capillary-like structures in vitro. Cell 58:933.

Haynes, W. L., Proia, A. D., and Klintworth, G. K., 1989, Effect of inhibitors of arachidonic acid metabolism on corneal neovascularization in the rat. Invest. Ophthalmol. Visual Sci. 30:1588.

Hogg, P.J., Stenflo, J., and Mosher, D. F., 1992, Thrombospondin is a slow tight-binding inhibitor of plasmin. Biochemistry 31:265.

Huang, H-J.S., Yee, J-K., Shew, J-Y., Chen, P-L., Bookstein, R., Friedman, T., Lee, E.Y-H.P., and Lee, W-H., 1988, Suppression of the neoplastic phenotype by replacement of the RB gene in human cancer cells. Science 242:1563.

Ingber, D., and Folkman, J., 1988, Inhibition of angiogenesis through modulation of collagen metabolism. Lab. Invest. 59:44.

Inoue, K., Korenaga, H., Tahnaka, N. G., Sakamoto, N., and Shizuo, K., 1988, The sulfated polysaccharide-peptidoglycan complex potently inhibits embryonic angiogenesis and tumor growth in the presence of cortisone acetate. Carbohydr. Res. 181:135.

Iruela-Arispe, M., Bornstein, P., and Sage, H., 1991, Thrombospondin exerts an antiangiogenic effect on cord formation by endothelial cells in vitro. Proc. Natl. Acad. Sci. USA 88:5026.

Jaffe, E. A., Ruggiero, J. T., and Falcone D. J., 1985, Monocytes and macrophages synthesize and secrete thrombospondin. Blood 65:79.

Klagsbrun, M., and D'Amore P.A., 1991, Regulators of angiogenesis. Annu. Rev. Physiol. 53: 217.

Klagsbrun, M., and Folkman, J., 1990, Peptide Growth Factors and Their Receptors II, in: "Angiogenesis. Handbook of Experimental Pharmacology, Vol., 95/II" , p. 549-586. M.B. Sporn and A.B. Roberts, (Eds.) Springer-Verlag, Berlin, Heildelberg, Germany.

Koch, A. E., Polverini, P. J., and Leibovich, S. J., 1986, Stimulation of neovascularization by human rheumatoid synovial tissue macrophages. Arth. Rheum. 29:471.

Koch, A. E.,Polverini, P.J., Kunkel, S. L., Harlow, L. A., DiPietro, L. A., Elner, V.M., Elner, S.G., and Strieter, R.M., 1992, Interleukin-8 (IL-8) as a macrophage-derived mediator of angiogenesis. Science 258:1798.

Langer, R., Brem, H., Falterman, K., Klein, M., and Folkman, J., 1976, Isolation of a cartilage factor that inhibits tumor neovascularization Science 193:70.

Lawler, J., 1986, The structural and functional properties of thrombospondin. Blood 67: 1197.

Lee, A., and Langer, R., 1983, Shark cartilage contains inhibitors of tumor angiogenesis. Science 221:1185.

Lutty, G. A., Thompson, D. C., Gallup, J. Y., Mello, R. J., Patz, A., and Fenselau, A., 1983, Vitreous: an inhibitor of retinal extract-induced neovascularization. Invest. Ophthalmol. Vis. Sci. 24:52.

Lynch, N. R., Castes, M., Astoin, M., and Salomon, J. C., 1978, Mechanisms of inhibition of tumor growth by asprin and indomethacin. Br. J. Cancer 38:503.

Maione, T. E., Gray, G. S., Petro, J., Hunt, A. J., Donner, A. L., Bauer, S. I., Carson, H. F., and Sharpe, R. J., 1990, Inhibition of angiogenesis by recombinant human platelt factor-4 and related peptides. Science 247:77.

Mansfield, P. J., Boxer, L. A., and Suchard, S. J., 1990, Thrombospondin stimulates motility of human neutrophils. J. Cell Biol. 111:3077.

Moroco, J.R., Solt, D.B., and Polverini, P.J., 1990, Sequential loss of suppressor genes for three specific functions during in vivo carcinogenesis. Lab. Invest. 63:298.

Moses, M.A., and Langer, R., 1991a, Inhibitors of angiogenesis. Biotechnology 9:630.

Moses, M. A., and Langer, R., 1991b, A metalloproteinase inhibitor as an inhibitor of neovascularization. J. Cell. Biochem. 47:230.

Nathan, C. F. (1987) Secreted products of macrophages. J. Clin. Invest. 79, 319-26.

Nickoloff, B. J., Mitra, R. S., Varani, J., Dixit, V. M., and Polverini, P. J. (submitted for publication) Psoriatic keratinocyte induced angiogenesis is inhibited by hrombospondin.

O'Shea, K. S., and Dixit, V. M., 1988, Unique distribution of the extracellular matrix component thrombospondin in the developing mouse embryo. J. Cell Biol. 107:2737.

Oikawa, T., Hirotani, K., Nakamura, O., Shudo, K., Hiragun, A., and Iwaguchi, T., 1989, A highly potent antiangiogenic activity of retinoids. Cancer Lett. 48:157

Orchard, P. J., Smith, C. M. II, Woods, W. G., Day, D. L., and Dehner, L. P., 1989, Treatment of hemangioendotheliomas with alpha interferon. The Lancet 2:565.

Pauli, B., Memoli, V., and Kuttner, K., 1981, Regulation of tumor invasion by cartilage-derived anti-invasive factor *in vitro* . J. Natl. Cancer Inst. 67:65.

Peterson, H-I., 1986, Tumor angiogenesis inhibition by prostaglandin synthetase inhibitors. Anticancer Res. 6:251.

Polverini, P. J., 1989, Macrophage-induced Angiogenesis: A review; in, "Macrophage-Derived Cell Regulatory Factors", C. Sorg, ed., S. Karger, Basel, pp. 54

Rastinejad, F., Polverini, P. J. and Bouck N. P., 1989, Regulation of the activity of a new inhibitor of angiogenesis by a cancer suppressor gene. Cell 56:345.

Robin, J. B., Regis-Pacheco, L. F., Kash, R. L., and Schanzlin, D. J., 1985, The histopathology of of corneal neovascularization, inhibitor effects. Arch Ophthalmol. 103:284.

Sage, H., and Bornstein, P., 1982, Endothelial cells from umbilical vein and a hemangioendothelioma secrete basement membrane largely to the exclusion of interstitial procollagens. Arteriosclerosis 2:27.

Sage, H., and Bornstein, P., 1991, Extracellular proteins that modulate cell-matrix interactions. J Biol. Chem. 266:14831.

Savill, J., Hogg, N., and Haslett, C., 1991, Macrophage vitronectin receptor, CD36, and thrombospondin cooperate in recognition of neutrophils undergoing programmed cell death. Chest 99:65.

Shapiro, R., and Valee, B. L., 1987, Human placental ribonuclease inhibitor abolishes both angiogenic and ribonucleolytic activities of angiogenin. Proc. Natl. Acad. Sci. 84:2238.

Sidky, Y. A., and Borden, E. C., 1987, Inhibition of angiogenesis by interferons: effects on tumor- and lymphocyte-induced vascular responses. Cancer Res. 47:5155.

Silverstein, R. L., and Nachman, R. L., 1987, Thrombospondin binds to monocytes and macrophages and mediates platelet-monocyte adhesion. J. Clin. Invest. 79:867.

Strieter, R. M., Kunkel, S. L., Elner, V. M., Martonyi, C. L., Koch, A. E., Polverini, P. J., and Elner, S. G., 1992, Interleukin-8: a corneal factor that induces neovascularization. Am. J. Pathol. 141:1279.

Tannock, I. F., and Hayashi, S., 1972, The proliferation of capillary endothelial cells. Cancer Res. 32:77.

Taylor, C. W., and Weiss, J. B., 1985, Partial purification of a 5.7k glycoprotein from bovine vitreous which inhibits both angiogenesis and collagenase activity. Biochem and Biophys. Res. Commun. 133:911.

Taylor, S., and Folkman, J., 1982, Protamine ia an inhibitor of angiogenesis. Nature 297: 307.

Thomas, K.A., 1988, Transforming potential of fibroblast growth factor genes. Trends in Biochem. Sci. 13:327.

Tolsma, S.S., Volpert, O.V., Good, D.J., Frazier, W.A., Polverini, P.J., and Bouck, N., 1993, Peptides derived from two separate domains of the matrix protein thrombospondin-1 have anti-angiogenic activity, J. Cell Biol. in press.

West, D. C., and Kumar, S., 1989, Hyaluronan and Angiogenesis, in, The Biology of Hyaluronan (Ciba Foundation Symp.) Wiley, Chichester, pp 143.

White, C. W., Wolf, S. J., Korones, D. N., Sondheimer, H. M., Tosi, M. F., and Yu, A., 1991, Treatment of childhood angiomatous diseases with recombinant interferon alpha. J. Pediatr. 118:59.

Woltering, E. A., Barrie, R., O'Dorisio, T. M., Arce, D., Ure, T., Cramer, A., Holmes, D., Robertson, J., and Fassler, J., 1991, Somatostatin analogues inhibit angiogenesis in the chick chorioallantoic membrane. J. Surg. Res. 50:245.

Ziche, M., Jones, J., and Gullino, P. M., 1982, Role of prostaglandin E1 and copper in angiogenesis. J. Natl. Cancer Inst. 69:475.

HUMAN BASIC FIBROBLAST GROWTH FACTOR:

STRUCTURE-FUNCTION RELATIONSHIP OF AN ANGIOGENIC MOLECULE

Marco Presta, Marco Rusnati, Anna Gualandris,
Patrizia Dell'Era, Chiara Urbinati, Daniela Coltrini,
Elena Tanghetti, and Mirella Belleri

Unit of General Pathology and Immunology
Department of Biomedical Sciences and Biotechnology
School of Medicine
University of Brescia
25123 Brescia, Italy

INTRODUCTION

Basic fibroblast growth factor (bFGF) belongs to the family of the heparin-binding growth factors which includes also acidic FGF and five other gene products (Basilico and Moscatelli, 1992). bFGF exerts various biological activities *in vitro* and *in vivo* on different cell types. In particular, bFGF is an angiogenic molecule that induces a set of complex, co-ordinated responses in cultured endothelial cells, including cell proliferation, chemotaxis, and protease production (Presta et al., 1986). The identification of the functional domains of bFGF appears to be of pivotal importance for the development of drugs aimed to stimulate or to inhibit angiogenesis in various pathological conditions. In the present paper we will summarize findings from different laboratories on the structure-function relationship of this angiogenic growth factor.

bFGF is a single chain, nonglycosylated protein. It consists of 155 amino acids (Figure 1), even though a number of N-terminal truncated forms of bFGF, generated by proteases released during the extraction procedure, have been isolated from different sources (Moscatelli et al., 1988). In the present paper the amino acid numbering 1-155 will be used for full length bFGF, even though amino acid numbering 1-146 can be encountered in the scientific literature, where the first residue corresponds to residue Pro-9.

X-RAY CRYSTALLOGRAPHY

The three-dimensional structure of the 146-residue form of human bFGF has been determined by x-ray crystallography (Eriksson et al., 1991; Zhang et al., 1991). The overall structure of bFGF can be described as a trigonal pyramid with a fold similar to that reported for interleukin-1α, interleukin-1β, and soybean trypsin inhibitor. The overall folding of bFGF is also similar to that of the actin-binding protein hisactophilin, even though the primary sequence of these two proteins are unrelated (Habazetti et al., 1992). bFGF is composed entirely of antiparallel adjacent hydrogen-bonded β-strands which form a barrel closed by the amino-terminal and by the carboxyl-terminal strands (Figure 1). When functional studies of bFGF are considered (see below), analysis of the 3-D structure suggests that the receptor-binding region and the heparin-binding region of bFGF are adjacent but separate domains on the β-barrel. This is consistent with the results of inhibition studies obtained with neutralizing antibodies (Kurokawa et al., 1989) and with the model proposed for FGF-receptor binding which invokes the formation of a trimolecular complex among bFGF, tyrosine- kinase transmembrane glycoprotein, and heparan sulphate proteoglycan (Klagsbrun and Baird, 1991). Indeed, bFGF, soluble heparin, and soluble FGF receptor can form a ternary complex *in vitro* (Ornitz et al., 1992).

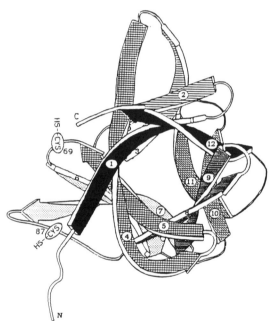

[1]MAAGSITTLPALPEDGGSG[20]

AFPPGHFKDPKRLYCKNGGF[40]

FLRIHPDGRVDGVREKSDPH[60]

IKLQLQAEERGVVSIKGVCA[80]

NRYLAMKEDGRLLASKCVTD[100]

ECFFFERLESNNYNTYRSRK[120]

YTSWYVALKRTGQYKLGSKT[140]

GPGQKAILFLPMSAKS[155]

Figure 1. Amino acid sequence and three-dimensional structure of bFGF. The twelve β-strands are depicted with flattened arrows and have been numbered starting from the N-terminus.

RECEPTOR-BINDING DOMAIN

bFGF binds to high affinity tyrosine-kinase receptors. Cloning of FGF receptors have lead to the discovery of a family of structurally related receptor molecules which differently interact with the various members of the heparin-binding growth factor family. Four distinct FGF receptor genes have been identified which can generate multiple isoforms by alternative splicing. Moreover, several studies have shown that different FGF receptors are also functionally different. Thus, an enormous degree of structural and functional complexity exists in the interaction of FGF receptors with their ligands (Johnson and Williams, 1993).

Surprisingly, very few studies have been published on the characterization of the receptor-binding domain of bFGF. Baird et al. (1988) proposed two domains of bFGF capable of interacting with the FGF receptor. The first domain is related to amino acid sequence 33-77. Indeed, Baird and co-workers have shown that a synthetic peptide corresponding to residues 33-77 exerts a limited partial bFGF agonist and antagonist activity. This peptide also inhibits the binding of bFGF to its receptors in different cell types, even though shorter fragments were inactive. However, in the same work the related peptides bFGF(39-59) and bFGF(41-62) showed a limited but statistically significant bFGF agonist and/or antagonist activity without affecting FGF receptor binding. Thus, the experimental evidences point to a possible biological activity of peptides related to bFGF region 33-77 which may not require binding to FGF receptor. This hypothesis seems to be confirmed by our findings with the synthetic peptide bFGF(38-61) (see below).

The second domain proposed by Baird and co-workers can be reduced to an active core corresponding to residues bFGF(115-124). This domain shows several features of a receptor binding peptide sequence. Indeed, peptides related to this sequence mimic the effect of bFGF in 3T3 cells and inhibit the binding of bFGF to its cell membrane receptor on different cell types (Schubert et al., 1987; Baird et al., 1988; Walicke et al., 1989). X-ray crystallography indicates that residues 115-124 form a distorted antiparallel β-turn on the surface of the molecule between the 9th β-strand and the 10th β-strand. This "strand-turn-strand" motif is partially conserved in interleukin-1β. As stated above, the putative receptor-binding and heparin-binding regions of bFGF are spatially separated and are located on different faces of the bFGF molecule.

Utilizing a different experimental approach, Seno et al. (1990) have hypothesized that the essential part for receptor binding is present between residues Asp-50 and Ser-109. This conclusion was drawn from experiments in which progressive deletions were performed at the amino-terminus or at the carboxyl-terminus of the bFGF molecule. Amino-terminal truncations up to residue Asp-50 did not affect bFGF activity. On the contrary, carboxyl-terminal truncations had a dramatic effect on the activity of the growth factor: truncations up to residue Ser-122 result in bFGF molecules which show a significant, even though reduced, mitogenic activity (ranging from 1% to 35% of the potency of the wild type molecule); truncation up to residues Ser-109 or Asn-113 results instead in bFGF mutants with very limited activity (less than 0.1% of the wild type bFGF). Thus, an interpretation of these data different from that proposed by the authors would suggest that the core of the receptor binding region of bFGF is included within residues Asn-113 and Ser-122, in good agreement with the domain bFGF(115-124) previously described.

For a better definition of the amino acid residues involved in the interaction of bFGF with its receptor, Presta et al. (1992) characterized the biological properties of two bFGF mutants in which basic residues Lys-128/Arg-129 (M6A- bFGF) or Arg-118/Lys-119/Lys-128/Arg-129 (M6B-bFGF) were substituted with neutral glutamine residues by site-directed mutagenesis (Figure 2).

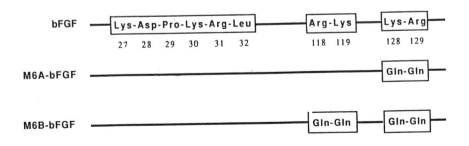

Figure 2. Schematic representation of M6A-bFGF and M6B-bFGF mutants. Basic amino acid clusters are shown within boxes.

Both mutants showed a receptor-binding capacity and mitogenic activity for cultured endothelial cells similar to that exerted by the wild type molecule, even though the latter mutant showed a reduced capacity to bind to immobilized heparin, to induce urokinase-type plasminogen activator production in endothelial cells, and to induce angiogenesis in a mouse-sponge model (Table 1). These results were unexpected since residues Arg-118 and Lys-119 are part of the core region of the putative receptor binding domain bFGF(115-124) and belong to the consensus sequence for *in vitro* protein kinase A-mediated phosphorylation of bFGF (Feige et al., 1989). Moreover, residues Lys-128 and Arg-129 belong to the cationic amino acid cluster involved in the heparin-binding activity of bFGF (see below). Thus, further mutagenesis studies are required for an unambiguous characterization of the amino acid residues responsible for the interaction of bFGF with its receptor.

HEPARIN-BINDING DOMAIN

bFGF is a highly positively charged molecule which binds to negatively charged heparan sulfate proteoglycans present on the cell surface and in the extracellular matrix (ECM) (Baskin et al., 1989). bFGF has been isolated from ECM produced *in vitro* and from basement membranes synthesized *in vivo* (Vlodavsky et al., 1987; Hageman et al., 1991). Also, the capacity of bFGF to bind to proteoglycans hampers its diffusion in the extracellular environment (Flaumenhaft et al., 1990). Interaction of bFGF with cell surface proteoglycans is required for the binding of bFGF to its cell membrane receptors (Yayon et al., 1991) and the capacity of bFGF to stimulate cell proliferation and plasminogen activator production in endothelial cells requires the presence of matrix-bound bFGF (Flaumenhaft et al., 1989; Presta et al., 1989). Moreover, ECM-bound bFGF represents a reservoir of biologically active bFGF that becomes available to target cells after enzymatic mobilization (Vlodavsky et al., 1991). bFGF binds not only to immobilized proteoglycans but also to glycosaminoglycans (GAGs) in solution. Interaction of bFGF with free heparin increases the radius of diffusion of the growth factor in the extracellular environment (Flaumenhaft et al., 1990) and protects bFGF from inactivation (Gospodarowicz and Cheng, 1986) and proteolytic degradation (Sommer and Rifkin, 1989).

The capacity of heparin to protect bFGF from trypsin cleavage has been ascribed to a possible interaction of the positively charged lysine and arginine residues with the negative sulfate groups of the GAG. Indeed, more than 20 basic residues are exposed on the surface of bFGF. In particular, x-ray crystallography has shown that a cluster of basic residues including Lys-35, Arg-53, Lys-128, Arg-129, Lys-134, Lys-138, and Lys-144 is present on the surface of bFGF in a region able to interact with a sulfate ion (Zhang et al., 1991; Eriksson et al., 1991). This cluster has been hypothesized to represent the heparin-binding domain of bFGF.

We have studied different recombinant bFGF mutants in which basic residues have been mutated or deleted. As described above, M6A-bFGF and M6B-bFGF were obtained by substitution with neutral glutamine residues of the basic residues Lys-128/Arg-129 or Arg-118/Lys-119/Lys-128/Arg-129, respectively. Two more mutants were obtained by deletion of the basic amino-terminal region bFGF(27-32) (M1-bFGF) or by substitution with glutamine residues of the basic residues Lys-27/Lys-30/Arg-31 (M1Q-bFGF). A schematic representation of the two latter mutants is shown in Figure 3. Thus, to identify the basic amino acid residues involved in the interaction of bFGF with sulfated GAGs, we evaluated the capacity of heparin to protect the different bFGF mutants from trypsin digestion (Coltrini et al., 1993). The results have demonstrated that substitution of the basic residues investigated with neutral glutamine residues does not significantly affect the capacity of heparin to protect bFGF from trypsin digestion. This was observed both when mutations were performed in the amino-terminal region bFGF(27-32) and in the carboxyl-terminal region bFGF(118-129). Our results suggest that the mutagenized amino acids are not essential for bFGF-heparin interaction, the minimal charge required for heparin binding being provided by the residual basic amino acids belonging to the same cationic cluster.

While neutralization of basic amino acids in the region bFGF(27-32) does not affect heparin protection, this is abolished when the full region is deleted, as it occurs in M1-bFGF (Coltrini et al., 1993). This indicates that M1-bFGF is present in a protein conformation which does not allow a proper interaction with heparin able to prevent tryptic digestion, as observed for heat-denatured wild-type bFGF (Sommer and Rifkin, 1989). To this respect, it is interesting to note that also the truncation of more than six amino acid residues from the carboxyl-terminus, but not from the amino-terminus, reduces the affinity of bFGF for heparin (Seno et al., 1990), suggesting that the affinity of bFGF for sulfated GAGs depends significantly on its carboxyl-terminal structure. On the other hand, the same truncations appeared to modify dramatically the tertiary structure of bFGF, reducing its solubility and its biological activity. Similar results were obtained in our laboratory when conservative single amino acid substitutions were performed in regions of the bFGF molecule corresponding to different loops connecting adjacent β-strands (Gualandris and Rusnati, unpublished observations). These findings suggest that any mutation of the bFGF molecule which alters the 3-D structure of the growth factor may result in a bFGF mutant with a reduced affinity for heparin, indicating that site-directed mutagenesis may not represent an useful approach for the characterization of the heparin-binding domain of bFGF.

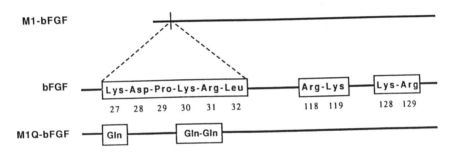

Figure 3. Schematic representation of M1-bFGF and M1Q-bFGF mutants. Basic amino acid clusters are shown within boxes.

Asp-Gly-Arg-CONTAINING DOMAINS

Experiments performed in our laboratory (Presta et al., 1991) have identified two regions in the primary structure of human bFGF which appear to be involved in the modulation of the mitogenic activity exerted by bFGF in endothelial cells and which are distinct from the putative receptor-binding region bFGF(115-124). These regions correspond to amino acid residues 38-61 and 82-101 (Figure 4). The identification of these functional domains was based on the following experimental evidences. A) The synthetic peptides bFGF(38-61) and bFGF(82-101) induce cell proliferation when administered to cultured endothelial cells. B) The two peptides exert a partial bFGF antagonist activity when assayed on endothelial cells. Inhibition of the mitogenic activity of bFGF is specific, since the two peptides do not inhibit endothelial cell proliferation induced by different mitogens. C) Polyclonal affinity-purified antibodies raised against the two peptides completely quench the mitogenic activity of human recombinant bFGF while antibodies raised against other bFGF fragments do not affect bFGF activity.

Interestingly, both bioactive peptides contain an Asp-Gly-Arg (DGR) sequence at positions bFGF(46-48) and bFGF(88-90), respectively (Figure 4). This sequence represents the inverted sequence of the widespread recognition signal RGD. RGD sequence is specifically recognized by the family of the integrin receptors, molecules that bind a variety of adhesive proteins like fibronectin, fibrinogen, laminin, vitronectin, and collagen (Ruoslathi and Pierschbacher, 1986). Short peptides containing the RGD or the DGR sequence have been shown to inhibit the interaction of integrins with their ligands (Akiyama and Yamada, 1985; Ruoslathi and Pierschbacher, 1987; Hynes, 1987).

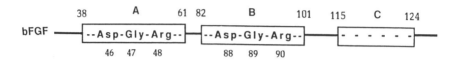

Figure 4. Schematic representation of the bFGF molecule. Putative "activation" domains A and B, each containing a DGR sequence, and the putative receptor binding domain C are shown within boxes. See text for further details.

On this basis, we hypothesized that the two DGR sequences may represent the core sequences responsible for the biological properties of the two bioactive bFGF fragments and may be involved in the modulation of the mitogenic activity exerted by the whole bFGF molecule on cultured endothelial cells. Indeed, we demonstrated that short DGR- and RGD-containing peptides inhibit in a specific manner the mitogenic activity of bFGF, as well as that exerted by the two bFGF fragments. Thus, our data strongly suggest that DGR sequences may play a role in mediating the mitogenic activity of bFGF. Interestingly, X-ray crystallography demonstrates that both DGR sequences are exposed on the surface of the bFGF molecule with a similar conformation and are localized in loops connecting two adjacent β-strands.

Even though peptides bFGF(38-61) and bFGF(82-101) exert a bFGF agonist and antagonist activity, they do not affect the binding of bFGF to its low and high affinity binding sites. Also, monospecific anti-peptide antibodies quench the mitogenic activity of bFGF without inhibiting the binding to its receptors. Furthermore, DGR- and RGD-containing peptides do not affect the interaction of bFGF with its low and high affinity binding sites (Presta et al., 1991). Thus, despite the biological activity exerted by peptides related to amino acid sequences bFGF(38-61) and bFGF(82-101), these sequences do not represent regions of bFGF responsible for the binding of bFGF to its plasma membrane receptor. This is in keeping with the identification of the sequence bFGF(115-124) as the core sequence of the receptor-binding region of bFGF (see above). Thus, it seems possible to hypothesize that different regions of the bFGF molecule are responsible for the biological activity of bFGF, as already suggested for EGF (Katsuura and Tanaka, 1989), transforming growth factor-α (Nestor et al., 1985), and glucagon (Krstenansky et al., 1986). Residues 115-124 represent the core of the receptor binding domain of bFGF while residues 33-77 and 82-101 represent possible "activation" domains, both with a DGR core sequence, not directly involved in the binding of bFGF to its receptor (Figure 4). This hypothesis would explain why peptides related to the sequence bFGF(115-124) inhibit receptor binding, while peptides bFGF(38-61) and bFGF(82-101), as well as short DGR- and RGD-containing peptides, exert a bFGF antagonist activity without affecting bFGF receptor binding.

The identification of residues 33-77 and 82-101 as possible "activation" domains of bFGF rises the question of the site(s) of interaction of these amino acid regions with the cell. The capacity of short DGR- or RGD-containing peptides to specifically antagonize the activity of peptides bFGF(38-61) and bFGF(82-101) and of the whole bFGF molecule points to the identification of the DGR sequences as the core sequences of the two "activation" domains. On this basis, it is tempting to hypothesize that bFGF may interact *via* its DGR sequences with an integrin-like molecule (Figure 5), and that this interaction may affect the mitogenic activity exerted by bFGF in endothelial cells. Several experimental evidences support this hypothesis. A) The structural requirements for the inhibition of bFGF activity by RGD-containing peptides mimic that observed for ligand/integrin interaction. B) bFGF promotes substrate adhesion of PC12 cells and of endothelial cells (Schubert et al., 1987; Baird et al., 1988). C) Peptides bFGF(38-61) and bFGF(82-101), as well as bFGF, promote the adhesion of endothelial cells to non-adhesive plastic and this capacity can be inhibited by short RGD-containing peptides and by anti-vitronectin receptor antibodies (Rusnati et al., manuscript in preparation). Moreover, extracellular matrix components have been shown to affect the mitogenic activity of bFGF in endothelial cells by modulating cell shape and intracellular pH (Ingber et al., 1987; Ingber and Folkman, 1989; Ingber, 1990; Ingber et al., 1990). It is interesting to note that the cell-adhesive proteins fibronectin, laminin, and collagen have been shown to interact with bFGF *in vitro*, affecting its phosphorylation by protein kinase A (Feige et al., 1989), and that a synthetic fragment of EGF has revealed a possible functional interrelationship between EGF and laminin (Eppstein et al., 1989). These findings shed a new light on the nature of the mechanisms that are responsible for the well-known modulation of the activity of soluble growth factors by extracellular matrix (Gospodarowicz et al., 1978; Ingber et al., 1987; Ingber and Folkman, 1988, 1989).

AMINO-TERMINAL EXTENSION AND NUCLEAR TRASLOCATION DOMAIN

The single copy human bFGF gene encodes for the 155 amino acid 18 kD form of bFGF and for three co-expressed isoforms, with an apparent molecular weight of 24 kD, 22.5 kD, and 22 kD. All bFGF isoforms are co-translated from a single mRNA transcript. 18 kD-bFGF is translationally initiated at a classic AUG codon while the high molecular weight bFGF isoforms (HMW-bFGFs) are initiated at novel CUG codons (Florkiewicz and Sommer, 1989). It has been hypothesized that HMW-bFGFs and 18 kD-bFGF may differ for intracellular distribution and function (Quarto et al., 1991a; Couderc et al., 1991).

Several laboratories have addressed this question by studying the intracellular localization of the various isoforms of bFGF in transfected and nontransfected cells (Renko et al., 1990; Dell'Era et al., 1990; Tessler and Neufeld, 1991; Florkiewicz et al., 1991; Bugler et al., 1991). The data demonstrate that HMW-bFGFs are mainly localized in the cell nucleus. Nuclear targeting of these bFGF isoforms appears to be mediated by amino acid sequences present in their N-terminal extension (Bugler et al., 1991; Quarto et al., 1991b) which are characterized by several repeats of the sequence Gly-Arg-Gly. These sequences contain methylated arginine residues which may play a role in directing nuclear translocation (Burgess et al., 1991).

Intracellular localization of 18 kD-bFGF is less well defined. Undoubtedly, a major portion of intracellular 18 kD- bFGF is localized in the cytoplasm (Renko et al., 1990). However, some reports indicate that also this form of bFGF is localized in the cell nucleus in a significant amount. For instance, more than 30% of endogenous 18 kD-bFGF is localized in the nucleus of fetal bovine aortic endothelial cells (Dell'Era et al., 1991). Also, intact BHK-21 cells and COS-1 cells transfected with human bFGF cDNA coding for 18 kD-bFGF show a strong immunoreactivity in the cell nucleus when probed with anti-bFGF antibodies (Tessler and Neufeld, 1990; Florkiewicz et al., 1991). Moreover, it has been shown that exogenously added 18 kD-bFGF is internalized and accumulates into the nucleus of cultured bovine aortic endothelial cells and of fetal rat astrocytes and hippocampal neurons (Bouche et al., 1987; Baldin et al., 1990; Walicke and Baird, 1991).

When we investigated the subcellular distribution of the different isoforms of bFGF in NIH 3T3 cells transfected with bFGF cDNAs that express 18 kD-bFGF only or all bFGF isoforms, we observed that a significant amount of 18 kD-bFGF localizes into the nucleus of transfected cells and that this form of bFGF, as well as HMW-bFGFs, interacts strongly with nuclear chromatin during interphase and the mitotic cycle (Gualandris et al., 1993). These observations suggest the possibility that not only HMW-bFGFs but also 18 kD-bFGF may possess one or more nuclear translocation sequences that are recognized under distinctive experimental conditions.

The sequence bFGF(27-32) Lys-Asp-Pro-Lys-Arg-Leu is in good agreement with a consensus sequence for nuclear translocation (Chelsky et al., 1989) and it is partially conserved in the corresponding basic amino acid sequence 21- 27 of aFGF (Imamura et al., 1992). Interestingly, the basic region present in the aFGF molecule has been suggested to be a nuclear translocation sequence for the growth factor and to play a role in mediating its biological activity (Imamura et al., 1990; Imamura et al., 1992). Deletion of amino acid residues 27-32 in the bFGF molecule results in the mutant M1-bFGF (Figure 3). As summarized in Table 1, this mutant shows a reduced capacity to induce plasminogen activator production in endothelial cells without alteration of its receptor binding capacity and mitogenic activity (Isacchi et al., 1992). Unfortunately, any attempt to define the intracellular localization of this protein in mammalian cells transfected with the corresponding mutagenized bFGF cDNA has been hampered by the very low levels of expression of this bFGF mutant (Gualandris, unpublished observations).

On this basis, we have characterized the biological activity (Table 1) and subcellular localization of a second bFGF mutant (M1Q-bFGF) in which basic residues Lys-27, Lys-30, and Arg-31, belonging to this putative nuclear translocation sequence, were substituted with neutral glutamine residues by site-directed mutagenesis (Figure 3). However, both immunocytochemical and cell fractionation data demonstrate that M1Q-bFGF, like the wild-type protein, accumulates into the nucleus of transfected COS-1 cells (Presta et al., manuscript in preparation). Thus, at variance with what suggested for aFGF, our data do not support the hypothesis that bFGF(27-32) may represent a nuclear translocation and/or retention sequence for 18 kD-bFGF.

Table 1. Biological properties of bFGF mutants.

	M1-bFGF	M1Q-bFGF	M6A-bFGF	M6B-bFGF
In vitro:				
-binding to immobilized heparin	++	++	++	+
-heparin-dependent trypsin resistance	-	++	++	++
On cultured endothelial cells:				
-binding to high affinity receptors	+ (++)	++ (++)	++ (++)	++ (++)
-mitogenic activity	+ (++)	++ (++)	++ (++)	++ (++)
-chemotactic activity	nd (++)	+ (+)	++ (++)	+ (+)
-plasminogen activator-inducing activity	- (-)	- (++)	++ (++)	- (++)
In vivo:				
-angiogenic activity in mouse-sponge model	- (-)	nd (nd)	- (-)	- (-)

++:potency equal to bFGF; + :potency 3-10 times lower than bFGF; - :potency 30-100 times lower than bFGF. Data in parenthesis refer to the potency of the molecule in the presence of soluble heparin. See text, Isacchi et al. (1992), and Presta et al. (1992) for further details.

These findings rise the question of the mechanism(s) responsible for the nuclear accumulation of 18 kD-bFGF observed both in transient and in stable transfectants producing high levels of the protein. The presence of a non-canonical nuclear translocation sequence in 18 kD-bFGF other than the sequence investigated can be hypothesized. Another possibility is that 18 kD-bFGF might diffuse into the nucleus through the nuclear membrane or after its distruption during the mitotic cycle (Silver, 1991). Diffused bFGF will then bind to chromatin and it will be retained in the nuclear compartment during interphase. Also, the existence of a facilitated nuclear transport for 18 kD-bFGF, as observed for other low molecular weight nuclear proteins (Breeuwer and Goldfarb, 1990), can not be ruled out. Further experiments will be necessary to clarify this point.

CONCLUSIONS

Figure 5 represents a working hypothesis for the study of the structure/function relationship of bFGF. Different domains appear to be involved in the interaction of bFGF with tyrosine kinase receptor (FGF-R), heparan sulfate proteoglycan (HSPG), and integrin-like molecules (I). These interactions may be responsible for different biological responses elicited by the growth factor in target cells, including endothelial cells. Further functional and structural studies are required for a full understanding of the biology of bFGF.

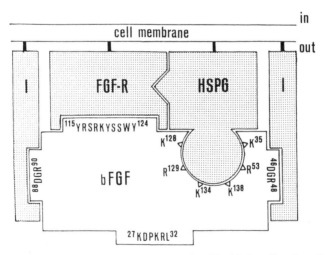

Figure 5. Schematic representation of the interaction of bFGF with the cell surface. See text for details.

ACKNOWLEDGMENTS

Figure 1 was kindly provided by P. Caccia, Farmitalia-Carlo Erba, Milan, Italy. Research in the authors' laboratory was supported by grants from Associazione Italiana per la Ricerca sul Cancro (AIRC) and from C.N.R. (Progetto Finalizzato Biotecnologie, Sottoprogetto Biofarmaci) to M. Presta. The bFGF mutants described in Table 1 were provided by Farmitalia-Carlo Erba, Milan, Italy.

REFERENCES

Akiyama, S.K., and Yamada, K.M., 1985, Synthetic peptides competitively inhibit both direct binding to fibroblasts and functional biological assays for the purified cell- binding domain of fibronectin, *J. Biol. Chem.*, 260:10402.

Baird, A., Schubert, D., Ling, N., and Guillemin, R., 1988, Receptor- and heparin-binding domains of basic fibroblast growth factor, *Proc. Natl. Acad. Sci. U.S.A.* 85:2324.

Baldin, V., Roman, A.M., Bosc-Bierne, I., Amalric, F., and Bouche, G., 1990, Translocation of bFGF to the nucleus is G_1 phase cell cycle specific in bovine aortic endothelial cells, *EMBO J.* 9:1511.

Basilico, C., and Moscatelli, D., 1992, The FGF family of growth factors and oncogenes, *Adv. Cancer Res.* 59:115.

Baskin, P., Doctrow, S., Klagsbrun, M., Svahn, C.M., Folkman, J. and Vlodavsky, I., 1989, Basic fibroblast growth factor binds to subendothelial extracellular matrix and is released by heparitinase and heparin-like molecules, *Biochemistry* 28:1737.

Bouche, G., Gas, N., Prats, H., Baldin, V., Tauber, J.P., Teissié, J., and Amalric, F., 1987, Basic fibroblast growth factor enters the nucleolus and stimulates the transcription of ribosomal genes in ABAE cells undergoing G_0-G_1 transition, *Proc. Natl. Acad. Sci. U.S.A.* 84:6770.

Breeuwer, M., and Goldfarb, D.S., 1990, Facilitated nuclear transport of histone H1 and other small nucleophilic proteins, *Cell* 60:999.

Bugler,B., Amalric, F., and Prats, H., 1991, Alternative initiation of translation determines cytoplasmic or nuclear localization of basic fibroblast growth factor, *Mol. Cell. Biol.* 11:573.

Burgess, W.H., Bizik, J., Mehlman, T., Quarto, N., and Rifkin, D.B., 1991, Direct evidence for methylation of arginine residues in high molecular weight forms of basic fibroblast growth factor, *Cell Regul.* 2:87.

Chelsky, D., Ralph, R., and Jonak, G., 1989, Sequence requirements for synthetic peptide-mediated translocation to the nucleus, *Mol. Cell. Biol.* 9:2487.

Coltrini, D., Rusnati, M., Zoppetti, G., Oreste, P., Iascchi, A., Caccia, P., Bergonzoni, L., and Presta, M., 1993, Biochemical bases of the interaction of human basic fibroblast growth factor with glycosaminoglycans: new insights from trypsin digestion studies, *Eur. J. Biochem.* in press.

Couderc, B., Prats, H., Bayard, F., and Amalric, F., 1991, Potential oncogenic effects of basic fibroblast growth factor requires cooperation between CUG and AUG-initiated forms, *Cell Regul.* 2:709.

Dell'Era, P., Presta, M., and Ragnotti, G., 1991, Nuclear localization of endogenous basic fibroblast growth factor in cultured endothelial cells, *Exp. Cell Res.* 192:505.

Eppstein, D.A., Marsh, Y.V., Schryver, B.B., and Bertics, P.J., 1989, Inhibition of epidermal growth factor/trasforming growth factor-α stimulated cell growth by a synthetic peptide, *J. Cell. Physiol.* 141:420.

Eriksson, A.E., Cousens, L.S., Weaver, L.H., Matthews, B.W., 1991, Three-dimensional structure of human basic fibroblast growth factor, *Proc. Natl. Acad. Sci. U.S.A.* 88:3441.

Feige, J.J., Bradley, J.D., Fryburg, K., Farris, J., Cousens, L.C., Barr, P.J., and Baird, A., 1989, Differential effects of heparin, fibronectin, and laminin on the phosphorylation of basic fibroblast growth factor by protein kinase C and the catalytic subunit of protein kinase A, *J. Cell Biol.* 109:3105.

Flaumenhaft, R., Moscatelli, D., Saksela, O., and Rifkin, D.B., 1989, Role of extracellular matrix in the action of basic fibroblast growth factor: matrix as a source of growth factor for long-term stimulation of plasminogen activator production and DNA synthesis, *J. Cell. Physiol.* 140:75.

Flaumenhaft, R., Moscatelli, D., and Rifkin, D.B., 1990, Heparin and heparan sulphate increase the radius of diffusion and action of basic fibroblast growth factor, *J. Cell Biol.* 111:1651.

Florkiewicz, R.Z., and Sommer, A., 1989, Human basic fibroblast growth factor gene encodes four polypeptides: three initiate translation from non-AUG codons. *Proc. Natl. Acad. Sci. U.S.A.* 86:3978.

Florkiewicz, R.Z., Baird, A., and Gonzalez, A.M., 1991, Multiple forms of bFGF: differential nuclear and cell surface localization, *Growth Factors* 4:265.

Gospodarowicz, D., Greenburg, G., and Birdwell, R.C., 1978, Determination of cellular shape by the extracellular matrix and its correlation with the control of cellular growth, *Cancer Res.* 38:4155.

Gospodarowicz, D., and Cheng, J., 1986, Heparin protects basic and acidic FGF from inactivation, *J. Cell. Physiol.* 128:475.

Gualandris, A., Coltrini, D., Bergonzoni, L., Isacchi, A., Tenca, S., Ginelli, B., and Presta, M., 1993, The N-terminal extension of high molecular weight forms of basic fibroblast growth factor (bFGF) is not essential for the binding of bFGF to nuclear chromatin in transfected NIH 3T3 cells, *Growth Factors* 8:49.

Habazetti, J., Gondol, D., Wiltscheck, R., Otlewsky, J., Schleicher, M. and Holak, T.A., 1992, Structure of hisactophilin is similar to interleukin-1β and fibroblast growth factor, *Nature* 359:855.

Hageman, G.S., Kirchoff-Rempe, M.A., Lewis, G.P., Fisher, S.K., and Anderson, D.H., 1991, Sequestration of basic fibroblast growth factor in the primate retinal interphotoreceptor matrix, *Proc. Natl. Acad. Sci. U.S.A.* 88:6706.

Hynes, R.O., 1987, Integrins: a family of cell surface receptors, *Cell* 48:549.

Imamura, T., Englera, K., Zhan, X., Tokita, Y., Forough, R., Roeder, D., Jackson, A., Maier, J.A.M., Hla, T., and Maciag, T., 1990, *Science* 249:1567.

Imamura, T., Tokita, Y., and Mitsui, Y.,Recovery of mitogenic activity of a growth factor mutant with a nuclear translocation sequence, 1992, Identification of a heparin-binding growth factor-1 nuclear translocation sequence by deletion mutation analysis, *J. Biol. Chem.* 267:5676.

Ingber, D.E., Madri, J.A., and Folkman, J., 1987, Endothelial growth factors and extracellular matrix regulate DNA synthesis through modulation of cell and nuclear expansion, *In Vitro Cell. Dev. Biol.* 23:387.

Ingber, D.E., and Folkman, J., 1988, Inhibition of angiogenesis through modulation of collagen metabolism, *Lab. Invest.*, 59:41.

Ingber, D.E., and Folkman. J., 1989, Mechanochemical switching between growth and differentiation during fibroblast growth factor-stimulated angiogenesis in vitro: role of extracellular matrix, *J. Cell Biol.* 109:317.

Ingber, D.E., 1990, Fibronectin controls capillary endothelial   cell growth by modulating cell shape, *Proc. Natl. Acad. Sci. U.S.A.* 87:3579.

Ingber, D.E., Prusty, D., Frangioni, J.V., Cragoe, E.J.Jr., Lechene, C., and Schwartz, M.A., 1990, Control of intracellular pH and growth by fibronectin in capillary endothelial cells, *J. Cell Biol.* 110:1803.

Isacchi, A., Statuto, M., Chiesa, R., Bergonzoni, L., Rusnati, M., Sarmientos, P., Ragnotti, G., and Presta, M., 1991, A 6-amino acid deletion in basic fibroblast growth factor dissociates its mitogenic activity from its plasminogen activator-inducing capacity, *Proc. Natl. Acad. Sci. U.S.A.* 88:2628.

Johnson, D.E., and Williams, L.T., 1993, Structural and functional diversity in the FGF receptor multigene family, *Adv. Cancer Res.* 60:1.

Katsuura, M., and Tanaka, S., 1989, Topographic analysis of human epidermal growth factor by monospecific antibodies and synthetic peptides, *J. Biochem.* 106:87.

Klagsbrun, M., and Baird, A., 1991, A dual receptor system is required for basic fibroblast growth factor activity, *Cell* 67:229.

Krstenansky, J., Trivedi, D., and Hruby, V.J., 1986, Importance of the 10-13 region of glucagon for its receptor interactions and activation of adenylate cyclase, *Biochemistry* 25:3833.

Kurokawa, M., Doctrow, S.R., and Klagsbrun, M., 1989, Neutralizing antibodies inhibit the binding of fibroblast growth factor to its receptor but not to heparin, *J. Biol. Chem.*, 264:7686.

Moscatelli, D., Joseph-Silverstein, J., Presta. M., and Rifkin, D.B., 1988, Multiple forms of an angiogenesis factor: basic fibroblast growth factor, *Biochimie* 70:83.

Nestor, J.J., Newman, S.R., DeLustro, B., Todaro, G.J., and Schreiber, A.B., 1985, A synthetic fragment of rat transforming growth factor-α with receptor binding and antigenic properties, *Biochem. Biophys. Res. Commun.* 129:226

Ornitz, D.M., Yayon, A., Flanagan, J.G., Svahn, C.M., Levi, E., and Leder, P, 1992, Heparin is required for cell-free binding of fibroblast growth factor to a soluble receptor and for mitogenesis in whole cells, *Mol. Cell. Biol.* 12:240.

Presta, M., Moscatelli, D., Joseph-Silverstein, J., and Rifkin, D.B., 1986, Purification from a human hepatoma cell line of a basic fibroblast growth factor-like molecule that stimulates capillary endothelial cell plasminogen activator production, DNA synthesis, and migration, *Mol. Cell. Biol.* 6:4060.

Presta, M., Maier, J.A.M., Rusnati, M., and Ragnotti, G., 1989, Basic fibroblast growth factor is released from endothelial extracellular matrix in a biologically active form, *J. Cell. Physiol.* 140:68.

Presta, M., Rusnati, M., Urbinati, C., Sommer, A., and Ragnotti, G., 1991, Biologically active synthetic fragments of human basic fibroblast growth factor (bFGF): identification of two asp-gly-arg-containing domains involved in the mitogenic activity of bFGF in endothelial cells, *J. Cell. Physiol.* 149:524.

Presta, M., Statuto, M., Isacchi, A., Caccia, P., Pozzi, A., Gualandris, A., Rusnati, M., Bergonzoni, L., and Sarmientos, P., 1992, Structure-function relationship of basic fibroblast growth factor: site-directed mutagenesis of a putative heparin-binding and receptor-binding region, *Biochem. Biophys. Res. Commun.* 185:1098.

Quarto, N., Talarico, D., Florkiewicz, R., and Rifkin, D.B., 1991a, Selective expression of high molecular weight basic fibroblast growth factor confers a unique phenotype to NIH 3T3 cells, *Cell Regul.* 2:699.

Quarto, N., Finger, F.P., and Rifkin, D.B., 1991b, The NH$_2$- terminal extension of high molecular weight bFGF is a nuclear targeting signal, *J. Cell. Physiol.* 147:311.

Renko, M., Quarto, N., Morimoto, T., and Rifkin, D.B., 1990, Nuclear and cytoplasmic localization of different basic fibroblast growth factor species, *J. Cell. Physiol.* 144:108.

Ruoslathi, E., and Piershbacher, M.D., 1986, Arg-Gly-Asp: a versatile cell recognition signal, Cell 44:517.

Ruoslathi, E., and Piershbacher, M.D., 1987, New perspectives in cell adhesion: RGD and integrins, *Science* 238:493.

Schubert, D., Ling, N., and Baird, A., 1987, Multiple influences of a heparin-binding growth factor on neuronal development, *J. Cell Biol.* 104:635.

Seno, M., Sasada, R., Kurokawa, T., and Igarashi, K., 1990, Carboxyl-terminal structure of basic fibroblast growth factor significantly contributes to its affinity for heparin, *Eur. J. Biochem.* 188:239.

Silver, P.A., 1991, How proteins enter the nucleus, *Cell* 64:489.

Sommer, A., and Rifkin, D.B., 1989, Interaction of heparin with human basic fibroblast growth factor: protection of the angiogenic protein from proteolytic degradation by a glycosaminoglycan, *J. Cell. Physiol.* 138:215.

Tessler, S., and Neufeld, G., 1990, Basic fibroblast growth factor accumulates in the nuclei of various bFGF-producing cell types, *J. Cell. Physiol.* 145:310.

Vlodavsky, J., Folkman, J., Sullivan, R., Fridman, R., IshaiMichaeli, R., Sasse, J., and Klagsbrun, M., 1987, Endothelial cell-derived basic fibroblast growth factor: synthesis and deposition into subendothelial extracellular matrix, *Proc. Acad. Sci. U.S.A.* 84:2292.

Vlodavsky, I., Bar-Shavit, R., Ishai-Michaeli, R., Bashkin, P., and Fuks, Z., 1991, Extracellular sequestration and release of fibroblast growth factor: a regulatory mechanism? *Trends Biochem. Sci.* 16:268.

Walicke, P.A., Feige J. J., and Baird, A., 1989, Characterization of the neuronal receptor for basic fibroblast growth factor and comparison to receptors on mesenchymal cells, *J. Biol. Chem.* 264:4120.

Walicke, P.A., and Baird, A., 1991, Internalization and processing of basic fibroblast growth factor by neurons and astrocytes, *J. Neurosci.* 11:2249.

Yayon, A., Klagsbrun, M., Esko, J.D., Leder, P., and Ornitz, D.M., 1991, Cell surface, heparin-like molecules are required for binding of basic fibroblast growth factor to its high affinity receptor, *Cell* 64:841.

Zhang, J., Cousens, L.S., Barr, P.J., and Sprang, S.R., 1991, Three-dimensional structure of human basic fibroblast growth factor, a structural homolog to interleukin-1β, *Proc. Natl. Acad. Sci. U.S.A.* 88:3446.

ANGIOGENESIS MODELS IDENTIFY FACTORS WHICH REGULATE ENDOTHELIAL CELL DIFFERENTIATION

D.S. Grant, D. Morales, M.C. Cid, H. K. Kleinman

Laboratory. of Developmental Biology, National Institute of Dental Research, NIH, Bldg. 30, Bethesda, MD, 20892- USA

INTRODUCTION

During development, blood vessels are rapidly formed from angioblasts in the mesoderm. This is followed by extensive branching of the vessels to supply the needs of the growing tissue. In the adult, however, vessels form a complex and stable network providing a transit-way for blood cells, oxygen, growth factors, hormones, etc., to all tissues. Usually changes in the vasculature occur only in response to injury stimulation, or cyclic changes in the female reproductive system. When induced to become angiogenic, the endothelial and smooth muscle cells enter a migratory and proliferative phase resulting in a reorganization of the vessel wall and the formation of new blood vessels. This involves a cascade of sequential events leading to endothelial cell invasion into the underlying stroma, and the formation of new blood vessels leading away from the parent vessel towards the stimulus [1, 2].

The exact mechanism of angiogenesis is not known, although several factors have been shown to induce angiogenesis in vivo. Folkman has demonstrated that the addition of corticosteroids or heparin can directly modulate the angiogenic process[2, 3]. Several growth factors have also been identified including fibroblast growth factor (FGF), tumor necrosis factors (TNF)[4], transforming growth factor (TGF alpha and beta)[5], platelet derived growth factor (PDGF)[6], vascular endothelial cell growth factor (VEGF)[7], and hepatocyte growth factor (HGF)[8, 9]. All these factors are able to stimulate angiogenesis in vivo and can affect endothelial cells either directly in the serum, the basement membrane or through unclear mechanisms, through the extravascular stroma (Figure 1). It has also been shown recently that the substances produced by the endothelial and muscle cells themselves can regulate their differentiation and function in an autocrine manner (Figure 1).

A greater understanding of the action of these and other factors on endothelial cells has been derived from experimental studies in which angiogenic models have been developed and used to determine the specific role of each factor [10-17]. Several of these models involve the stimulation of existing endothelial cells in situ whereas other models have been developed which mimic some or all of the events occurring in vivo in in vitro assay systems. In this paper we will describe several models, including two new models our lab has developed that are rapid and can accurately determine factors which have an effect on endothelial cells. In addition, we present recent data on female homones which promote angiogenesis.

ANGIOGENESIS MODELS

Angiogenesis can be studied in vivo and in vitro. Most of the recent discoveries have been possible with the use of in vivo methods [17-19] While these in vivo methods relate directly to many events occurring during normal blood vessel formation, they have several drawbacks.

First, in vivo angiogenic models are often difficult to control and it is hard to discern if the action of the stimulant is direct or the result of a secondary product of inflammatory cells. The most widely used model is the chicken chorioallantoic membrane assay and it has been used primarily to show capillary inhibition by numerous substances such as fumagillin, cortisone and a laminin-derived synthetic peptide, YIGSR [20], located on the B1 chain of the basement membrane glycoprotein laminin. Drawbacks to this assay, are that it takes up to four days to complete and there can be great variability between chicks during the preparation of the eggs. The rabbit eye model of angiogenesis provides an avascular system that can be used to test the stimulatory or inhibitory effect of angiogenic compounds. This assay, however, is expensive since it employs the use of many rabbits. Other in vivo models use a subcutaneous implantation of a Goretex sponge or polyvinyl disk containing bFGF into mice[17]. There is a certain amount of variability between all of these methods and thus they require large number of tests samples to obtain reliable results.

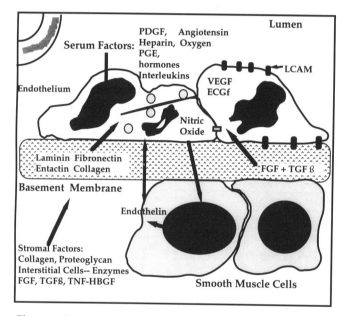

Figure 1. Diagram of factors regulating endothelial cell behavior.

The mechanisms involved in endothelial cell attachment or vessel formation are difficult to examine in vitro due to the complex relationship of endothelial cells to both soluble and insoluble factors in situ. Endothelial cells from various sources have been isolated and maintained in culture[21]. If the cells are permitted to become superconfluent without change of the culture medium, structures form above the monolayer which resemble capillary-like vessels [14, 22]. This differentiation can be accelerated by the additon of a matrix such as either gelatin fibrin or collagen to the surface of the plastic [22-24]. The induction of capillary formation can also be initiated by the addition of phorbol esters, and by incubating the cells with a collagen I gel [14, 25]. These are good models of angiogenesis but they usually require three or more days for vessels to form, and in many cases only half the cells participate in the differentiation process (Table 1). Others have shown that if cultured endothelial cells are incubated in a fibrin clot or with collagen IV, vessel formation will occur within a week [26]. With these systems, capillary-like structures are observed but the response is generally slow, requiring several days to weeks(Table 1), and in some cases, the vessels are inside out, secreting basement membrane material and interstitial collagen into the lumen.

Tube formation on Matrigel:

The stability and integrity of blood vessels is maintained by many factors in the blood, the surrounding tissues and by an important extracellular matrix, the basement membrane, which is closely associated with the endothelium of the vessels (Figure 1). Basement membranes are composed of an organized network of collagen (type IV), heparan sulfate proteoglycan, and glycoproteins such as entactin, fibronectin, and laminin. Laminin is one of the most important and abundant substances in basement membranes. It has a direct role in cell attachment, migration, and induction of the differentiated phenotype of many cells. We have examined and defined the role(s) of laminin and its specific cell-binding sites at the biochemical level using an in vitro angiogenic model [20, 27]. This model involves the differentiation of cultured endothelial cells on a laminin-rich reconstituted basement membrane matrix, Matrigel, into capillary-like structures.

Table 1. Comparison of processing time in different angiogenesis models.

Angiogenesis Models

In Vivo:

Tissue	Time taken
Chick chorioallantoic membrane	4-8 days
Mouse Gortex sponge implant	1-2 weeks
Rabbit cornea	2-5 days
Polyvinyl disc implant in mouse	14 days
Matrigel plug in mouse	3-14 days

In Vitro:

Substratum	Time taken
Plastic-low serum	3-6 weeks
Gelatin or Fibronectin	4-8 weeks
Collagen I gel	2-4 weeks
Fibrin Clot	4-7 days
Sprouts from aortic rings	5-11 days
Adipose tissue in collagen gel	5-12 days
Basement membrane collagen	4 days
Laminin gel	1 day
Matrigel	18 hours

Previous work in our laboratory has shown that endothelial cells attach, migrate, and assemble on a laminin-rich reconstituted matrix, Matrigel, to form tube-like structures resembling capillaries within 18 hours [15, 20, 27]. Matrigel is a mixture of basement membrane components and growth factors[28], extracted from the Englebreth-Holm-Swarm (EHS) tumor and has been found to induce and/or maintain the differentiation of a wide variety of cells [29]. The components are extracted with 2.0M urea and then dialyzed into a physiological buffer. At 4°C, the components remain in solution but polymerize when warmed to 24-37°C. The gelled extract has the appearance of authentic basement membrane in the electron microscope. Human umbilical vein endothelial cells HUVEC (as well as microvascular cells) when seeded onto a 16mm-well, coated with Matrigel, rapidly attach to the matrix and begin to migrate (Figure 2). Work in our lab has shown that there is significant remodeling of the Matrigel during the first 4 hours of migration and organization of the cells. Within a period of 8-10 hours, capillary-like structures are already apparent (Figure 2). When these structures are cut in cross-section and viewed in the electron microscope, they are cylindrical with a lumen [27, 30].

The cells within the tubes maintain the ability to bind acetylated-LDL [15] and continue to produce VonWillebrand factor (unpublished observations). One distinct difference in tube formation on Matrigel that makes it different from other in vitro angiogenesis assays is that the proliferation of endothelial cells is reduced. Therefore Matrigel induces endothelial cells differentiation , i.e. the completing steps of angiogenesis during which cells are slowing down in migration and proliferation and forming vascular strucures. In addition, the Matrigel in vitro model does not include other cells (smooth muscle and pericytes) which are associated with normal vessel formation in vivo.

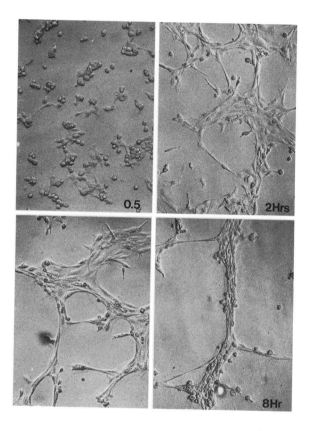

Figure 2. HUVEC cultured on a thin layer of Matrigel over time. The cells attach within 0.5 hours, and then organize into tubes over the period of 8 hours.

The Matrigel in vitro model system has been used to define the cellular interactions with the basement membrane and to identify several intracellular events occurring during capillary vessel formation. The cells use multiple receptors to interact with basement membrane components, through multiple receptors. Protein synthesis, an intact cytoskeleton, and the activities of protein kinase C and collagenase IV are all required for tube formation [27]. Using this model system, both inhibitors and stimulators of angiogenesis can also be screened prior to in vivo testing.

In vivo vessel formation in Matrigel

An in vivo angiogenesis model employing subcutaneously injected Matrigel enriched with an exogeneously added angiogenic factor has been developed to alleviate some of the limitations of the in vitro Matrigel model[12, 31]. Since Matrigel is a liquid at 4°C, it can be easily injected beneath the skin, and rapidly forms a solid plug of matrix as it warms up to the mouse's body temperature (Figure 3).

These plugs are easily retrievable by peeling back the mouse skin, even months after the injection. After the plugs have been fixed in formaldehyde, histological sections illustrate the cells that have migrated into the Matrigel. If Matrigel alone is injected, and the plug is examined 14 days later, very few cells are observed in the matrix. When bFGF or other

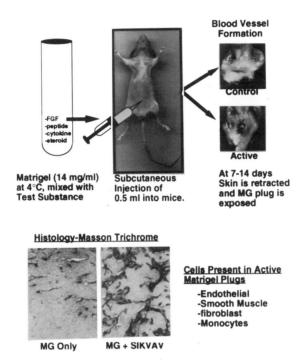

Figure 3. Diagram of the Matrigel mouse angiogenesis model, and resulting histological sections of supplemented and unsupplemented samples.

angiogenic factors are mixed with the Matrigel, then injected into mice, vessels form in the plug within 14 days (Figure 3) [12]. The time frame of vessel formation is different for each compound and some substances may require indirect action of, i.e., inflammatory cells to induce angiogenesis, whereas other compounds (bFGF) appears to act directly on the surrounding vasculature to induce vessel formation. The major advantage of using this in vivo Matrigel angiogenic assay is that, all events occurring in angiogenesis (i.e., adhesion, migration and tube formation) are present in this assay, and the newly forming vessels can easily be identified and quantified in histological sections.

We have used the in vivo Matrigel angiogenic assay to re-examine substances tested using the in vitro tube formation assays. In most cases substances which promoted tube formation in vitro also stimulated or enhanced angiogenesis in the Matrigel in vivo assay (Table 2). Conversely substances that inhibited tube formation such as inhibitors of collagen synthesis or collagenase activity (TIMPs) also blocked vessel formation in the in vivo assay. Comparison of the in vivo Matrigel model to other more established models such as the chicken chorioallantoic membrane and the rabbit cornea, showed a similar correlation of action with this new assay.

Examining the ability of substances to modulate endothelial cell tube formation in vitro, can serve as a rapid method for identifying new angiogenic or anti-angiogenic components. Conversely, this in vitro method can be used to evaluate the endothelial cell components that are necessary for the cells to migrate on the Matrigel, and then align and form tubes. We are currently investigating the proteins that are involved in early events of tube formation. Preliminary results indicate that there are several genes that are turned on during the first 4 hrs of tube formation. Thus, this in vitro model provides us with information as to the products produced by the endothelial cells as well as the exogenous substances that affect their differentiation. The in vivo Matrigel plug method described above can then be used to confirm the activity of a substances tested in vitro. These methods may be more useful than other models, in that they reduce the lengthy and expensive animal testing such as the rabbit eye model assay for angiogenesis.

Table 2. Comparison of the Matrigel in vitro and in vivo methods to other angiogenic models.

Compound	Matrigel In Vitro /In Vivo	Other Assays In Vivo
YIGSR	inhibit/ ND	- inhibit in the CAM and rabbit eye model
RGD	inhibit/ ND	-ND
SIKVAV	promote/stimulate	-stimulate: sponge and CAM models
bFGF	promote/stimulate	-stimulates in CAM
TGFß	promote/stimulate	-stimulate in rabbit ear
alpha IFN	promote/stimulate	-inhibit angiomas
gamma IFN	inhibit/ ND	ND
Scatter Factor HGF	promote/stimulate	-stimulate in rabbit eye
Haptoglobin	promote/stimulate	-stimulate in sponge disk
Phorbol Esters	promote/ ND	-stimulate in CAM
Inhibitors of collagen synthesis	inhibit/inhibit	-inhibit in CAM

The role of gonadal steroids in angiogenesis.

As an example of the applicability of the two Matrigel models, we have employed these methods to examine the role of estrogen in angiogenesis. There are several pathological conditions that are thought to be exacerbated by the presence of increased estrogen. For example, certain chronic inflammatory diseases, in particular Takayasu's arteritis, occur most often in women of child-bearing age[32]. These diseases are associated with endothelial cell proliferation and extensive blood vessel collateralization [33]. Normally, the only tissues in the body where angiogenesis is actively occurring (cyclically) are in the uterus and ovaries. Furthermore, a recent investigation has demonstrated that the presence of excised pieces of the endometrium (in active luteal phase) can elicit an angiogenic response in the cornea of the rabbit eye [34]. Thus, it is quite likely that angiogenesis may be under control of steroid hormones present in the female reproductive tissues.

We have examined the effect of estradiol on endothelial cell behavior in vitro. When endothelial cells are cultured in the presence of increasing amounts of estradiol, the cells have increased attachment to laminin (Table 3). In Boyden chamber migration assays, endothelial cells exposed to estradiol demonstrated increased migration through Millipore filters in response to a chemoattractant (Table 3). In wounding experiments, where a confluent monolayer is scratched with a 2mm rubber policeman, the endothelial cells also demonstrated increased cell migration and wound closure (not shown). These studies suggest that estrogen can affect the biologic activity of endothelial cells by increasing their ability to migrate and adhere to the substratum.

The in vitro Matrigel assay was also used to examine the effect of estradiol on endothelial cell differentiation. HUVEC preincubated with estradiol at 1 ng/ml showed almost three-fold increase in tube formation (Table 3). This increase in tube formation was also dose dependent; with the maximal stimulation occurring at 1 ng/ml, well within physiological range.

Table 3. The effect of estradiol on HUVEC adhesion, migration and tube formation.

Estradiol Dose (ng/ml)	Adhesion to Laminin (relative cell area)	Migration (# cells/field)	Tube formation (area/well)
0	1.0 ± 0.08	69.9 ± 3.8	2.2 ± 0.22
0.1	1.39 ± 0.04	106.9 ± 3.1	4.5 ± 0.29
0.5	ND	110.0 ± 2.4	5.4 ± 0.22
1	1.73 ± 0.06	131.4 ± 3.9	6.0 ± 0.32
2	2.14 ± 0.06	165.4 ± 6.2	6.0 ± 0.52

This enhancement of tube formation was reexamined in the in vivo Matrigel model. Matrigel with or without bFGF was co-injected into normal and ovariectomized mice. The gross appearance of the plugs in mice injected with Matrigel alone show a clear plug area with very little vasculature, whereas with bFGF (50 ng/ml) showed a bright red plug full of vessels (Figure 4). Matrigel plugs containing bFGF cultured in ovariectomized mice showed less blood vessels than control mice. When estradiol was added back in the form of a slow release pellet, the plug was bright red and had many associated blood vessels. The factor VIII immunostained histological sections corresponded to the gross morphology (Figure 4). Estradiol also increased the number of mature arteries and veins . In addition, quantitation of the vessels area in these histological specimens demonstrated a two fold increase in total vessel area in plugs containing both added estradiol and bFGF as compared to bFGF alone.

Summary

Angiogenesis can be regulated by many different types of compounds. Several in vitro and in vivo assays exist for detecting angiogenic activity. Many of the established assays are problematic due to difficulty in quantitation, prolonged (days or weeks) assay time, technical difficulties in reproducibility and the need for many animals. We describe here a new, quick and easy in vitro assay which is reliably quantitative and an in vivo assay which is technically easy and reproducibly quantitative. Comparing the activities of known angiogenic compounds in these assays with established assays demonstrates the usefulness of these new assays. In addition, we present data on the angiogenic activity of estrogens using these new assays. The verification of the two new assays should promote their use and allow for a better understanding of the mechanisms involved in the angiogenic process.

Matrigel Plug gross morphology and Histology

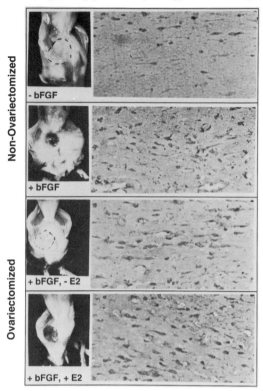

Figure 4. The mouse Matrigel assay. The gross morphology of the plugs on the left and the factor VIII immunostained histology on the right. Mice (normal (-NOv) and ovariextomized (-OV)) were injected with Matrigel with or without bFGF.

REFERENCES:

1.   Form, D.M., B.M. Pratt, and J.A. Madri, Endothelial cell proliferation during angiogenesis. In vitro modulation by basement membrane components. Lab Invest, 55: 521-30 (1986).

2.   Folkman, J., The role of angiogenesis in tumor growth. Semin Cancer Biol, 3: 65-71 (1992).

3.   Folkman, J. and Y. Shing, Control of angiogenesis by heparin and other sulfated polysaccharides. Adv Exp Med Biol, 313: 355-64 (1992).

4.   Sato, N., H. Nariuchi, N. Tsuruoka, T. Nishihara, J.G. Beitz, P. Calabresi, and A.J. Frackelton, Actions of TNF and IFN-gamma on angiogenesis in vitro. J Invest Dermatol, (1990).

5.   Pepper, M.S., D. Belin, R. Montesano, L. Orci, and J.D. Vassalli, Transforming growth

factor-beta 1 modulates basic fibroblast growth factor-induced proteolytic and angiogenic properties of endothelial cells in vitro. J Cell Biol, 111: 743-55 (1990).

6.   Heldin, C.H., K. Usuki, and K. Miyazono, Platelet-derived endothelial cell growth factor. J Cell Biochem, 47: 208-10 (1991).

7.   Kim, K.J., B. Li, K. Houck, J. Winer, and N. Ferrara, The vascular endothelial growth factor proteins: identification of biologically relevant regions by neutralizing monoclonal antibodies. Growth Factors, 7: 53-64 (1992).

8.   Grant, D.S., H.K. Kleinman, I.D. Goldberg, M.M. Bhargava, B.J. Nickoloff, J.L. Kinsella, P. Polverini, and E.M. Rosen, Scatter factor induces blood vessel formation in vivo. Proc Natl Acad Sci U S A, 90: 1937-41 (1993).

9.   Bussolino, F., R.M. Di, M. Ziche, E. Bocchietto, M. Olivero, L. Naldini, G. Gaudino, L. Tamagnone, A. Coffer, and P.M. Comoglio, Hepatocyte growth factor is a potent angiogenic factor which stimulates endothelial cell motility and growth. J Cell Biol, 119: 629-41 (1992).

10.   Wright, P.S., D.D. Cross, J.A. Miller, W.D. Jones, and A.J. Bitonti, Inhibition of angiogenesis in vitro and in ovo with an inhibitor of cellular protein kinases, MDL 27032. J Cell Physiol, 152: 448-57 (1992).

11.   Wilting, J. and B. Christ, A morphological study of the rabbit corneal assay. Anat Anz, 174: 549-56 (1992).

12.   Passaniti, A., R.M. Taylor, R. Pili, Y. Guo, P.V. Long, J.A. Haney, R.R. Pauly, D.S. Grant, and G.R. Martin, A simple, quantitative method for assessing angiogenesis and antiangiogenic agents using reconstituted basement membrane, heparin, and fibroblast growth factor. Lab Invest, 67: 519-28 (1992).

13.   Nicosia, R.F., P. Belser, E. Bonanno, and J. Diven, Regulation of angiogenesis in vitro by collagen metabolism. In Vitro Cell Dev Biol, (1991).

14.   Montesano, R., M.S. Pepper, J.D. Vassalli, and L. Orci, Modulation of angiogenesis in vitro. Exs, 61: 129-36 (1992).

15.   Lawley, T.J. and Y. Kubota, Induction of morphologic differentiation of endothelial cells in culture. J Invest Dermatol, 93S: 59S-61S (1989).

16.   Knighton, D.R., V.D. Fiegel, and G.D. Phillips, The assay of angiogenesis. Prog Clin Biol Res, 365: 291-9 (1991).

17.   Auerbach, R., W. Auerbach, and I. Polakowski, Assays for angiogenesis: a review. Pharmacol Ther, 51: 1-11 (1991).

18.   Takigawa, M., Y. Nishida, F. Suzuki, J. Kishi, K. Yamashita, and T. Hayakawa, Induction of angiogenesis in chick yolk-sac membrane by polyamines and its inhibition by tissue inhibitors of metalloproteinases (TIMP and TIMP-2). Biochem Biophys Res Commun, 171: 1264-71 (1990).

19.   Wilting, J., B. Christ, and H.A. Weich, The effects of growth factors on the day 13 chorioallantoic membrane (CAM): a study of VEGF165 and PDGF-BB. Anat Embryol (Berl), 186: 251-7 (1992).

20.   Grant, D.S., J.L. Kinsella, R. Fridman, R. Auerbach, B.A. Piasecki, Y. Yamada, M. Zain, and H.K. Kleinman, Interaction of endothelial cells with a laminin A chain peptide (SIKVAV) in vitro and induction of angiogenic behavior in vivo. J Cell Physiol, 153: 614-25 (1992).

21.   Jaffe, E.A., R.L. Nachman, C.G. Becker, and C.R. Minick, Culture of human endothelial cells derived from umbilical veins-identification by morphological and immunological criteria. J. Clin. Invest., 52: 2745-2756 (1973).

22.   Madri, J.A., L. Bell, M. Marx, J.R. Merwin, C. Basson, and C. Prinz, Effects of soluble factors and extracellular matrix components on vascular cell behavior in vitro and in vivo: models of de-endothelialization and repair. J Cell Biochem, 45: 123-30 (1991).

23.   Liu, H.M., D.L. Wang, and C.Y. Liu, Interactions between fibrin, collagen and endothelial cells in angiogenesis. Adv Exp Med Biol, 281: 319-31 (1990).

24.   Ingber, D., Extracellular matrix and cell shape: potential control points for inhibition of angiogenesis. J Cell Biochem, 47: 236-41 (1991).

25.   Kinsella, J.L., D.S. Grant, B.S. Weeks, and H.K. Kleinman, Protein kinase C regulates endothelial cell tube formation on basement membrane matrix, Matrigel. Exp Cell Res, 199: 56-62 (1992).

26.   Madri, J.A. and S.K. Williams, Capillary Endothelial Cell Cultures: Phenotypic Modulation by Matrix Components. J. Cell Biol., 97: 153-165 (1983).

27.   Grant, D.S., H.K. Kleinman, and G.R. Martin, The role of basement membranes in vascular development. Ann N Y Acad Sci, 588: 61-72 (1990).

28.   Vukicevic, S., H. Kleinman, F.P. Luyten, A.B. Roberts, N.S. Roche, and A.H. Reddi, Identification of multiple active growth factors in basement membrane Matrigel suggests caution in interpretation of cellular activity related to extracellular matrix components. Exp. Cell Res., 202: 1-8 (1992).

29.   Kleinman, H.K., J. Graf, Y. Iwamoto, G.T. Kitten, R.C. Ogle, M. Sasaki, Y. Yamada, G.R. Martin, and L. Luckenbill-Edds, Role of basement membranes in cell differentiation. Ann. N. Y. Acad. of Sci., 513: 134-145 (1987).

30.   Grant, D.S., K.-I. Tashiro, B. Segui-Real, Y. Yamada, G.R. Martin, and H.K. Kleinman, Two different laminin domains mediate the differentiation of human endothelial cells into capillary-like structures in vitro. Cell, 58: 933-943 (1989).

31.   Kibbey, M.C., D.S. Grant, and H.K. Kleinman, Role of the SIKVAV site of laminin in promotion of angiogenesis and tumor growth: an in vivo Matrigel model. J Natl Cancer Inst, 84: 1633-8 (1992).

32.   Shelhamer, J.H., D.J. Volkman, J.E. Parillo, T.J. Lawley, M.R. Johnston, and A.S. Fauci, Takayasu's arteritis and its therapy. Ann. Intern. Med., 103: 121-126 (1985).

33.   Ahmed, S.A., W.J. Penhale, and N. Talal, Sex hormones, immune responses, and autoimmune diseases. Mechanisms of sex hormones action. Am. J. Pathol., 121: 531-551 (1985).

34.   Torry, R.J. and B.J. Rongish, Angiogenesis in the uterus: potential regulation and relation to tumor angiogenesis. Am J Reprod Immunol, 27: 171-9 (1992).

EMA-1, A NOVEL ENDOTHELIAL CELL SURFACE MOLECULE THAT IS PREFERENTIALLY EXPRESSED BY MIGRATING ENDOTHELIAL CELLS

Hellmut G. Augustin-Voss[1] and Bendicht U. Pauli[2]

[1]Cell Biology Laboratory, Dept. of Gynecology
  and Obstetrics, University of Göttingen,
  3400 Göttingen, Germany
[2]Cancer Biology Laboratories, Dept. of Pathology,
  Cornell University College of Veterinary Medicine,
  Ithaca, NY 14853, USA

INTRODUCTION

The vascular endothelium forms a continuous sheet of squamous epithelium that lines the inside of all vessels. Extending throughout the body, it is a tissue with considerable biological potential. Many of the complex functions of endothelial cells appear to involve specific structural and chemical domains of the lumenal cell surface.[9,10] Structural differences characterized by the formation of continuous, fenestrated and discontinuous endothelia serve the need for different transendothelial perfusion rates in various organs, while the biochemical heterogeneity of endothelial cells is closely linked to their role in inflammation, immunity, and neoplasia. Constitutively expressed and inducible endothelial cell surface molecules have been identified that play critical roles in the regulation of lymphocyte and leukocyte trafficking, and in the metastatic colonization of select organ sites.[10,14] Distinct biochemical cell surface properties have recently been associated with the migratory phenotype of endothelial cells.[1] The lumenal surface of migrating endothelial cells expresses a distinct pattern of hyperglycosylation and specific migration-associated cell surface molecules. These findings suggested that endothelial cells express a distinct phenotype during angiogenesis and/or reendothelialization.

In the present study, we describe the production of monoclonal antibodies that preferentially recognize migrating endothelial cells using lumenal endothelial cell membrane vesicles from growth-arrested and migrating cultures in a passive/active immunization protocol. An endothelial cell migration-associated antigen (EMA-1) is identified by mAb 6F8 and characterized in some detail. It defines an endothelial cell surface molecule with an apparent molecular weight of 110 kDa that is preferentially expressed at the leading edge of migrating endothelial cells.

Angiogenesis: Molecular Biology, Clinical Aspects
Edited by M.E. Maragoudakis *et al*, Plenum Press, New York 1994

MATERIALS AND METHODS

Migration Assays

Two different assays were used to study the migration of bovine aortic endothelial cells (BAEC). In order to probe individual, migrating BAEC for monoclonal antibody (mAb) staining, BAEC were grown to confluence within the confines of rectangular silicon templates after which they were released from growth arrest and allowed to migrate for 24 h by removing the silicon templates.[2] The second assay was a circular scraping technique that was designed to produce large numbers of simultaneously migrating BAEC.[1] It was used for the screening of hybridoma supernatants by ELISA as well as for the biochemical analysis of migrating BAEC.

Production of Monoclonal Antibodies

Monoclonal antibodies which preferentially recognize migrating endothelial cells were produced following a passive/active immunization protocol using outside out endothelial membrane vesicles as antigen.[15] Outside-out endothelial cell membrane vesicles were harvested from confluent (arrested) and circularly scraped (migrating) BAEC monolayers by incubating them with a solution of 100 mM paraformaldehyde, 2 mM DTT, 1 mM $CaCl_2$, and 0.5 mM $MgCl_2$ in PBS.[8,15] Mice were passively immunized by intravenous injection of syngeneic mouse antiserum directed against vesicles from confluent monolayers. Five minutes later, the mice were actively immunized intraperitoneally with vesicles obtained from circulary scraped BAEC monolayers emulsified in adjuvant. Mice were boostered after 4 weeks by passive/active immunization. Three days before fusion of splenocytes with myeloma cells, mice were injected intrasplenically with vesicles from migrating BAEC monolayers. Fusion procedures, cloning of hybridomas, and immunoglobulin subclass determination were performed following standard protocols. Hybridoma supernatants were screened for relevant mAbs against migrating endothelial cells using a differential ELISA comparing binding intensities of arrested BAEC to that of migrating BAEC.

Immunohistochemistry and Immunoprecipitation

Serial sections were prepared from organs of healthy adult cattle. Tissue specimen were fixed with formaldehyde and embedded in paraffin. Deparaffinized sections were stained by an indirect immunoperoxidase technique using hybridoma supernatants as primary antibody and biotinylated anti-mouse Ig as secondary antibody. Antibody binding was visualized with strepatavidin-peroxidase and the peroxidase substrate AEC. For immunoprecipitation of the mAb 6F8 antigen, surface proteins of migrating endothelial cells were labeled by lactoperoxidase-catalyzed iodination. Labeled cell lysates were incubated with hybridoma supernatant and Ag-Ab complexes were precipitated with anti-mouse-IgM sepharose. Precipitates were resuspended in electrophoresis sample buffer and resolved by SDS-PAGE.

RESULTS

Monoclonal Antibody Production

Three fusions were performed following a passive/active immunization protocol in order to produce monoclonal antibodies that would preferentially recognize migrating endothelial cells. More than 1500 hybridoma supernatants were tested by differential ELISAs that compared the binding intensitiesy of confluent, growth-arrested BAEC monolayers to that of migrating or subconfluent BAEC monolayers. On average, the absorbance values of migrating and subconfluent monolayers were 62.8% of the absorbance values of confluent monolayers, reflecting differences in cell number and exposed cell surface of the different cell populations. Of the hybridomas tested, 51 had higher absorbance values in the migrating/subconfluent monolayers than in the confluent monolayers, i.e., the binding intensity was at least 1.6 times upregulated on the migrating/subconfluent cells. One hybridoma (6F8) was selected that yielded a 1.7 times higher absorbance value in subconfluent and circularly scraped BAEC monolayers than in confluent BAEC monolayers. Hybridoma 6F8 was subcloned by limiting dilution and found to produce antibodies of the IgM isotype. The putative antigen recognized by this antibody was termed endothelial cell migration-associated antigen-1 (EMA-1).

EMA-1 Cytochemistry

As suggested by the ELISA data, the EMA-1 antibody bound to both confluent and migrating BAEC. However, expression of EMA-1 was strongly upregulated on migrating endothelial cells (Fig. 1A). Binding was most intense on cell processes and at the rim of migrating endothelial cells (Fig.1B). In contrast, binding to confluent endothelial cell monolayers was characterized by a weak, diffuse evenly distributed surface staining of the cells.

EMA-1 Tissue Distribution

The tissue distribution of EMA-1 was determined by indirect immunohisto-chemistry. Stained tissues were the lungs, liver, kidney, heart, skeletal muscle, brain, small and large intestine, skin, ovary, uterus, spleen, and lymph node. Anti-EMA-1 monoclonal antibody 6F8 selectively recognized endothelial cells of varying vessel calibers (Table 1). The antibody bound consistently to different caliber arteries and

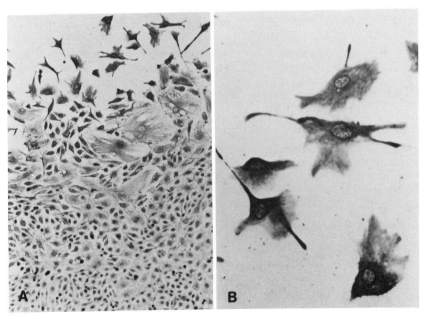

Figure 1. EMA-1 cytochemistry. Direct comparison of migrating and confluent endothelial cells reveals a prominent upregulation of the binding of anti-EMA-1 mAb 6F8 to the cells at the migrating front (A). Higher magnification of the cells at the migrating front demonatrates the preferential expression of EMA-1 at cell processes and at the leading edge of migrating endothelial cells (B).

veins in all organs except skeletal muscle. Arterioles, capillaries, and venules were only stained in the ovary (rete ovary), the intestine (submucosal vessels), and the skin. Ovaries and uteri, normally expressing various degrees of angiogenic activity, contain EMA-1 positive sprouting endothelial cells in the proximity of the growing follicle and in the mucosa of the uterus. This observation suggests that EMA-1 is expressed *in vivo* in areas of neovasculatization.

Table 1. Tissue distribution of EMA-1

	Arteries	Arterioles	Capillaries	Venules	Veins
Lung	+/++	-	-	-/+	+/++
Liver	++	-	-	-/+	++
Kidney	+/++	+	-/+	+	++
Heart	-/+	-	-	-/+	+
Skeletal muscle	-	-	-	-	-/+
Brain	-/+	-	-	-	-/+
Intestine	+++	++	+	+	+++
Skin	+++	++	+	+	+++
Ovary	++	+	+/++	+/++	+++
Uterus	++	++	+/++	++	++
Spleen	+	-	-	-	+
Lymph node	+/++	-	-	-	+/++

Biochemical Analysis of EMA-1

Anti-EMA-1 monoclonal antibody 6F8 was used to characterize EMA-1. Immunoprecipitation and subsequent gel electrophoresis under reducing conditions of surface iodinated endothelial cell surface molecules yielded a single autoradiographic band with an apparent molecular weight of 110-kDa. The molecular weight of this band was estimated from its electrophoretic mobility which was slightly faster than that of ß-galactosidase (119-kDa). EMA-1 could also be detected by Western blot analysis of whole cell lysates. The 110 kDa band could not be immunoprecipitated after N-glycanase treatment of lysates of surface iodinated endothelial cells, suggesting that the 6F8 antibody recognizes a carbohydrate antigen on endothelial cells.

DISCUSSION

In the present study we describe an endothelial cell surface molecule (EMA-1) that is strongly expressed by migrating BAEC *in vitro*. EMA-1 has been identified and isolated by immunoaffinity procedures using anti EMA-1 mAb 6F8. The monoclonal antibody has been selected from a battery of hybridomas that preferentially bound to migrating endothelial cells. Classical immunization and hybridoma techniques have identified several endothelial cell specific molecules.[5,7,11,12,13] Nevertheless, few endothelial cell specific molecular probes are available considering the extensively characterized heterogeneity of different endothelial cell populations.[3,6,9] In the present study, hybridomas were obtained following a passive/active immunization protocol. This procedure was used in our laboratory to produce several mAbs against endothelial cell surface determinants of distinct vascular branches in distinct organs.[14,15] Mice were actively immunized with endothelial cell membrane vesicles from migrating BAEC (specific vesicles), immediately following passive immunization with syngeneic polyclonal antiserum directed against an antigenically closely related vesicle population obtained from confluent BAEC monolayers (unspecific vesicles). The procedure greatly improves the probability of generating mAbs against minor and/or weakly immunogenic determinants.[15] The effectiveness of the technique was potentiated by the use of membrane vesicles instead of whole cells in the immunization.

EMA-1 is an endothelial cell-specific, but not migration-specific molecule. It is expressed constitutively on growth-arrested endothelia of arteries and veins in almost all organs, but a strong staining reaction of capillary and venular endothelia with anti-EMA-1 mAb 6F8 is restricted to the ovary and the uterus. Interestingly, these organs undergo cyclic remodeling processes and, thus, contain numerous sprouting capillaries with migrating endothelial cells. Similarly, minor EMA-1 staining reactions with endothelia in skin and intestinal microvessels may be associated with focal inflammatory and repair processes regularly observed in these organs. The reasons for the selective expression of EMA-1 on growth-arrested large vessel and migrating capillary and venular endothelial cells are unclear and need further investigation.

Monoclonal antibodies against migrating endothelial cells, such as the anti-EMA-1 mAb, will be usefull tools to better understand the molecular events during endothelial cells migration and particularly for the analysis of the functional relevance of carbohydrate moieties on endothelial cells during angiogenesis. Subset specific endothelial cell surface molecules might also serve as probes to target specific subpopulations of endothelial cells for diagnostic or therapeutic purposes, such as the targeting of migrating endothelial cells to inhibit pathologic angiogenesis.[4,7,9]

REFERENCES

1. H. G. Augustin-Voss and B. U. Pauli, Migrating endothelial cells are distinctly hyperglycosylated and express specific migration-associated cell surface glycoproteins, *J. Cell Biol.* 119:483 (1992).
2. H. G. Augustin-Voss and B. U. Pauli, Quantitative analysis of autocrine-regulated, matrix-induced, and tumor cell-stimulated endothelial cell migration using a silicon template compartmentalization technique, *Exp. Cell Res.* 198:221 (1992).
3. P. N. Belloni and G. L. Nicolson, Differential expression of cell surface glycoproteins on various organ-derived microvascular endothelia and endothelial cell cultures, *J. Cell. Physiol.* 136:398 (1988).
4. M. S. F. Clarke and D. C. West, The identification of proliferation and tumor-induced proteins in human endothelial cells: a possible target for tumor therapy, *Electrophoresis* 12:500 (1991).
5. S. Goerdt, L. J. Walsh, G. F. Murphy, and J. S. Pober, Identification of a novel high molecular weight protein preferentially expressed by sinusoidal endothelial cells in normal human tissues, *J. Cell. Biol.* 113:1425 (1991).
6. F. Gumkowski, G. Kaminska, M. Kaminski, L. W. Morrissey, and R. Auerbach, Heterogeneity of mouse vascular endothelium. In vitro studies of lymphatic, large blood vessel, and microvascular endothelial cells. *Blood Vessels* 24:11 (1987).
7. H. H. Hagemeier, E. Vollmer, S. Goerdt, K. Schulze-Osthoff, and C. Sorg, A monoclonal antibody reacting with endothelial cells of budding vessels in tumors and inflammatory tissues, and non reactive with normal adult tissues, *Int. J. Cancer* 38:481 (1986).
8. R. C. Johnson, H. G. Augustin-Voss, D. Zhu, and B. U. Pauli, Preferential binding of lung-derived endothelial membrane vesicles to lung metastatic tumor cells, *Cancer Res.* 51:394 (1992).
9. S. A. McCarthy, I. Kuzu, K. C. Gatter, and R. Bicknell, Heterogeneity of the endothelial cell and its role in organ preference of tumour metastasis, *Trends Pharmacol. Sci.* 12:462 (1991).
10. B. U. Pauli, H. G. Augustin-Voss, M. E. El-Sabban, R. C. Johnson, and D. A. Hammer, Organ preference of metastasis: the role of organ-specifically modulated endothelial cells, *Cancer Metastas. Rev.* 9:175 (1990).
11. M. C. Rorvik, D. P. Allison, J. A. Hotchkiss, H. P. Witschi, and S. J. Kennel, Antibodies to mouse lung endothelium, *J. Histochem. Cytochem.* 36:741 (1988).
12. D. J. Ruiter, R. O. Schlingemann, F. J. R. Rietveld, and R. M. W. de Waal, Monoclonal antibody-defined human endothelial cell antigens as vascular markers, *J. Invest. Dermatol.* 93:25S (1989).
13. H. Seulberger, F. Lottspeich, and W. Risau, The inducible blood-brain barrier specific molecule HT7 is a novel immunoglobulin-like cell surface glycoprotein, *EMBO J.* 9:2151 (1990).
14. D. Zhu, C. F. Cheng, and B. U. Pauli, Mediation of lung metastasis of murine melanomas by a lung-specific endothelial cell adhesion molecule, *Proc. Natl. Acad. Sci.* 88:9568 (1991).
15. D. Zhu and B. U. Pauli, Generation of monoclonal antibodies directed against organ-specific endothelial cell surface determinants, *J. Histochem. Cytochem.* 39:1137 (1991).

EARLY RESPONSE GENES IN ENDOTHELIAL CELLS

Timothy Hla

Department of Molecular Biology
Holland Laboratory
American Red Cross
15601 Crabbs Branch Way
Rockville, Maryland 20855

SUMMARY

The vascular endothelial cells initiate angiogenesis when stimulated by growth factors and cytokines. While the cellular activities of the angiogenic factors such as the fibroblast growth factors (FGF), tumor necrosis factor (TNFα) and type-ß transforming growth factor are well-characterized, molecular mechanisms involved in the different phases of angiogenesis, namely, migration, proliferation and differentiation are not well-understood. Protein kinase-C pathway is involved in the regulation of angiogenesis because the tumor promoter phorbol myristic acetate (PMA) is a potent inhibitor of FGF-induced endothelial cell proliferation and an inducer of differentiation into capillary-like tubules. Because immediate-early (IE) genes have been shown to be involved in critical regulatory events, we have cloned and characterized PMA-inducible IE genes from human umbilical vein endothelial cells (HUVEC). Collagenase type-I and a novel gene termed edg-1 were isolated as abundant PMA-inducible transcripts. The structure of edg-1 suggests that it encodes a novel G-protein coupled receptor. Furthermore, an isotype of cyclooxygenase (Cox) enzyme, Cox-2, was also induced as an IE gene in HUVEC. The expression of immunoreactive Cox isotypes in vivo correlates with the angiogenesis that occurs in chronic inflammatory diseases such as rheumatoid arthritis (RA). Because prostaglandins induce inflammation and angiogenesis, exaggerated and persistent expression of the Cox-2 may be important in maintaining the inflammatory disease phenotype. Regulated induction of IE genes such as edg-1, collagenase type I and Cox-2 may be important in physiological events that require angiogenesis; however, exaggerated and dysregulated expression of IE genes may result in enhanced angiogenesis, a characteristic of chronic inflammatory diseases and solid tumor growth.

INTRODUCTION

Angiogenesis, also known as new blood vessel formation, is initiated by endothelial cells and involves the orderly migration, proliferation and differentiation of endothelial cells into new capillary channels (1). Enhanced angiogenesis is thought to be necessary for solid tumor growth and chronic inflammatory disease progression (2). Cultured vascular endothelial cells express many of the characteristics of endothelium *in vivo* and are used as a model system to study angiogenesis (3). Thus, the FGF family of polypeptide mitogens were identified as potent inducers of endothelial cell proliferation and migration (4). In contrast, the cytokines IL-1, TNFα and TGFß inhibit the growth of endothelial cells (5-7). Phorbol myristic acetate (PMA), a potent activator of protein kinase C, also inhibit the FGF-stimulated growth of endothelial cells (8). Prolonged treatment of endothelial cells grown on collagen or fibrin matrices with PMA results in the induction of differentiation into capillary-like tubules (9). We utilized the PMA-treated human umbilical vein endothelial cells (HUVEC) as a model system to investigate the transcriptional events involved in angiogenesis.

Immediate-early (IE) genes are rapidly induced at the level of transcription by various cellular stimuli (10). Growth factor-inducible nuclear IE genes, such as c-fos and c-jun, are involved in the critical regulatory events of cell-cycle traverse (11). Thus, microinjection of specific antibodies against c-fos inhibits DNA synthesis and cellular transformation (12). Secreted IE genes such as JE and KC are thought to act as cytokines that communicate with neighboring cells (13). For example, monocyte chemoattractants such as MAP-1 and MCP-1 are produced by activated fibroblasts (14). We assumed that the characterization of PMA-inducible IE genes in HUVEC may provide some novel insights into the phenomenon of angiogenesis *in vitro* .

We utilized the differential and subtractive hybridization techniques (15,16) to isolate IE genes from HUVEC. We also developed a novel technique, termed polymerase chain reaction-amplified subtractive hybridization (PCRASH) to enhance the sensitivity of the classical subtraction methods (17). While the exhaustive analysis of the PMA-inducible IE genes were not done, it is noteworthy that the repertoire of IE genes in HUVEC is different from that in fibroblasts (11). It is perhaps not surprising that differentiated cells may utilize different sets of inducible genes.

COLLAGENASE TYPE-I

Treatment of quiescent HUVEC with PMA induced the 2 kb mRNA for collagenase type-I. The induction was apparent within 30 min of PMA addition and was sustained for at least 24 h. The addition of cycloheximide, a protein synthesis inhibitor did not affect the PMA induction. Collagenase type-I represents a major PMA-inducible transcript in endothelial cells; it is estimated to be 0.1% of total mRNA population in HUVEC cells. The degradation of interstitial-type collagens and the modification of extracellular matrix is thought to be critical for the angiogenic response (1). Thus, rapid and sustained upregulation of this mRNA may be an important event in angiogenesis. Indeed, previous work by Moscatelli et al. have provided a biological basis since collagenase inhibitors block angiogenesis in *vitro* (18). Moreover, Montesano et al. have documented the coordinate regulation of proteases and protease inhibitors in the *in vitro* angiogenesis of endothelial cells (19).

ENDOTHELIAL DIFFERENTIATION GENE (EDG)-1

An abundant PMA-inducible IE gene, termed edg-1 was cloned by differential hybridization techniques (20). The 3 kb mRNA was rapidly induced by PMA but declines after 4 h. In contrast to collagenase type I, the edg-1 mRNA is super-induced in the presence of cycloheximide. This is probably due to the inhibition of degradation of the short-lived edg-1 mRNA ($T_{1/2}$ ~ 60 min). Nuclear run-on analysis indicated that the edg-1 gene transcription is induced by PMA. In addition, FGF-1 also induced the edg-1 mRNA, albeit with a different kinetics. The nucleotide sequence of the 2.7 kb cDNA clone for edg-1 was determined and was found to encode a polypeptide of 380 amino acids. Consistent with this deduction, *in vitro* translation of the edg-1 mRNA results in the synthesis of a polypeptide of approximately 40 kd. The hydrophobicity plot of the edg-1 polypeptide is shown in figure 1. A characteristic 7 transmembrane motif was seen, suggesting that the edg-1 polypeptide is a member of the G-protein-coupled receptor (GPR) superfamily. Indeed, significant sequence similarity exists between the edg-1 polypeptide and other members of the GPR family such as ß₂-Adrenergic receptor, Substance P receptor, cannabinoid receptor and the mas oncogene/ angiotensin receptor. Significantly, sequence motifs that are invariant in almost all members of the GPR superfamily are also present in edg-1, suggesting that it is a novel GPR. The proposed model of edg-1 is shown in figure 2. While the ligand and the signalling mechanisms of edg-1 are not known at present, expression of edg-1 in Cos cells results in the depression of basal cAMP levels. Thus edg-1 may interact with the G_i family of trimeric G-proteins. Moreover, ectopic expression of the edg-1 cDNA in NIH 3T3 cells results in the formation of transformed foci. This observation suggested that either the edg-1 expression constitutively stimulates signalling pathways in 3T3 cells or alternatively, the ligand for edg-1 is either produced by the 3T3 cells or is present in the cell growth medium. High-level expression of the edg-1 mRNA is specific for endothelial cells; however, edg-1-related mRNAs are present at low levels in fibroblasts, smooth muscle cells and epithelioid cell lines. The GPR family of receptors are well-known for their ability to transduce signals from extracellular peptides, neurotransmitters and hormones (21). The cellular

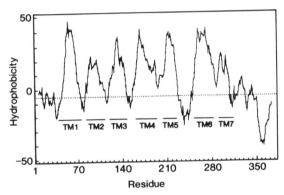

Figure 1. Hydrophobicity plot of the deduced amino acid sequence of the edg-1 cDNA.

responses that are regulated by the GPRs are generally thought to be rapid, such as ionic changes and transient changes in second messenger levels. Recent studies have implicated the role of the GPRs in the regulation of longer-term cellular responses such as migration, cell proliferation and differentiation of higher eukaryotes. For example, the neutrophil chemotaxis receptors for IL-8 and f-met-leu-phe were shown to be the members of the GPR superfamily (22). Ectopic expression of the receptors for angiotensin (mas oncogene), serotonin (5HT1c) and M1 muscarinic acetylcholine results in fibroblast transformation (23). In contrast, lower eukaryotes such as yeast and dictyostelium primarily utilize GPRs to regulate growth and differentiation pathways. In these organisms, GPRs are inducible genes and the enhanced expression of the receptors are required for growth arrest and differentiation. For example, inhibition of the expression of the cAMP receptor of dictyostelium by antisense RNA results in the blockage of cAMP-induced growth arrest and differentiation (24). Because edg-1 is an IE gene that is abundantly expressed in endothelial cells, edg-1-regulated signalling pathways may play an important role in angiogenesis.

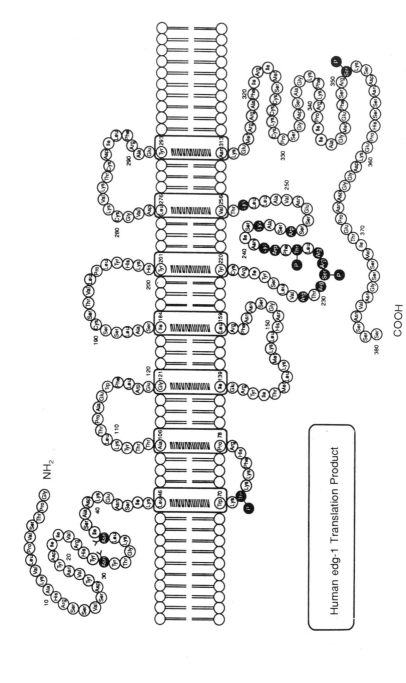

Figure 2. Structural model of the edg-1 polypeptide. Analogous to other GPRs, the N-terminus is assumed to be on the extracellular side of the plasma membrane. The shaded residues in the intracellular domains represent potential phosphorylation sites.

Human edg-1 Translation Product

CYCLOOXYGENASE (COX)-2

Prostaglandins (PGs) are oxygenated metabolites of arachidonic acid and are involved in the regulation of many biological processes including pain, inflammation, vascular tone, among others (25). The rate-limiting enzyme in the PG biosynthetic pathway is Cox, also known as PG H synthase (E.C.1.14.99.1) (25). PGs are produced rapidly in response to cellular stimuli and are secreted immediately to the extracellular milieu. They are then thought to act as short-lived biological response modifiers. PGE_1 and PGE_2 are potent inducers of angiogenesis in *vivo* , molecular mechanisms involved are not well-understood (26). Previous work has resulted in the cloning of the cDNA for the Cox enzyme from sheep seminal vesicles (27). While PMA induces the formation of PGs rapidly in a variety of cells, the kinetics of induction of PG synthesis does not correlate with that of the Cox mRNA levels. However, a homologous gene, termed Cox-2 is a PMA-inducible IE gene (28,29). The human Cox-2 cDNA possess significant sequence similarity to previously-isolated IE genes from 3T3 cells (TIS10) and CEF cells (CEF147) (28-30). The Cox-1 and -2 polypeptides are approximately similar in size (~ 69 kd) and are 61% similar in primary sequence. Amino acid residues that are critical for catalysis in the Cox-1 polypeptide (31) are also present in the Cox-2 protein, suggesting functional similarity. Expression of the Cox-1 and -2 cDNAs in Cos cells results in the elaboration of Cox activity, as measured by the conversion of radioactive arachidonic acid into PGs. The enzymatic activity of both Cox-1 and -2 proteins are blocked by indomethacin and ibuprofen. Interestingly, the mRNA levels for the Cox-2 were significantly lower in the transfected Cos cells, suggesting that the half-life of the Cox-2 mRNA is relatively short.

The mRNA for Cox-1 and -2 genes are expressed in a variety of human cells. For example, the Cox-1 mRNA is ubiquitously expressed in all human cells examined. In contrast, the mRNA for Cox-2 is detected in PMA-induced human endothelial, smooth muscle, monocytic cells and fibroblasts. Interestingly, the Cox-2 mRNA is not expressed in a variety of epithelioid cell lines such as A431 cells either treated or not with PMA (30). Previous studies on murine Cox-2 gene expression have also indicated that the Cox-2 gene expression is limited. The molecular mechanisms responsible for the tissue-specific expression of the Cox-2 gene is not understood. In HUVEC, cellular quiescence depresses the levels of Cox-2 and up-regulates the Cox-1 expression. This is probably because the growth factor FGF-1 induces the Cox-2 gene transiently as an IE gene whereas it represses the mRNA levels for Cox-1 (32). PMA, which is an inhibitor of HUVEC growth and a promoter of differentiation induces the Cox-2 mRNA as an IE gene and does not modulate the levels of the Cox-1 mRNA. Thus, the expression of the Cox-1 and -2 genes are under complex and dynamic control by angiogenic mediators *in vitro*. Because the products of the Cox pathway, for example, PGE_2 , induces angiogenesis and inflammatory responses, the modulation of the Cox-1 and -2 genes may be important *in vivo*.

In order to obtain *in vivo* correlates of Cox expression, we studied by immunohistochemistry, the expression of Cox isotypes in rheumatoid arthritis (RA). RA is a chronic inflammatory disease with a prominent angiogenic component (33). The affected individuals develop a classic inflammatory reaction in the various joint cavities, whereby mononuclear cell infiltration, fibroblast and synovial cell proliferation and enhanced angiogenesis develop. The proliferative rheumatoid synovium produces high levels of PGs, growth factors and proteases. In contrast to RA, osteoarthritis (OA) is characterized by a less aggressive development of the disease. Synovial tissues from the OA patients exhibit little angiogenesis and fibroblast proliferation. The expression of Cox proteins were studied on biopsy specimens of synovium from RA and OA patients by immuohistochemistry procedures using the polyclonal Cox antiserum, which reacts with both Cox-1 and -2 polypeptides. Strong immunostaining was observed in the cytoplasm of inflammatory mononuclear cells, synovial fibroblasts and endothelial cells of RA specimens. In contrast, the OA synovium stained less prominently. These data indicate that the extent and intensity of Cox immunoreactivity correlates with the inflammatory disease phenotype in RA (34).

The Cox-2 gene in endothelial cells represents a paradigm whereby transient induction is important in specific biological events such as cell growth or differentiation. Transient and regulated expression of the IE genes may play specific protective functions in a normal inflammatory scenario. However, exaggerated and dysregulated expression may contribute to the evolution of the chronic inflammatory diseases. Because of the crucial role that endothelial cells play in the tissue responses to inflammation, the study of IE genes in these cells may yield novel insights into the critical processes of chronic inflammatory diseases.

REFERENCES

1. Folkman, J. and C. Haudenschild (1980) *Nature* 288,551-556.
2. Folkman, J. (1985) *Adv. Cancer Res.* 43, 175-203.
3. Maciag, T. et al. (1982) *J. Cell Biol.* 94, 511-520.
4. Burgess, W.H. and T. Maciag (1989) *Ann. Rev. Biochem.* 58, 575-606.
5. Norioka, K. et al. (1987) *Biochem. Biophys. Res. Comm.* 145,969-975.
6. Frater-Schröder, M. et al. (1987) *Proc. Natl. Acad. Sci. USA* 84, 5277-5281.
7. Heimark, R. et al. (1986) *Science* 233, 1078-1080.
8. Doctrow, S.R. and J. Folkman (1987) *J. Cell Biol.* 104, 679-687.
9. Montesano, R. and Orci, L. (1985) *Cell* 42, 469-477.
10. Lau, L. and Nathans, D. (1987) *Proc. Natl. Acad. Sci., USA* 84, 1182-1186.
11. Herschman, R. (1991) *Ann. Rev. Biochem.* 60, 281-319.
12. Riabowol, K.T. et al. (1988) *Mol. Cell. Biol.* 8, 1670-1676.
13. Cochran, B. H. et al. (1983) *Cell* 33, 939-947.
14. Rollins, B. J. et al. (1988) *Proc. Natl. Acad. Sci., USA* 85, 3738-3742.
15. Almendral, J.M. et al. (1988) *Mol. Cell. Biol.* 8, 2140-2148.
16. Hla, T. and Maciag, T. (1990) *Biochem. Biophys. Res. Comm.* 167, 637-643.
17. Sargent, T.D. and Dawid, I. (1983) *Science* 222, 135-139.
18. Moscatelli, D. et al. (1980) *Cell* 20, 343-351.
19. Montesano, R. et al. (1990) *Cell* 62, 435-445.
20. Hla, T. and Maciag, T. (1990) *J. Biol. Chem.* 265, 9308-9313.
21. Lefkowitz, R.J. and Caron, M.G. (1988) *J. Biol. Chem.* 263, 4993-4996.
22. Murphy, P.M. and Tiffany, H.L. (1991) *Science* 253, 1280-1283.
23. Julius, D. et al. (1989) *Science* 244, 1057-1062.
24. Devreotes, P. (1989) *Science* 245, 1054-1058.
25. Needleman, P. et al. (1986) *Ann. Rev. Biochem.* 55, 69-102.
26. Ziche, M. et al. (1982) *J. Natl. Cancer Inst.* 69, 475-482.
27. DeWitt, D. and Smith, W.L. (1988) *Proc. Natl. Acad. Sci., USA* 85, 1412-1416.
28. Xie, W. et al. (1991) *Proc. Natl. Acad. Sci., USA* 88, 2692-2696.
29. Kujubu, D. et al. (1991) *J. Biol. Chem.* 266, 12866-12872.
30. Hla, T. and Neilson, K. (1992) *Proc. Natl. Acad. Sci., USA* 89, 7384-7388.
31. Shimokawa, T. and Smith, W.L. (1991) *J. Biol. Chem.* 266,6168-6173.
32. Hla, T. and Maciag, T. (1991) *J. Biol. Chem.* 266, 24059-24063.
33. Harris, E.D. (1990) *New Eng. Jour. Med.* 322, 1277-1289.
34. Sano, H. et al. (1992) *J. Clin. Invest.* 89, 97-108.

PlGF: PLACENTA DERIVED GROWTH FACTOR

Domenico Maglione* and Maria Graziella Persico

International Institute of Genetics and Biophysics, CNR, Via Marconi 10, 80125 Naples, Italy.
* Geymonat SpA Pharmaceutical industry,Via S. Anna 2, 03012 Anagni (FR), Italy

INTRODUCTION

The formation of new blood vessels from pre-existing ones, a phenomenon called angiogenesis, is a complicated process that involves the migration, proliferation and organization of endothelial cells. Angiogenesis is virtually absent in the healthy adult organism, restricted to a few conditions including wound healing and the formation of corpo luteum, endometrium, and placenta. In contrast, in certain pathological conditions, angiogenesis is dramatically enhanced, for example, in rheumatoid arthritis, psoriasis, diabetic retinopathy, hemangiomas and in solid tumors (Folkman and Klagsbrun, 1987; Klagsbrun and D'Amore, 1991).

A number of angiogenic growth factors have been purified and characterized (Klagsbrun and D'Amore, 1991). These growth factors appear to fall into two groups: (i) those that act directly on endothelial cells and (ii) those that act indirectly by inducing host cells to release specific endothelial growth factors (Risau, 1990; Yang and Moses, 1990). One member of the first group is vascular permeability factor/ vascular endothelial growth factor (VPF/VEGF), a dimeric protein isolated from the conditioned medium of various cell types. Besides angiogenic activity, VPF/VEGF displays the physiological function of increasing the permeability of the capillary vessels to various macromolecules (Keck et al., 1989; Connolly et al., 1989). A new factor has been recently isolated which shares biochemical and functional features with VPF/VEGF. This factor was named placenta growth factor (PlGF), because it was first isolated from a human placenta cDNA library. Due to the strong similarity with VPF/VEGF and its ability to stimulate the growth of endothelial cells, an angiogenic role for PlGF has been inferred (Maglione et al., 1991a). Topical informations on PlGF are presented in this article.

PlGF-1 AND PlGF-2: ISOLATION AND CHARACTERIZATION

Two different forms of PlGF have been isolated, PlGF-1 and PlGF-2, which are generated by an alternative splicing (see below). The cDNA coding for PlGF-1 was isolated from a human term-placenta library (Maglione et al., 1991a), while PlGF-2 cDNA was obtained from a human choriocarcinoma library (Maglione et al., 1993a). Comparison of the cDNA deduced amino acid sequences revealed that these two forms of PlGF are identical except for the presence of a highly basic 21 amino acid stretch at the carboxyl end of PlGF-2 (boxed region in Fig.1). They have a strong homology (53% identity) with the platelet-derived growth factor-like (PDGF-like) domain of VPF/VEGF. This homology reaches to 71% if both the identical and conservative amino acid changes are considered. Interestingly, there is a perfect conservation of distances among the 8 cysteines of PDGF-like domain (circled cysteines in Fig.1) between PlGFs and VPF/VEGF. Finally, one of the two putative N-glycosylation sites of PlGFs (double underlined in Fig.1) corresponds to the single site present in the VPF/VEGF (Maglione et al., 1991a).

The biochemical characterization of PlGF-1 and PlGF-2 was mainly performed using conditioned media from COS-1 cells transiently expressing the proteins. In this system the PlGF proteins were revealed by immuno-precipitation and/or immunoblot using antibodies obtained from rabbit serum or chicken egg yolk immunized with partially purified bacterial PlGF-1 (Maglione et al., 1991a). After applying these approaches, both PlGF-1 and -2 resulted secreted into culture medium. Using Von Heijene's algorithm (Von Heijene, 1986) we predicted four probable cleavage sites of the signal peptide responsible for PlGFs secretion (Maglione et al., 1991b) (arrows in Fig.1). The one which showed the best score (longest arrows in Fig.1) occurred between aminoacids 20 and 21 of the predicted amino acid sequences of PlGFs (Maglione et al., 1991a and unpublished results).

By using N-glycosidase F and tunicamycin, and reducing or unreducing conditions, we have demonstrated that both PlGFs are N-glycosylated and dimeric proteins with the monomers bound together by disulfite bonds (Maglione et al., 1991a and unpublished results).

We also have demonstrated that PlGF-2, but not PlGF-1, binds to (i) heparin in vitro and (ii) to heparin-like molecules of cell surface and/or extracellular matrix of COS-1 cells. Therefore, these unique properties of PlGF-2, due to the highly basic 21 amino acid stretch, mentioned above, are present only in PlGF-2 (unpublished results).

Human PlGF-1, from conditioned medium of transfected COS-1 cells, is able to stimulate, *in vitro*, the growth of calf pulmonary endothelial cells (Maglione et al., 1991a) and bovine adrenal cortex capillary endothelial cells (Maglione et al., 1991b). However, the level of observed stimulation is quite low, probably owing to the expression of PlGFs by endothelial cells themselves (see below).

M P V M R L F P C F L Q L L A G L A L P A V P P Q Q W A L S A G N̲ G S S E V E V − 40

V P F Q E V W G R S Y Ⓒ R A L E R L V D V V S E Y P S E V E H M F S P S Ⓒ V S L − 80

L R C T G Ⓒ Ⓒ G D E N L H Ⓒ V P V E T A N̲̲ V T M Q L L K I R S G D R P S Y V E L −120

T F S Q H V R Ⓒ E Ⓒ R P L R E K M K P E R [R R P K G R G K R R R E K Q R P T D C] −160

[H L] C G D A V P R R   −170

Figure 1. Amino acid sequence of the PlGFs.

The peptide encoded by exon 6, present only in PlGF-2, is boxed. The cysteines of PDGF-like domain are circled. The putative N-glycosylation site, present only in PlGFs, is single underlined, while the one conserved with VPF/VEGF is double underlined. Arrows indicate the putative cleavage sites of the signal peptide; the longest arrow indicates the most probable cleavage site.

PlGF GENE STRUCTURE AND ITS EXPRESSION

The gene coding for human PlGF maps on chromosome 14. The chromosome mapping was performed by probing a Southern blot of digested genomic DNAs prepared from 18 hamster-human somatic cell hybrids with a specific PlGF probe (Maglione et al., 1993a).

Comparison of cDNA and genomic PlGF sequences indicated that the PlGF gene presents 7 exons and 6 introns spanning an 8000-kb-long DNA interval (Maglione et al., 1993a). The mRNA coding for PlGF-2 contains the exon 6 that is spliced-out in the PlGF-1 mRNA. Table 1 describes the characteristics of these 7 exons and correlates them with the contents of the 9 exons of VPF/VEGF gene (Ferrara et al., 1991; Tisher et al., 1991). Table 1 also points out a strong similarity between the PlGF and VPF/VEGF gene structures which is consistent with a common evolutionary origin of these genes.

Expression of PlGF gene (see table 2) was assayed in several tissues and cell lines by northern blot analysis. However, to distinguish between the expression of PlGF-1 and PlGF-2, we also performed reverse PCRs using primers around the exon 6 (Maglione et al., 1993a). Highest expression of PlGF was seen in placenta at various stages and choriocarcinoma cell lines (JEG-3 and JAR). Low expression of PlGF was detected in human heart, brain, lung and skeletal muscle, while no expression was seen in human kidney, pancreas and HeLa cell line.

The majority of tissues and cell lines analized express both PlGF-1 and PlGF-2 transcripts. Only the PlGF-2 transcript was detected in invasive hydatiform mole and 8-week placenta, while PlGF-1 is the only form expressed in two human tumors (colon and mammary carcinomas).

Interestingly, PlGF is expressed, at least in culture, by endothelial cells derived from 3 types of tissues: human umbilical vein (HUVEC), bovine aortic (BAEC) and calf pulmonary artery (CPA-47) (unpublished results). From these data it is possible to suppose an autocrine role for PlGF on endothelial cells.

TRANSLATION OF PlGFs IS SEVERELY AFFECTED BY A SMALL OPEN READING FRAME

Regulation of PlGF gene expression is unknown. However, we have presented evidence indicating that expression of this gene could be regulated at translational level by a small open reading frame (ORF). This small ORF is localized within a region upstream to the coding region of PlGF mRNAs and it potentially codes for a peptide of 15 amino acids. Elimination of this element by point mutations in its initiator codon or by deletion, increases the PlGF synthesis in reticulocyte and wheat germ lysates, as well as in CV-1 cells, about 7-fold with respect to wild type. Furthermore, the small ORF may be an important regulatory element, because it is conserved also in bovines (Maglione et al., 1993b).

Even if little is known regarding the physiological conditions and the tissue(s) where the potential translational regulation of PlGF gene could take place, we have supposed that this could happen under hypoxia conditions

Table 1. Comparison between contents of PlGF and VPF/VEGF exons.

Exons of VPF/VEGF gene	Contents		Exons of PlGF gene
	VPF/VEGF gene	PlGF gene	
1	5' UTR and majority of the signal peptide	5' UTR and majority of the signal peptide	1
2	Peptide unrelated to PlGF exon 2	Peptide unrelated to VPF/VEGF exon 2	2
3	First six cysteines of the PDGF-like domain	First six cysteines of the PDGF-like domain	3
4	Last two cysteines of the PDGF-like domain	Last two cysteines of the PDGF-like domain	4
5	Terminal region of the PDGF-like domain	Terminal region of the PDGF-like domain	5
6	Higly basic region present in the C-terminus of VEGF189 and VEGF206	Higly basic region present only in the C-terminus of PlGF-2	6
7	Region present only in the C-terminus of VEGF206	Absent	–
8	Region present only in the C-terminus of VEGF165	Absent	–
9	Last amino acids of VPF/VEGF proteins and 3' UTR	Last amino acids of PlGF proteins and 3' UTR	7

UTR = untranslated region

Table 2. Expression of PlGF in tissues and cell lines.

Tissues/cell lines	PlGFs	Form 1	Form 2
8-week human placenta	+	--	yes
13-week human placenta	+	yes	yes
Human term placenta	+	yes	yes
Human placenta cells	+	yes	yes
Human amnion cells	+	yes	yes
Human heart	low	nd	nd
Human brain	low	nd	nd
Human lung	low	nd	nd
Human skeletal muscle	low	nd	nd
Human kidney	--		
HeLa cell line	--		
Human pancreas	--		
Hydatiform mole (non invasive)	+	yes	yes
Hydatiform mole (invasive)	+	--	yes
Choriocarcinoma cell line(JEG-3)	+	yes	yes
Hepatoma cell line (Hep-G2)	+	nd	nd
Human coloncarcinoma	low	yes	--
Human mammary carcinoma	low	yes	--
HUVEC	+	yes	yes
BAEC	+	nd	nd
CPA-47	+	nd	nd

+ = high level expression of PlGF.
low = low level expression of PlGF.
nd = non determined.
-- = no expression.
HUVEC = human umbilical vein endothelial cells.
BAEC = bovine aorta endothelial cells.
CPA-47 = calf pulmonary artery endothelial cell line.

(Maglione et al., 1993a). In fact, it is know that hypoxia induces an angiogenic response in many systems analyzed (Strik et al., 1991, and Shweiki et al., 1992).

CONCLUSIONS AND PERSPECTIVES

The present knowledge regarding PlGF indicates that it has a strong similitude with VPF/VEGF at various levels. The same kind of similitude was found between PDGF-A and -B (Westermark and Heldin, 1991), which are able to form a A-B heterodimer. Therefore, we have hypothesized that a heterodimeric PlGF-VPF form may exist in the cells in which they are expressed simultaneously (Maglione et al., 1993b). The finding of this heterodimer could open several interesting new lines of research. For instance, the PlGF-VPF heterodimer could binds to a new receptor(s), now unknown, inducing new cellular and physiological effect(s). In addition, at most eight sub-types of PlGF-VPF heterodimers could exist that represent all combinations between the two forms of PlGF (1 and 2) and four forms of VPF/VEGF (121, 165, 189 and 206). They could be compartmentalized on heparan sulfate proteoglycans to various degrees owing to the differential heparin binding capacity of the single monomers.

Acknowledgements

This work was supported by grants from the Progetto Finalizzato "Applicazioni Cliniche della Ricerca Oncologica", CNR.

REFERENCES

Connolly, D.T., Heuvelman, D.M., Nelson, R., Olander, J.V., Eppley, B.L., Delfino, J.J., Siegel, N.R., Leimgruber, R.M., and Feder, J., 1989, Tumor vascular permeability factor stimulates endothelial cell growth and angiogenesis, *J. Clin. Invest.* 84: 1470.

Ferrara, N., Houck, K.A., Jakeman, L.B., Winer, J., and Leung, D.W., 1991, The vascular endothelial growth factor family of polypeptides, *J. Cell. Biochem.*, 47: 211.

Folkman, J., and Klagsbrun, M., 1987, Angiogenic factors, *Science*. 235: 442.

Keck, P.J., Hauser, S.D., Krivi, G., Sanzo, K., Warren, T., Feder, J., and Connolly, D.T., 1989, Vascular permeability factor, an endothelial cell mitogen related to PDGF, *Science* . 246: 1309.

Klagsbrun, M. and D'Amore, P.A., 1991, Regulators of angiogenesis, *Annu. Rev. Physiol.* 53: 217.

Maglione, D., Guerriero, V., Viglietto, G., Delli Bovi, P., and Persico, M.G., 1991a, Isolation of human placenta cDNA coding for a protein related to the vascular permeability factor, *Proc. Natl. Acad. Sci.*, 88: 9267.

Maglione, D., Guerriero, V., Viglietto, G., Risau, W., Delli Bovi, P., and Persico, M.G., 1991b, PlGF: a new gene coding for a novel human vascular permeability related protein, *in* "Growth Factors of the Vascular and Nervous Systems," C. Lenfant, R. Paoletti, A. Albertini, eds, S. Karger, Basel.

Maglione, D., Guerriero, V., Viglietto, G., Ferraro, M.G., Aprelikova, O., Alitalo, K., Del Vecchio, S., Lei, K.J., Yang Chou, J., and Persico, M.G., 1993a, Two alternative mRNAs coding for the angiogenic factor, placenta growth factor (PlGF), are transcribed from a single gene of chromosome 14, *Oncogene*, 8: 925.

Maglione, D., Guerriero, V., Rambaldi, M., Russo, G., and Persico, M.G., 1993b, Translation of the placenta growth factor mRNA is severely affected by a small open reading frame localized in the 5' untranslated region, *Growth Factors*. 8: 141.

Risau, W., 1990, Angiogenic growth factors, *Prog. Growth Factor Res.* 2: 71.

Shweiki, D., Itin, A., Soffer, D., and Keshet, E., 1992, Vascular endothelial growth factor induce by hypoxia may mediate hypoxia-initiated angiogenesis, *Nature.* 259: 843.

Strick, D.M., Waycaster, R.L., Montani, J.P., Gay, W.J., and Adair, T.H., 1991, Morphometric measurements of chorioallantoic membrane vascularity: effects of hypoxia and hyperoxia, *Am. J. Physiol.* 260: H1385.

Tischer, E., Mitchell, R., Hartman, T., Silva, M., Gospodarowicz, Fiddes, J.C., and Abrahams, J.A., 1991, The human gene for vascular endothelial growth factor, *J. Biol. Chem.* 266: 11947.

Von Heijene, G., 1986, A new method for predicting signal sequence cleavage sites., Nucl. *Acids Res.*, 14: 4683.

Westermark, B., and Heldin, C.H., 1991, Platelet-derived growth factor in autocrine transformation,*Cancer Res.* 51: 5087.

Yang, E.Y. and Moses, H.L., 1991, Trasforming growth factor beta1-induce change in cell migration, proliferation, and angiogenesis in the chorioallantoic membrane, *J. Cell. Biol.*, 111: 731.

EXPRESSION OF THE M-SUBUNIT OF LAMININ CORRELATES WITH INCREASED CELL ADHESION AND METASTATIC PROPENSITY

Wheamei Jenq, Shi-Jun Wu, and Nicholas A. Kefalides

Connective Tissue Research Institute and Department of Medicine, University of Pennsylvania, and University City Science Center, Philadelphia, Pennsylvania 19104

INTRODUCTION

The presence of several isoforms of laminin in a variety of tissues and species has been reported. S-laminin, a B1 chain homologue of 190 kDa, has been identified in association with the synaptic cleft of the neuromuscular junction (Hunter, et al., 1989), while O'Rear (1992) has reported a novel laminin B1 chain variant (B1-2) in the avian eye, and Marinkovich et al. (1992) reported another variant, K-laminin in human amniotic fluid.

Previous studies in this laboratory demonstrated the presence of laminin M or "M" subunit of about 300 kDa, which was shown to be absent in EHS tumor laminin or in laminin synthesized by some neoplastic cell lines (Ohno et al., 1983, 1986; Jenq et al., 1993). Paulsson & Saladin (1989) isolated from mouse heart a Mr 300,000 polypeptide which is not antigenically related to the A or B chains of laminin. In addition, a laminin like protein, merosin, has been identified in human placenta, striated muscle, peripheral nerve, and schwannoma cells (Leivo & Engvall, 1988; Ehrig et al., 1990; Engvall et al., 1992). The role of these laminin variants in cells and tissues have not been studied. Pozzatti et al. (1986) demonstrated that the ras oncogene transformed early passage rat embryo cells (4R) showed a ten-fold increase in metastatic propensity than cells (RE4) doubly transformed with ras oncogene plus the adenovirus type 2 E1a gene. RE4 cells, which are highly tumorigenic but show little or no metastatic propensity, do not produce type IV collagenase. However, cells transformed with the ras oncogene alone secrete high levels of collagenase and exhibit high metastatic ability.

In our current studies we wanted to examine the nature of the laminin molecules synthesized by the 4R and RE4 cells and determine whether alteration in subunit composition correlates with changes in biological activity. Laminins isolated from conditioned media of 4R and RE4 cells were subjected to SDS-PAGE and immunoblot analyses. The steady-state mRNA level of the laminin "M" subunit in both cell lines was also measured by slot blot and Northern blot hybridization. The data show that the highly metastatic 4R cells produce a three to four-fold excess of the "M" subunit protein compared to the highly tumorigenic but not metastatic RE4 cells. In addition, we found that the mRNA for the "M" subunit is ten-fold higher in 4R than in RE4 cells. The laminin isolated from the medium of the 4R cells has a higher cell attachment promoting activity than that of RE4 cells. It is suggested that the tumorigenicity of RE4 cells correlates with the lack of expression of the "M" subunit, whereas the metastatic propensity of 4R cells and the cell adhesion promoting activity of their laminin correlate with expression of the laminin "M" subunit.

MATERIALS AND METHODS

Cell Cultures

Ras oncogene transformed (4R) and ras & E1a co-transformed (RE4) rat embryonic fibroblast cell lines were obtained from Dr. William G. Stetler-Stevenson, National Cancer Institute, Bethesda, MD. 4R and RE4 cells were grown in DMEM with 10% heat-inactivated fetal bovine serum, 100 µU/ml penicillin and 100 µg/ml streptomycin (Pozzatti et al., 1986). HT-1080 cells were grown in DMEM supplemented with 10% fetal bovine serum and 25 µg/ml of gentamicin (Brown et al., 1990b).

Purification of Laminin from Serum-free Conditioned Media

Laminin was isolated and purified from the clear supernatant of serum-free conditioned medium by the combined procedures reported previously (Dixit, 1985; Paulsson & Saladin, 1989; Brown et al., 1990a; Jenq et al.,1993). The clear supernatant was passed (0.5 ml/min. flow rate) through a Heparin-Agarose affinity column (6.0 x 1.8 cm), equilibrated with 0.05 M Tris-HCl/0.15 M NaCl (pH 7.4), containing 2.5 mM EDTA and 1 mM each of PMSF and NEM. The unbound material was washed off with the same buffer. All the Tris buffers used were at pH 7.4 and 0.05 M in concentration. The bound material was eluted with 1 M NaCl/Tris buffer and the eluted peak was pooled. The pooled fractions were dialyzed against 0.15 M NaCl/Tris buffer before applying to a DEAE-Sephacel column (5.0 x 2.0 cm). The column was eluted with Tris buffer containing 0.3 M NaCl, and the peak fractions were pooled, dialyzed and concentrated against 0.15 M NaCl/ Tris buffer to a final volume of 2 ml. The concentrated solution was fractionated onto a Sepharose CL-6B column and eluted with 0.15 M NaCl/Tris buffer. Fractions which reacted by ELISA with the polyclonal anti-EHS laminin antibody were pooled, dialyzed and concentrated against 0.15 M NaCl/Tris buffer to a final volume of 1 ml.

Purification of Native Laminin from Human Placenta

The purification protocol used here was based on the procedure described by Jenq et al. (1993). Briefly, the tissue was cut, homogenized, and extracted in solution I (20 mM-Tris/1M-NaCl/10 mM-EDTA/0.3 mM-PMSF/10 mM-NEM/1 mM-NaN$_3$/0.14 mM-hydroxymercuribenzoate, pH 7.4) containing 1.8% Triton X-100. After centrifugation at 4 °C for 30 min. at 20,000 xg, NaCl was added to the supernatant to 5 M followed by centrifugation at 35,000 xg. The resulting pellet was dissolved in solution II (50 mM-Tris/0.15M-NaCl/2.5 mM-EDTA/1 mM-PMSF/1 mM-NEM, pH 7.4), and the clear supernatant was passed through a Sepharose CL-6B gel filtration column. The eluted fractions that reacted with the EHS laminin antibody were pooled and re-chromatographed through a DEAE-Sephacel column equilibrated with solution II. Elution proceeded with solution III (50 mM-Tris/0.3 M-NaCl/2.5 mM-EDTA/1 mM-PMSF/1 mM-NEM, pH 7.4) and the peak fractions were pooled, dialyzed and concentrated against solution II to obtain the purified placenta laminin.

SDS-Polyacrylamide Gel Electrophoresis and Immunoblot

Gel electrophoresis of the laminins was performed on the 5% separating gel with a 3% stacking gel according to the method of Laemmli (1970). Proteins were stained with Coomassie Brilliant Blue R. Electrophoretic transfer of proteins from the gel onto a nitrocellulose membrane was carried out in a semi-dry Sartoblot II unit and laminin was detected by reacting with anti-EHS or anti-human placenta laminin antibody. The Peroxidase Vectastain Elite kit was used for color formation in the immunoblotting experiments.

Metabolic Labeling and Immunoprecipitation

The immunoprecipitation procedure used was that of Aratani & Kitagawa (1988) with modifications. 4R and RE4 cells at 85% confluence were labeled with 100µCi/ml of trans-labeled ^{35}S-methionine for 18 hr, in the methionine free DMEM media supplemented with 15 mM HEPES. After lysis of the cell layers, the cell lysates were diluted and then immunoprecipitated with antibodies to laminin. The pellet material was analyzed by SDS-gel electrophoresis and visualized by fluorography.

Cell Attachment and Spreading

Quantitation of the human fibrosarcoma HT-1080 cells attached to substratum-bound laminin was achieved following the procedure reported by Turner et al. (1987) and Rao & Kefalides (1990). The 6-well plastic plates were coated with 2.0 ml of 0.1 M Tris buffer, pH 9.5, containing increasing amount of laminins to be assayed for cell adhesion activity. HT-1080 cells at about 85 % confluence were labeled with ^{35}S-methionine (1 μ Ci/ml) in growth medium for 18-20 hr. The cells were detached with 0.1% trypsin solution and washed successively with 10-ml volume of growth medium, attachment medium without BSA, and then attachment medium. The cell pellet was suspended in attachment medium after the last wash. One hundred microliters of this cell suspension containing 1.0 - 1.2 x 10^6 cells was seeded to each well, shaken for 5 sec, and incubated at 37 °C for 1 hr. The unattached cells were removed with three 2.0-ml washes, and each wash agitated at 80 rpm for 30 sec. Attached cells were then lysed for 30 min. in 2.0 ml per well of 1% Triton X-100 in PBS. Percent attachment was defined as : [radioactivity extracted from attached cells]/ [radioactivity in cells added to assay] x 100. In preliminary experiments, we observed a linear relationship between cell number and radioactivity (data not shown). For photo-taking, after removing the unattached cells, 2 ml of HBSS was added to the attached cells in each well.

Poly A-RNA Isolation and Slot Blot Hybridization

Poly A-RNA from 4R and RE4 cells were isolated using a kit from Invitrogen (Cat. No. K1593-02) following the manufacturer's instruction manual. Briefly, about 1 x 10^8 cells were homogenized in 15 ml lysis buffer which contains 0.02 vol. of RNase protein degrader. Incubation was carried out in a 45 °C water bath for 1 hr to digest proteins and to release the poly A-RNA. Oligo(dT) cellulose was then added to the lysate, and the mixture was centrifuged at 2,000 xg. The oligo(dT) cellulose pellet was transferred into a disposable column, and the poly A-RNA was eluted off the oligo(dT) cellulose with elution buffer. For concentrating poly A-RNA, 0.1 vol. of 2 M Na-acetate and 2.5 vol. of 100% ethanol were added to the eluent and kept at -20 °C overnight. After centrifugation at 16,000 xg for 15 min., the poly A-RNA pellet was then washed with 75% ethanol and recentrifuged. The slot blot hybridization procedure was based on the method of Dyson (1991). The probe used was a γ^{32}P-ATP (6,000 Ci/mMole) 5' end-labeled oligomer which is specific to a human merosin cDNA sequence (Gene bank data base, accession no. M59832M32076).

RESULTS

Purification of Laminins from Conditioned Medium and Placenta

The material, isolated from serum free-conditioned media of 4R and RE4 cultures, was eluted through a Heparin-Agarose affinity column, dialyzed and chromatographed on a DEAE-Sephacel column. A major peak, eluted with 0.3 M NaCl, was applied to a Sepharose CL-6B column. The gel filtration pattern is shown in Figure 1. The fractions which contained the material reacting by ELISA with the laminin antiserum, were pooled, concentrated, and dialyzed to obtain the purified laminin. In addition, the laminin extract from the human placenta was further purified by chromatography as described previously (Jenq et al., 1993).

Immunoprecipitation, SDS-PAGE, and Immunoblot of Laminin

The ^{35}S-methionine labeled and immunoprecipitated 4R and RE4 cell lysates were electrophoresed in a SDS-polyacrylamide gel under reducing conditions. Figure 2 is the fluorogram which indicated that there is a three to four-fold excess of the laminin "M" subunit in 4R cells compared to RE4 cells. SDS-PAGE, under reducing conditions, and immunoblot of the purified laminins from both 4R and RE4 conditioned media verified the above findings, and the "M" subunit was lacking in EHS laminin (data not shown).

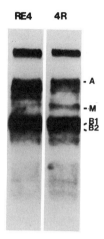

Figure 1. Sepharose CL-6B gel filtration of the pooled laminin peak from the ion exchange chromatography. The optical absorption curve at 280 nm is shown by the open circles. The solid circles indicate the peak obtained by ELISA using rabbit anti-laminin antibody. V_o and V_t designate the void and total volume of the column, respectively. The bar indicates the anti-laminin antibody reactive peak.

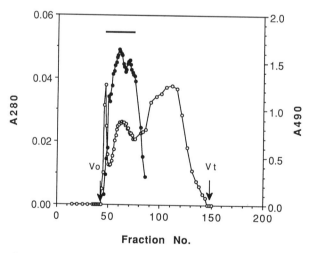

Figure 2. Mobility of subunits from the antibody-precipitated laminin complexes from 4R and RE4 cells on SDS-PAGE under reducing conditions. Radiolabeled cell lysates were immunoprecipitated with the purified polyclonal anti-EHS laminin antibody and separated by SDS-PAGE.

Cell adhesion on Laminin Substrata

Both 4R and RE4 laminins, purified from the serum-free conditioned media, were tested along with human placenta laminin and EHS mouse tumor laminin for their ability to promote the adhesion of human fibrosarcoma HT-1080 cells. Various concentrations of laminin in 2 ml of coating buffer were coated to each well. The coating efficiency of EHS laminin was 92%, whereas both 4R and RE4 laminins showed about 90% coating efficiency. The data in Figure 3 clearly show that, laminin synthesized by 4R cells or isolated from placenta (at a concentration of 250 ng/2 ml) promotes the attachment of HT-1080 cells more efficiently (68% and 63%, respectively) than RE4 or EHS laminin (39% and 26%, respectively). The data are consistent with the observation made earlier that RE4 cells synthesize little or no "M" subunit and that EHS laminin is devoid of the "M" subunit. Figure 4 is a micrograph showing the higher degree of cell attachment and further spreading when 4R laminin was used as a substrate at 0.2 μg and 0.5 μg per well as compared to RE4 laminin at the same concentrations.

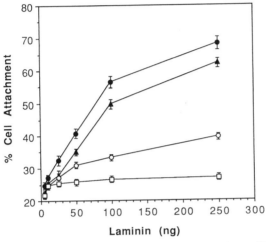

Figure 4. Adhesion of human fibrosarcoma HT-1080 cells on 4R and RE4 laminins coated onto the plastic at two concentrations, 0.2 μg and 0.5 μg per well. The increased adhesion and spreading of HT-1080 cells onto the 4R laminin is evident. Micrographs were taken after 1 hr incubation. Bar = 100 μm.

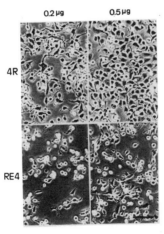

Figure 3. Concentration effect of laminin from EHS tumor, human placenta, 4R, and RE4 cells on the extent of cell attachment using human fibrosarcoma HT-1080 cells. Data points for laminin from EHS are represented by open squares, from placenta by solid triangles, from 4R by solid circles, and RE4 by open circles. Each point represents the average measurement from triplicate wells. Error bars indicate standard error of the mean.

Comparative Analysis of mRNA for Laminin M Subunit

The steady state level of the laminin "M" subunit mRNA was measured in confluent cultures of 4R and RE4 cells. The message level of "M" was detected by slot blot M" subunit. The autoradiogram (Figure 5) revealed that the mRNA level of the laminin "M" subunit is ten-fold higher in 4R cells than in RE4 cells. Analysis of laminin "M" mRNA by hybridization to

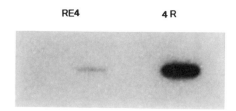

Figure 5. Slot blot hybridization of poly A-RNA (0.1 μg/slot) isolated from 4R and RE4 cells. The 5' end-labeled oligomers, specific for human merosin cDNA, were the probes. The hybridized laminin "M" mRNA from both 4R and RE4 cells is shown.

the synthetic oligomer probes which are specific to the merosin or laminin "Northern blot hybridization confirmed the data obtained by slot blot hybridization (data not shown). We have also tested the steady state mRNA level of the house-keeping "actin" gene and found the same amount of actin mRNA, in terms of band density, in both 4R and RE4 cells (data not shown).

DISCUSSION

Our data show that the subunit composition of laminin varies in normal and neoplastic cells or tissues, and that this correlates with the biological behavior of cells which synthesize it. We have shown that purified laminin from human placenta and from the culture medium of the highly metastatic 4R cells contain substantial amounts of the "M"subunit. This property correlates with a higher cell adhesion promoting activity of laminin made by these cells, compared to the RE4 laminin, which contains little "M" subunit, or the EHS laminin, which lacks the "M" subunit. The ras oncogene transformed 4R cells are highly metastatic but have a low tumorigenicity, whereas the ras oncogene and viral E1a gene co-transfected RE4 cells are highly tumorigenic but have little metastatic propensity (Pozzatti et al., 1986). Based on our findings, the expression of the "M" subunit of laminin seems to correlate proportionably with the degree of cell adhesion and metastatic propensity. We had reported earlier (Jenq et al., 1993) that laminin purified from the culture media of two related mouse epithelial cell lines, B82 and its tumorigenic derivative B82HT, showed quantitative differences in the subunit composition, namely the presence of more "M" subunit, and a higher content of "M" mRNA in B82 than in B82HT cells. In addition, laminin from B82 cells promoted adhesion of HT-1080 cells to a greater extent than laminin from B82HT cells. A previous report by Rao & Kefalides (1991), using similar mouse cell lines, A9 and A9HT, also indicated the higher amount of laminin "M" subunit in A9 cells than in A9HT cells. Although, the mechanism by which various activated ras oncogenes can induce the complex metastatic phenotype in suitable recipient cells (Thorgeirsson et al., 1985; Bernstein and Weinberg, 1985; Muschel et al., 1985; Bradley et al., 1986; Pozzatti et al., 1986) remains unclear, we speculate that the ras oncogene alone, transfected into the proper recipient cells, may increase the expression of laminin "M". On the other hand, simultaneous introduction of the viral E1a gene into the same cell suppresses its expression. Garbisa et al.(1987) showed that c-Ha-ras oncogene alone, transfected into early passage rat embryo fibroblasts, induced these cells to secrete high levels of type IV collagenolytic metalloproteinase and to concomitantly exhibit a high increase of spontaneous metastasis in nude mice. They also indicated that co-transfection with c-Ha-ras plus the adenovirus type 2 E1a gene into nontumorigenic and nonmetastatic NIH 3T3 cells yields cells which are highly tumorigenic but nonmetastatic and fail to produce type IV collagenase. It is of interest to note that Chambers et al. (1993), reported that the c-Ha-ras induced metastatic ability in NIH 3T3 cells increased the expression of osteopontin, a phosphoprotein which has some adhesive property. The present studies show that the subunit composition of immunoprecipitated laminin from 4R and RE4 cell lysates agrees with the subunit composition determined by immunoblot of laminins purified from the culture media of 4R and RE4 cells, namely a three to four-fold increase in the "M" subunit of the 4R laminin. Furthermore, the steady-state mRNA level of the "M" subunit was shown to be ten-fold higher in 4R cells compared to RE4 cells. The data indicate that the steady-state level of the "M" mRNA is markedly depressed in the doubly transfected RE4 cells, which correlates with the decreased expression of the "M" subunit of laminin as well as the decreased ability of this laminin to promote HT-1080 cell adhesion. It is tempting to speculate that for tumor cells to develop the metastatic phenotype, they must also synthesize and secrete the isoform of laminin which contains substantial amount of the "M" subunit.

SUMMARY

In the present study we have isolated the laminin synthesized by two cells (4R and RE4) and examined their subunit structure as well as their cell adhesion promoting activity. The ras oncogene transformed 4R cells are highly metastatic, whereas the RE4 cells, which are co-transfected with the ras oncogene plus the adenovirus type 2 E1a gene, are highly tumorigenic but are not metastatic. Analysis of laminin by SDS-PAGE and immunoblot from the 4R and RE4 culture media, using an antibody to placenta laminin, demonstrated that the metastatic 4R cells produced a three to four-fold excess of the laminin "M" subunit compared to RE4 cells. SDS-PAGE of immunoprecipitated ^{35}S-methionine labeled cell lysate, using an antibody against EHS laminin, confirmed the above findings. In adhesion studies, using HT-1080 cells, laminin (250 ng/well) from 4R cells had a greater adhesion promoting activity (68%) than RE4 laminin (39%). Similarly, laminin isolated from human placenta, which expresses both the A, B1 & B2 and M, B1 & B2 isoforms, had a higher cell adhesion promoting activity (63%) than EHS laminin (26%), which lacks the "M" subunit. The mRNA level for "M" in the 4R cells was significantly (ten-fold) higher compared to the RE4 cells. The data indicate that tumorigenicity correlates with lack of expression of the laminin "M" subunit in RE4 cells, whereas expression of the "M" subunit correlates with both the metastatic propensity and increased cell adhesion promoting activity of the laminin isoform which contains the "M" subunit.

ACKNOWELEDGEMENTS

The authors wish to express their appreciation to Dr. William G. Stetler-Stevenson for providing us 4R, RE4, and HT-1080 cells, to Zahra Ziaie for critically reading the manuscript. This study was supported in part by NIH Grants AR 20553, HL-29492, and AR-07490.

REFERENCES

Aratani, Y., & Kitagawa, Y. (1988) Enhanced synthesis and secretion of type IV collagen and entactin during adipose conversion of 3T3-L1 cells and production of unorthodox laminin complex. *J. Biol. Chem. 263*, 16163-16169.

Bernstein, S. C., & Weinberg, R. A. (1985) Expression of the metastatic phenotype in cells transfected with human metastatic tumor DNA. *Proc. Natl. Acad. Sci. USA 82*, 1726-1730.

Bradley, M. O., Kraynak, A. R., Storer, R. D., & Gibbs, J. B. (1986) Experimental metastasis in nude mice of NIH 3T3 cells containing various ras genes. *Proc. Natl. Acad. Sci. USA 83*, 5277-5281.

Brown, J. C., Spragg, J. H., Wheeler, G. N., & Taylor, P. W. (1990a) Identification of the B1 and B2 subunits of human placental laminin and rat parietal-yolk sac laminin using antisera specific for murine laminin-ß-galactosidase fusion proteins. *Biochem. J. 270,* 463-468.

Brown, P. D., Levy, A. T., Margulies, I. M. K., Liotta, L. A., & Stetler-Stevenson, W. G. (1990b) Independent expression and cellular processing of Mr 72,000 type IV collagenase and interstitial collagenase in human tumorigenic cell lines. *Cancer Res. 50,* 6284-6191.

Chambers, A. F., Hota, C., & Prince, C. W. (1993) Adhesion of metastatic, ras-transformed NIH 3T3 cells to osteopontin, fibronectin, and laminin. *Cancer Res. 53,* 701-706.

Dixit, S. N. (1985) Isolation, purification and characterization of intact and pepsin-derived fragments of laminin from human placenta. *Connect. Tissue. Res. 14,* 31-40.

Dyson, N. J. (1990) Immobilization of nucleic acids and hybridization analysis, *in : Essential Molecular Biology , A Practical approach, Vol. 2* (Brown, T. A., Ed.) pp 111-156, IRL Press at Oxford University Press, Oxford.

Ehrig, K., Leivo, I., Argraves, R. E., & Engvall, E. (1990) Merosin, a tissue-specific basement membrane protein, is a laminin-like protein. *Proc. Natl. Acad. Sci. USA 87,* 3264-3268.

Engvall, E., Earwicker, D., Day, A., Muir, D., Manthorpe, M., & Paulsson, M. (1992) Merosin promotes cell attachment and neurite outgrowth and is a component of the neurite-promoting factor of RN22 Schwannoma cells. *Exp. Cell. Res. 198,* 115-123.

Garbisa, S., Pozzatti, R., Muschel, R. J., Saffiotti, U., Ballin, M., Goldfarb, R. H., Khoury, G., & Liotta, L. A. (1987) Secretion of type IV collagenolytic protease and metastatic phenotype : Induction by transfection with c-Ha-ras but not c-Ha-ras plus Ad 2-E1a. *Cancer Res. 47,* 1523-1528.

Hunter, D. D., Shah, V., Merlia, J. P., & Sanes, J. R. (1989) A laminin-like adhesive protein concentrated in the synaptic cleft of the neuromuscular junction. *Nature 338,* 229-234.

Jenq, W., Wu, S. J., & Kefalides, N. A. (1993) Adhesion promoting property of laminin from normal tissue and from a tumorigenic cell line. *Connect. Tissue. Res.* In press.

Laemmli, U. K. (1970) Cleavage of structural proteins during the assembly of the head of bacteriophage T4. *Nature 227,* 680-685.

Leivo, I., & Engvall, E. (1988) Merosin, A protein specific for basement membrane of Schwann cells, striated muscle, and trophoblast, is expressed late in nerve and muscle development. *Proc. Natl. Acad. Sci. USA 85,* 1544-1548.

Marinkovich, M. P., Lunstrum, G. P., Keene, D. R., & Burgeson, R. E. (1992) The dermal-epidermal junction of human skin contains a novel laminin variant. *J. Cell Biol. 119,* 695-703.

Muschel, R., William, J. E., Lowy, D. R., & Liotta, L. A. (1985) Harvey ras induction of metastatic potential depends upon oncogene activation and type of recipient cells. *Am. J. Pathol. 121,* 1-8.

Ohno, M., Martinez-Hernandez, A., Ohno, N., & Kefalides, N. A. (1983) Isolation of laminin from human placental basement membranes: Amnion, chorion and chorionic microvessels. *Biochem. Biophys. Res. Commun. 112,* 1091-1098.

Ohno, M., Martinez-Hernandez, A., Ohno, N., & Kefalides, N. A. (1986) Laminin M is found in placental basement membranes, but not in basement membranes of neoplastic origin. *Connect. Tissue. Res. 15,* 199-207.

O'Rear, J. J. (1992) A novel laminin B1 chain variant in avian eye. *J. Biol. Chem. 267,* 20555-20557.

Paulsson, M., & Saladin, K. (1989) Mouse heart laminin, purification of the native protein and structural comparison with Engelbreth-Holm Swarn tumor laminin. *J. Biol. Chem. 264,* 18726-18732.

Pozzatti, R., Muschel, R., Williams, J., Padmanabhan, R., Howard, B., Liotta, L. A., & Khoury, G. (1986) Primary rat embryo cells transformed by one or two oncogenes show different metastatic potentials. *Science 232,* 223-227.

Rao, N., Brinker, J. M., & Kefalides, N. A. (1991) Changes in the subunit composition of laminin during the increased tumorigenesis of mouse A9 cells. *Connect. Tissue. Res. 25,* 321-329.

Rao, C. N., & Kefalides, N. A. (1990) Identification and characterization of a 43-kilodalton laminin fragment from the "A" chain (long arm) with high-affinity heparin binding and mammary epithelial cell adhesion-spreading activities. *Biochemistry 29,* 6769-6777.

Thorgeirsson, U. P., Turpeenniemi-Hujanen, T., William, J. E., Westin, E. H., Heilman, C. A., Talmadge, J. E., & Liotta, L. A. (1985) NIH/3T3 cells transfected with human tumor DNA containing activated ras oncogenes express the metastatic phenotype in nude mice. *Mol. Cell Biol.* *5* , 259-262.

Turner, D. C., Filter, L. A., & Carbonetto, S. (1987) Magnesium dependent attachment and neurite outgrowth by PC12 cells on collagen and laminin substrata. *Dev. Biol.* *121,* 510-525.

THE PROLIFERATIVE EFFECT OF SUBSTANCE P ON CAPILLARY ENDOTHELIAL CELLS IS MEDIATED BY NITRIC OXIDE

M. Ziche, L. Morbidelli, L. Mantelli, E. Masini, F. Ledda, #H.J. Granger, and*C.A. Maggi

Dept. of Pharmacology, University of Florence, Viale Morgagni 65, 50134 Florence, Italy,

#Dept of Medical Physiology, Texas A&M University, College Station, Texas, USA

*Pharmacology Dept., A. Menarini Pharmaceuticals, Florence, Italy

INTRODUCTION

Under physiological circumstances the ability of a given tissue to produce a neovascular response is under strict control, i.e. the balance between factors and events favoring or interfering with neovascular growth is in equilibrium. Modification of the normal cell/cell and cell/matrix boundaries, which maintain tubular morphology of the endothelium at the postcapillary level, is a prerequisite for capillary endothelial cell mobilization and proliferation to constitute a neovascular response.

Vasoactive peptides released by peripheral endings of sensory neurons or locally produced during inflammatory reaction, exert their primary action at the postcapillary level causing a reduction in cell/cell and cell/matrix interaction and an increment of vascular permeability. SP has been proposed as a main mediator of neurogenic inflammation produced when noxious stimuli of various origin activate the peripheral endings of primary sensory neurons, causing vasodilation and increased vascular permeability (1). The vasorelaxant response to SP is dependent on the presence of endothelium and is mediated by NK1 receptors (2, 3). Endothelium-dependent relaxation has been clearly demonstrated to be caused by

Angiogenesis: Molecular Biology, Clinical Aspects
Edited by M.E. Maragoudakis *et al*, Plenum Press, New York 1994

an endothelium-derived relaxing factor identified as nitric oxide (NO) (4, 5). We have reported that SP promotes angiogenesis in vivo and endothelial cell growth and mobilization in vitro and that these effects are mediated by NK1 receptors (6, 7). Recently we have also reported that NO-donor drugs promote endothelial cells proliferation in vitro (8).

In the present study we have evaluated the effect of NO production on the proliferative response induced by SP in cultured endothelial cells isolated from coronary postcapillary venules of bovine origin (CVEC) (9). These cells were selected because postcapillary venules are believed to be a main site for the action of SP in increasing vascular permeability and are actively involved in the angiogenesis process (10, 11). In order to evaluate a possible role of NO production in the proliferative effect of SP, the effect of NO-synthase inhibition on DNA synthesis promoted by SP in CVEC was investigated. The ability of CVEC to be responsive to NO action was evaluated by measuring the cyclic GMP levels following exposure to SP.

METHODS AND RESULTS

DNA synthesis

Coronary venular endothelial cells (CVEC) were isolated from bovine heart as previously described (9). The endothelial identity of the cells was determined by immunofluorescence detection of factor VIII-related antigen and by electron microscopy. Cells were cloned and each clone was subcultured up to a maximum of 25 passages. Passages between 15 and 20 were used in these experiments. DNA synthesis was quantified by [^{3}H]-thymidine incorporation of subconfluent cell monolayers seeded onto 24 multiwell plates (10). After 48 h in serum-free media (0.1% FCS), media was removed and cells were incubated with increasing concentrations of the agents for 24 h and pulsed for 1 h with 0.5 μCi [^{3}H]-thymidine per well. Exposure of serum-deprived subconfluent monolayers of CVEC to SP induced a concentration-dependent activation of DNA synthesis as measured by [^{3}H]-thymidine incorporation (6). Maximal biological activity was obtained in capillary endothelial cells at concentration of 10 nM SP. The NO-generating drug sodium nitroprusside (SNP) (13) also induced DNA synthesis and maximal effect was observed at 10 μM (8). To evaluate the role played by NO in promoting SP induced proliferation, cells were treated for 1h with N$^{\omega}$-monomethyl-L-arginine (L-NMMA) to inhibit NO-synthase (14) before exposure to SP. Pretreatment with 100 μM L-NMMA did not affect basal DNA synthesis, but reduced significantly thymidine incorporation produced by SP (Table 1). Similar results have been obtained with other NO synthase inhibitors.

Cyclic GMP

NO released from endothelial cells, promotes its effect by activation of cyclic GMP production on the target cell. Therefore to evaluate NO production we measured cyclic GMP accumulation in CVEC. Cells were plated in 100 mm diameter dishes and allowed to grow to 90% confluence (5×10^6 final cell number). Stimulation was carried out in PBS with calcium and magnesium. Cell monolayers were pretreated for 30 min with 10 μM indomethacin and 50 μM IBMX. Levels of cyclic GMP were measured in the aqueous phase of TCA cell extracts following the procedure previously reported (11). The cyclic GMP content of each dish was expressed as fmol/mg protein/10^6 cells. The ability of CVEC to respond to NO by increasing cyclic GMP levels was assessed by exposing CVEC to SNP. After 30 min incubation with IBMX (50 μM) and indomethacin (10 μM), to block phosphodiesterase and cyclo-oxygenase activity respectively the cells were exposed to SP and SNP for 5 min. Cyclic GMP was doubled when CVEC were exposed to 10 μM of SNP. Following 5 min exposure to 10 nM SP cyclic GMP levels increased by 61 %. Treatment of cell monolayers with L-NMMA reduced the basal production of cyclic GMP by 66.5 %. L-NMMA treatment produced a 52% reduction in the cyclic GMP elevation obtained after SP stimulation. The cyclic GMP elevation produced by exogenous administration of NO, obtained with SNP, was not affected by L-NMMA treatment (Table 1).

Table 1. Effect of substance P and of sodium nitroprusside on DNA synthesis and cyclic GMP production in postcapillary endothelial cells.

	DNA synthesis	Cyclic GMP
	[^{3}H]-Thymidine incorporation (% of basal)	fmol/mg protein/10^6 cells
Basal	100	110.6±8
+L-NMMA	110±5	49.0±9 *
SP 10 nM	130±4 *	210.6±15 **
+L-NMMA	97±2.8 #	121.2±16 @
SNP 10 μM	144±8 *	246.0±9 **
+ L-NMMA	–	276.4±21

In DNA synthesis experiments L-NMMA (100 μM) was added 1 h before the addition of SP. Cyclic GMP production was evaluated in endothelial cell monolayers exposed to test substances for 5 min. Cells were pretreated with L-NMMA for 60 min. Data are means ± sem of 5 esperiments run in duplicate. * P<0.01, ** P<0.001 vs basal. # P<0.05 and @ P<0.001 vs SP alone (Student's t test).

DISCUSSION

The increase in vascular permeability produced by SP occurs at the postcapillary level (1) and has been related to local NO production (15). The cells used in this study were obtained from the microvascular endothelium, which is the site of action of SP and of a large number of inflammatory and immune mediators actively involved in the healing process (10, 16). Postcapillary endothelial cells have a higher basal turnover and proliferate when exposed to several vasoactive peptides (6, 10).

In the vascular tree, NO is mainly released by endothelial cells and exerts its effect by activating guanylate cyclase on target cells thus favouring vasorelaxation (17). Endothelial cells from different sections of the circulation have been reported to possess the enzymes responsible for NO synthesis (4). However the physiological meaning of NO production at different level of the circulatory system is not yet clear. Cytokines and endotoxins activate an inducible NO-synthase in endothelial cells which causes release of NO (18) which, according to some authors is associated with endothelial cell damage (19). On cultured cells the growth-promoting effect of NO as well as the involvement of cyclic GMP in DNA synthesis activation is not univocal among the authors (19, 20, 21)

The data presented here show that in resting conditions CVEC spontaneously release NO as indicated by the basal levels of cyclic GMP detected. The significant reduction in cyclic GMP levels obtained following inhibition of NO production, induced by the NO-synthase inhibitor L-NMMA, indicates that basal cyclic GMP levels are specifically linked to NO production. This observation suggests that the spontaneous production of cyclic GMP measured in these cells involves an autocrine effect of NO.

Critical to our hypothesis was the demonstration that CVEC respond to NO by increasing cyclic GMP levels. Our results show that following stimulation induced by SNP which directly releases NO (13), cyclic GMP levels rise up to 2 fold basal levels. Likewise, SP induces a significant increase of cyclic GMP levels. L-NMMA treatment did not affect the cyclic GMP increase induced by SNP since, in this condition, NO is released exogenously and is independent of endothelial cell synthase. On the contrary L-NMMA significantly reduced cyclic GMP production induced by SP indicating NO production as mediator of the response to SP. Other reports have suggested that SP induced endothelium-dependent vasodilation and plasma extravasation in vivo could be modulated by NO production at the microvascular level (15). However the present one represents the first indication that cyclic GMP pathway can be directly activated by SP in postcapillary endothelial cells.

We have reported that NO release obtained with the NO generating drug SNP, promotes DNA synthesis of endothelial cells of the microcirculation (8). We have also reported that SP promotes proliferation of capillary endothelium (6, 7). The data reported here indicate that the proliferative effect induced by the vasoactive peptide SP is modulated by NO production. These observations propose a new mechanism through which endothelial cells could control their proliferation via an autocrine effect exerted by NO and guanylate cyclase activation. Recently on the chick chorioallantoic membrane (CAM) it has been reported that increased NO inhibits the physiological development of the CAM vascular network, while inhibition of the NO synthase activity was accompained by an increased neovascular response (22). Although in apparent contraddition, these results together with our data suggest a physiological role played by NO in the fine control of capillary proliferation with opposite actions in embrionic versus highly differentiated capillary endothelium.

The increasing availability of potent antagonists of peptide and non-peptide nature endowed with high selectivity for the three main types of SP receptors and the active research in the discovery of NO pharmacology, could offer a new approach in the understanding of the local control of the angiogenesis process and in the design of therapies for angiogenesis-dependent diseases.

Acknowledgments

This work was supported by funds from Italian Ministry for the University and for Scientific and Technological Research and by the National Research Council of Italy.

REFERENCES

1. F. Lembek, and P. Holzer, Substance P as a neurogenic mediator of antidromic vasodilation and neurogenic plasma extravasation, Naunyn-Schmiedeberg's Arch. Pharmacol. 310: 176-183 (1979).
2. P. D'Orleans-Juste, S. Dion, G. Drapeau, and D. Regoli, Different receptors are involved in the endothelium-mediated relaxation and the smooth muscle contraction of the rabbit pulmonary artery in response to substance P and related neurokinins, Eur. J. Pharmacol. 125: 37-44 (1985).
3. P. D'Orleans-Juste, S. Dion, J. Mizrahi, and D. Regoli, Effects of peptides and nonpeptides on isolated arterial smooth muscles: Role of endothelium, Eur. J. Pharmacol. 114: 9-21 (1985).
4. L.J. Ignarro, G.M. Buga, K.S. Wood, R.E. Byrns, and G. Chaudhuri, Endothelium-derived relaxing factor produced and released from artery and vein is nitric oxide. Proc. Natl. Acad. Sci. USA 84: 9265-9269 (1987).
5. R.M.J. Palmer, A.G. Ferridge, and S. Moncada, Nitric oxide release accounts for the biological activity of endothelium-derived relaxing factor, Nature 327: 524-526 (1987).

6. M. Ziche, L. Morbidelli, M. Pacini, P. Geppetti, G. Alessandri, and C.A. Maggi, Substance P stimulates neovascularization in vivo and proliferation of cultured endothelial cells, Microvasc. Res. 40: 264-268 (1990).

7. M. Ziche, L. Morbidelli, P. Geppetti, C.A. Maggi, and P. Dolara, Substance P induces migration of capillary endothelial cells: a novel NK1 selective receptor mediated activity, Life Sci. 48: PL7-PL11 (1991).

8. M. Ziche, L. Morbidelli, E. Masini, H.J. Granger, P. Geppetti, and F. Ledda, Nitric oxide promotes DNA synthesis and cyclic GMP formation in endothelial cells from postcapillary venules, Biochem. Biophys. Res. Commun. 192: 1198-1203 (1993).

9. M.E. Schelling, C.J. Meininger, J.R. Hawker, and H.J. Granger, Venular endothelial cells from bovine heart, Am. J. Physiol. 254: H1211-H1217 (1988).

10. L. Morbidelli, A. Parenti, H.J. Granger, M. Ziche, and F. Ledda, Activation of DNA synthesis and inositol-phosphate turnover in coronary venular endothelial cells exposed to bradykinin, Pharmacol. Res. 25: 150-151 (1992).

11. E. Masini, P.F. Mannaioni, A. Pistelli, D. Salvemini, and J. Vane, Impairment of the L-arginine-nitric oxide pathway in mast cells from spontaneously hypertensive rats, Biochem. Biophys. Res. Commun. 177: 1178-1182 (1991).

13. M. Feelish, and E.A. Noack, Correlation between nitric oxide formation during degradation of organic nitrates and activation of guanylate cyclase, Eur. J. Pharmacol. 139: 19-30 (1987).

14. R.M.J. Palmer, D.D. Rees, D.S. Ashton, and S. Moncada, L-arginine is the physiological precursor for the formation of nitric oxide in endothelium-dependent relaxation, Biochem. Biophys. Res. Commun. 153: 1251-1256 (1988).

15. S.R. Hughes, T.J. Williams, and S.D. Brain, Evidence that endogenous nitric oxide modulates oedema formation induced by substance P, Eur. J. Pharmacol. 191: 481-484 (1990).

16. R. M. Marks, W.R. Roche, M. Czerniecki, R. Penny, and D.S. Nelson, Mast cell granules cause proliferation of human microvascular endothelial cells, Lab. Invest. 55: 289-293 (1990).

17. L.J. Ignarro, R.G. Harbinson, K.S. Wood, and P.J. Kadowitz, Activation of purified guanylate cyclase by endothelium-derived relaxing factor from intrapulmunary artery and vein: Stimulation by acetylcholine, bradykinin and arachidonic acid, J. Pharmacol. Exp. Ther. 237: 893-900 (1986).

18. R.G. Kilbourn, and P. Belloni., Endothelial cell production of nitrogen oxides in response to interferon gamma in combination with tumor necrosis factor, interleukin-1, or endotoxin, J. Natl. Cancer Inst. 82: 772-776 (1990).

19. R.M.J. Palmer, L. Bridge, N.A. Foxwell, and S. Moncada, The role of nitric oxide in endothelial cell damage and its inhibition by glucocorticoids, Br. J. Pharmacol. 105: 11-12 (1992).

20. S.A. Moodie, and W. Martin, Effects of cyclic nucleotides and phorbol myristate acetate on proliferation of pig aortic endothelial cells, Br. J. Pharmacol. 102, 101-106 (1991).

21. U.C. Garg, and A. Hassid, Inhibition of rat mesangial cell mitogenesis by nitric oxide-generated vasodilation, Am J. Physiol. 257: F60-F66 (1989).

22. E. Pipili-Synetos, E. Sakkoula, and M.E. Maragoudakis, Nitric oxide is involved in the regulation of angiogenesis, Br. J. Pharmacol. 108: 855-857 (1993).

REACTIVATION OF METALLOPROTEINASE-INHIBITOR COMPLEX BY A LOW

MOLECULAR MASS ANGIOGENIC FACTOR

Mclaughlin B., and Weiss J.B.,

Wolfson Angiogenesis Unit
Department of Rheumatology,
Manchester University,
Hope Hospital,
Manchester M6 8HD
England

INTRODUCTION

It has now become widely accepted that there is involvement of neutral matrix metalloproteinases (MMPs),in angiogenesis. The initial event in angiogenesis was described by Ausprunk in 1979 as dissolution or degradation of the capillary basement membrane, which it was suggested would permit migration of microvessel endothelial cells through the dense collagenous stroma, enabling penetration of new microvessels. The control of matrix degradation via metalloproteinase activity must therefore be regarded as an important step in the promotion of angiogenesis.

Endothelial cell Stimulating Angiogenesis Factor (ESAF)

Treatment of Tumour Angiogenic Factor,(TAF),Folkman 1971, with an affinity column derived from an antibody to TAF,(Phillips and Kumar 1979) resulted in the activity,measured on the chick chorioallantoic membrane,(CAM) being found in a fraction which was freely dialysable,(Weiss et al 1979). Subsequently this low molecular weight fraction was shown to be mitogenic to bovine brain derived microvessel endothelial cells and the fraction was referred to as Endothelial cell Stimulating Angiogenesis Factor, (ESAF), (Schor et al 1980).

It became evident that ESAF was not primarily a tumour factor and it would be found in growth plate,(Brown et al 1987) pineal glands,(Taylor et al 1988a) as well as in the serum from patients with pathological conditions where angiogenesis is occurring.(see Chapter,J.B.Weiss,in this book.) ESAF is freely dialysable and has a molecular mass of less than 600 daltons, is soluble in water, ethanol and methanol, but not in more non-polar solvents such as ether,benzene and chloroform. It binds weakly to Diethyl amino ethyl (DEAE)cellulose, suggesting that it has a weak negative charge. ESAF is rapidly inactivated by acid but is moderately heat stable(50° C for 10 minutes).

It is important to stress that ESAF is not a protein,or peptide and has no intrinsic catalytic activity.

Biological tests for ESAF include,the Chick Chorioallantoic membrane(CAM),the chick yolk sac membrane(YSM),the rabbit cornea and sponge implants into rat skin.(Odedra and Weiss 1991). ESAF has been purified to a single homogeneous peak by reverse phase High Performance Liquid Chromatography,(HPLC) and this peak of activity has been found to be identical in extracts from all tissues and also from the same tissues of a variety of species.(Taylor et al 1987, Brown et al 1987, Elstow et al 1985).

We have shown previously that ESAF activates procollagenase, (Weiss et al 1984)) and is capable of reactivating collagenase from its complex with its inhibitor Tissue Inhibitor of Metalloproteinases,(TIMP-1)(McLaughlin et al 1991). Also we have recently described ESAF activation of gelatinase-A,and ESAF reactivation of the gelatinase-A, TIMP-2 complex.(Weiss and McLaughlin 1993.)

Table 1. The Three Major Interstitial Metalloproteinases

Enzyme		Substrate
Collagenase	MMP-1	Collagen Types I,III,II,VII,X
Gelatinase A	MMP-2	Collagen Types V,IV,VI,gelatin.
Stromelysin-1	MMP-3	Collagen Types III,IV,V, proteoglycan,lamanin, fibronectin

Neutral Matrix Metalloproteinases (MMPs)

Collagenase is one of the family of neutral metalloproteinases, which are all secreted as proenzymes, activated more or less instantaneously, and immediately inhibited by TIMPs. Activation of these proenzymes whether by proteolytic processing by plasmin, trypsin etc, or chemically by amino phenyl mercuric acetate,(APMA), which has no proteolytic activity results in loss of molecular mass. This is also the case with ESAF activation of the proenzymes,though as has been mentioned previously, ESAF is itself not catalytic. It is probable that a change in molecular structure following binding of ESAF results in autocatalytic processing of the proenzymes. Three of the major interstitial MMPs are Collagenase, Gelatinase-A,and Stromelysin-1 (Matrisian 1992.). As shown in Table.1, collagenase is capable of degrading the major interstitial collagens. Gelatinase A, as its name implies is capable of degrading gelatin but is more specific for type V collagen and the helical portion of type IV collagen involved in the basement membrane. Stromelysin-1 has the unusual ability to increase the proteolytic processing of procollagenase by plasmin or trypsin.

Recently we have shown that ESAF is also capable of activating prostromelysin-1.(Weiss and McLaughlin 1993). In addition we can show that gelatinase-A complexed with both TIMPs 1 and 2 can be reactivated by ESAF at physiological concentrations.(Table 2.) Two types of Tissue Inhibitors of metalloproteinases(TIMPs) have been described, TIMP-1 as a 28.5Kd glycoprotein, with no free sulphydryl groups(Cawston et al 1983) and TIMP-2, not a glycoprotein which may account for its lower molecular mass of 21 Kd. TIMP-2 has been described as having a possible preferential affinity for gelatinase-A ,(Stetler-Stevenson et al 1989) and has recently been shown to be capable of binding to the pro-form of gelatinase A.(Howard and Banda 1991.)

Table 2. Reactivation of the Gelatinase A-TIMP2 complex by ESAF

Gelatinase-A	50286 ± 106
Gelatinase-A + 1 unit of TIMP-2	$1762 \pm 35 = 96\%$ inhibition
Gelatinase-A + 1 unit of TIMP-2 +2 units of ESAF	$50047 \pm 59 = 99\%$ reactivation
values are cpm $\pm$ S.E.M.	

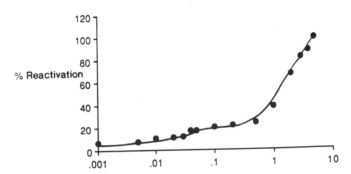

Figure 1. Reactivation of the Gelatinase-A,TIMP-2 Complex a range of ESAF units 0-10,(plotted on a log scale.)

Time curves for the activation of progelatinase-A was seen to give 50% activation by ESAF at 60 minutes and 100% activation at 90 minutes.(Weiss and McLaughlin 1993). Full reactivation of the gelatinase-A, TIMP-2 complex was achieved with 5 units of ESAF. (Fig 1.) A unit of ESAF is described as 100% activation of 1 unit of procollagenase.

DISCUSSION

ESAF is the only physiological factor capable of activating one of the major enzymes, progelatinase-A, involved in the degradation of the basement membrane and specifically involved in the cleavage of type IV collagen. ESAF is also the only physiological factor capable of reactivating the complexes formed between TIMPs 1 and 2 and collagenase, and gelatinase-A to produce a free active enzymes and free inhibitors. Reactivation of the enzyme-inhibitor complexes has been achieved by non-physiological means, but this only gives rise to a free active inhibitor (Cawston et al 1983). Activation of progelatinase A and indeed procollagenase and prostromelysin-1 by ESAF all resulted in an apparent loss in molecular mass as observed on gel electrophoresis.(Result not shown.) This would suggest that ESAF activation of the proMMPs causes a conformational change of the enzyme enabling autocatalytic activation to occur. The amounts of ESAF necessary to achieve these activations and reactivations are thought to be in the order of picogram quantities. It is now widely accepted that activation of the metalloproteinase system is a primary event in angiogenesis and several serine proteinases have been postulated for such activation. However none of these enzymes have the speed of activation achieved by ESAF. Also ESAF being a small molecule is easily removed from its site of activation. We consider that the role of this angiogenic factor in the activation and reactivation of the major enzymes involved in degradation of the basement membrane and the surrounding collagenous stroma is now patently very important in our understanding of the initial events in angiogenesis.

REFERENCES

Ausprunk, D., 1979 Tumour angiogenesis in: " Handbook of Inflammation" J.C. Houck ed.1, 318-351.

Brown, R.A., Taylor, C., McLaughlin, B., McFarland, C.D., Weiss, J.B., and Ali, S.Y., 1987, Epiphyseal growth plate cartilage and chondrocytes in mineralising culture produce a low molecular mass angiogenic procollagenase activator, *Bone Miner.* 3(2):143.

Cawston, T.E., Murphy, G., Mercer, E., Galloway, W.A., Hazelman, B.L., and Reynolds, J.J., 1983, Purification of an inhibitor of rabbit bone metalloproteinase, *Biochem. J.* 211:313.

Elstow, S.F., Schor, A.M., and Weiss, J.B., 1985, Bovine retinal angiogenesis factor is a small molecule (molecular mass <600), *Invest. Opthalmol. Vis.Sci.* 26:74.

Folkman, J., Merler, E., Abernathy, C., and Williams, G., 1971, Isolation of a tumour factor responsible for angiogenesis, *J.Exp.Med.* 133:275.

Howard, E.W., and Banda, M.J., 1991, Binding of tissue inhibitor of metalloproteinase 2 to two distinct sites on human 72Kda gelatinase, *J.Biol.Chem.* 266:17972.

Keegan, A., Hill, C.R., Kumar, S., Phillips, P., Schor, A.M., and Weiss, J.B., 1982, Purified tumour angiogenesis factor enhances proliferation of capillary but not aortic endothelial cells in vitro, *J. Cell Sci.* 55:261.

Matrisian, L.M., 1992, Matrix metalloproteinases, *Bioessays.* 14:445.

McLaughlin, B., Cawston, T.E., and Weiss, J.B., 1991, Activation of the matrix metalloproteinase inhibitor complex by a low molecular weight angiogenic factor, *Biochem.Biophys. Acta.* 1073:295.

Nagase, H., Ozata, Y., Suzuki, K., Enghild, J.J., and Salvesen G., 1991, Activation mechanisms of the precursors of matrix metalloproteinases, *Biochem.Soc.Trans.* 19:715.

Odedra, R., and Weiss, J.B., 1991, Low molecular weight angiogenic factors, *Pharmac.Ther.* 49:111.

Phillips, P., and Kumar, S., 1979, Tumour angiogenesis factor (TAF) and its neutralisation by a xenogenic antiserum, *Int.J.Cancer* 23:82.

Schor, A.M., Schor, S.L., Weiss, J.B., Brown, R.A., Kumar, S., and Phillips, P., 1980, Stimulation by a low molecular weight angiogenic factor of capillary endothelial cells in culture, *Brit.J.Cancer* 4:1790

Stetler-Stevenson, W.G., Krutzsch, H.C., and Liotta, L.A., 1989, Tissue inhibitor of metalloproteinase (TIMP-2), *J.Biol.Chem.* 264:17374.

Taylor, C.M., McLaughlin, B., and Weiss, J.B., 1987, Quantitation of endothelial cell stimulating angiogenesis factor (ESAF) in a range of tissues, *Int.J.Microcirc* 7:288

Taylor, C.M., McLaughlin, B., Weiss, J.B., and Smith, I. 1988a, Bovine and human pineal glands contain substantial quantities of endothelial cell stimulating angiogenesis factor, *J. Neural Transm.* 71:79.

Taylor, C.M., Weiss, J.B., McLaughlin, B., and Kissun, R.D. 1988b, Increased procollagenase activating angiogenic factor in the vitreous humour of oxygen treated kittens. *Br. Journal Opthalmol.* 72:2.

Weiss, J.B., Brown, R.A., Kumar, S., and Phillips, P., 1979, An angiogenic factor isolated from tumours: a potent low molecular weight compound, *Br.J. Cancer* 40:493.

Weiss, J.B., Hill, C.R., Davis, R.J., and McLaughlin, B., 1984, Activation of mammalian procollagenase and basement membrane degrading enzymes by a low molecular weight angiogenesis factor. *Agents and Actions* 15:107.

Weiss, J.B., and McLaughlin, B., 1993, Activation of gelatinase-A and reactivation of the gelatinase-A inhibitor complex by endothelial cell stimulating angiogenesis factor (ESAF), *J. Physiol.* 467:49.

INTERACTIONS BETWEEN OXIDIZED LIPOPROTEIN (a) [Lp(a)], MONOCYTES AND ENDOTHELIAL CELLS

Mahmoud Ragab, Periasamy Selvaraj, and Demetrios S. Sgoutas

Department of Pathology and Laboratory Medicine, Emory University School of Medicine, Atlanta, GA 30322

Address correspondence to: Demetrios Sgoutas, Department of Pathology and Laboratory Medicine, Room F-147, Emory University Hospital, 1364 , Clifton Road, N.E., Atlanta, GA 30322

SUMMARY

Lipoprotein (a) [Lp(a)] is a subclass of low density lipoproteins (LDL) whose elevated blood concentrations constitute an independent risk factor for coronary artery disease (CAD). Lipoprotein (a) derives its uniqueness from an appendage, apolipoprotein (a), which is linked to the apolipoprotein B-100 portion of LDL. Apolipoprotein (a) has been found to display a remarkable homology with plasminogen. This homology has lead to suggestions that there may be competition between Lp(a) and plasminogen for binding to fibrin, and to cell membranes. Consequently, much attention has been paid to interactions between fibrin, Lp(a), and plasmin and to the linkage of thrombosis and atherogenesis.

In contrast, this study draws attention at the cholesterol rich component of Lp(a) which is identical to LDL, investigates its oxidation and the role of oxidized Lp(a) in activating resident intimal macrophages. It provides evidence that oxidatively modified Lp(a), enhanced the expression of surface antigens on U937 cells and stimulated adhesion of U937 cells to cultured endothelial cells. These results suggest that oxidatively modified Lp(a) like oxidized LDL behaves as an inducer of differentiation, adhesion and activation of monocytes, leading to the hypothesis that oxidized Lp(a) may play an important role in atherogenesis, and CAD.

INTRODUCTION

Atherogenesis is generally regarded as an example of uncontrolled angiogenesis. The proliferative arteriosclerotic lesion could be considered the arterial response to sustained or episodic pathological vascular stimulation occuring over the years. Recent investigations on cells of the arteriosclerotic lesion (injured endothelial cells, macrophages, lymphocytes and proliferating smooth muscle cells) focused on growth factors, growth inhibitors, cytokines, autocrine and paracrine regulators collectively known as factors of "angiogenesis" [1, 2]. Studies on the progression of restenosis after angioplasty, in particular, focused on the in vitro interaction of lipoproteins with monocytes, endothelial cells and smooth muscle cells [3, 4]. To further delineate these interactions, we studied the effect of

Angiogenesis: Molecular Biology, Clinical Aspects
Edited by M.E. Maragoudakis *et al*, Plenum Press, New York 1994

oxidized and native lipoprotein (a) [Lp(a)] on monocyte activation and their adhesion to endothelial cells. The results indicated that pretreatment with oxidized Lp(a) in contrast to native Lp(a) increased the adhesive properties of monocytes. The results were comparable to those obtained with oxidized low density lipoproteins (LDL) [5].

Elevated levels of plasma Lp(a) are assumed to be an additional risk factor for atherosclerotic disease in man. Several clinical trials have shown a strong correlation between elevated plasma Lp(a) and cardiovascular disease [5, 6] including restenosis after angioplasty [7]. These studies suggested that Lp(a) is a much more atherogenic lipoprotein than LDL [8, 9]. Hence, it became important to understand how Lp(a) exerts its pathogenic effects. The chemical structure of Lp(a) differs from LDL by the presence of the glycoprotein apo(a). Because the aminoacid sequence of apo(a) is approximately 80% identical to that of plasminogen [8-10], it is possible that the pathophysiolocal properties of Lp(a) including its effects on thrombogenesis and fibrinolysis, are attributable to apo(a). Lp(a) binds to endothelial and macrophage cells and to extracellular components such as fibrin and inhibits cell associated plasminogen activation and plasmin generation [11, 12]. The resulting inhibition of clot lysis undoubtfully contributes to the process of atherosclerosis and the homology between apo(a) and plasminogen points to a linkage among lipoproteins, atherosclerosis and thrombosis [11, 12]. Furthermore, Lp(a) could be linked to neovascularization and angiogenesis because it has been shown that plasmin causes release of basic fibroblast growth factor (a well known angiogenetic factor) [4] from extracellular matrix .

Other efforts to understand the atherogenecity of Lp(a), on the other hand, have been focused at its cholesterol-rich moeity, which is identical to the lipid-rich moeity of LDL. This component, especially because of its attachment to the glycoprotein, apo(a), which is capable of binding with high affinity to many extracellular matrix components could be responsible for the local accumulation of cholesterol in damaged areas of the blood vessel wall [11, 13]. Recent *in vitro* studies have shown that endothelial cells, smooth muscle cells and macrophages release substances causing oxidative modification of LDL [14]. Other *in vitro* studies have shown that macrophages readily bind and internalize oxidized LDL by a specific scavenger receptor [15]. Evidence indicating that oxidized LDL exist *in vivo* has also come from the presence of oxidized forms of LDL in human atherosclerotic plaques [16].

It is likely, that Lp(a), because of its cholesterol-rich component could be equally susceptible to oxidation and could follow similar steps in the oxidative LDL pathway [16]. Uptake of oxidized Lp(a) by intimal macrophages resulting in the formation of macrophage-derived foam cells may explain the increased risk of coronary artery disease associated with high plasma levels of Lp(a). Histochemical methods have demonstrated that Lp(a) occurs in the intima of atherosclerotic lesions.

MATERIALS AND METHODS

Lipoprotein Separation. LDL (1.019 to 1.055 g/ml) and Lp(a) (1.055 to 1.090 g/ml) were isolated by sequential ultracentrifugation from each individual plasma sample obtained from donors with Lp(a) concentrations over 70 mg/dl. Lp(a) was purified by column chromatography over Biogel A-5m (Bio-Rad Laboratories, Richmond, CA) and lysine-sepharose (Pharmacia, Uppsala, Sweden) in order to exclude any LDL.

Modification of Lipoproteins. Lp(a) and LDL (8-10 mg/ml) were dialyzed (to remove EDTA) against a 200-fold volume of 0.001 M phosphate-buffered saline (PBS) (pH 7.4) for 48 hrs at 4°C in the dark. Lp(a) and LDL were oxidized by incubation with 10 uM copper chloride in PBS at 37°C. Details of the methodology are given elsewhere [17]. The kinetics of the reaction of the lipoproteins were followed by measuring the 234 nm absorption at different time intervals.

Cell Culture. The U937 cells were grown in RPMI 1640 medium supplemented with 10% vol/vol heat-inactivated fetal calf serum and gentamycin sulfate (50 ug/ml) and kept in an atmosphere of 5% carbon dioxide and 95% air. Endothelial cells were prepared from human umbilical veins.

Expression of Surface Antigens. Monoclonal fluorescein isothiocyanate -conjugated antibodies were purchased from Becton Dickinson. The cells were washed three times in PBS containing 0.1% bovine albumin and 0.02% sodium azide. The pellet was then gently suspended, the monoclonal antibodies were added to the cell suspension, and the suspension was covered and placed on ice for 30 min. There after the cells were washed twice in the same as above PBS. Cell viability was determined by tryptan blue dye exclusion before analysis. The expression of surface antigens was determined in a FACS Scan from Becton Dickinson.

Adhesion to Endothelial Cells. In order to study adhesion to endothelial cells, U937 cells were incubated for 24 hrs with native and oxidized Lp(a) (10 μg/ml), native and oxidized LDL (10 ug/ml) or phorbol-12-myristate (PMA). The cells were then counted and 1 ml of cell suspension (half a million cells per ml) was added to each well containing endothelial cells on glass coverslips. After 30 min of cocultivation at 37°C, each glass coverslip was washed six times in PBS and placed in a well containing a trypsin/EDTA solution to detach cells. Cells were suspended into single cells and counted in an electronic cell counter. The number of cells, after subtraction of a mean of endothelial cells from coverslips without the addition of U937 cells was regarded as the number of adherent U937 cells.

Analytical Procedures. Lipid peroxide levels were also determined by spectrophotometric assay as thiobarbituric acid-reactive substances (TBARS), and results were expressed as nmol malondialdehyde (MDA) equivalents /mg protein. Plasma Lp(a) concentration was determined by an ELISA [Macra Lp(a); Terumo Medical Corp., Elkton, MD]. The fatty acid composition of Lp(a) and LDL was determined by gaschromatography after transesterification with BF3/MeOH (20%) in benzene. Protein was determined on a Kodak Ektachem 700 analyzer (Eastman Kodak Co., Rochester, NY). Alpha-tocopherol levels in Lp(a) and LDL were determined by high performance liquid chromatography.

RESULTS

Figure 1. Shows results of the copper mediated oxidation of Lp(a) compared with that of LDL from the same donors (n=5). The onset of lipid peroxidation is preceded by a lag phase [17]. In all five cases the lag phase of Lp(a) isolated from the corresponding donors was shorter than the lag phase of LDL. Mean values expressed in min ± SD were 46.4 ± 6.5 for Lp(a) and 57.4 ± 7.8 for LDL, respectively. The amount of conjugated dienes formed during oxidation was calculated assuming a molar extinction coefficient of 29500 [17]. The rate of conjugated diene formation, expressed as nmol/mg. min ± SD was significantly (p <0.05) greater in the case of Lp(a) (0.65 ± 0.09) as compared to LDL (0.54 ± 0.08).

TBARS expressed as MDA equivalents in the above samples after 3 hrs oxidation were 36.4 ± 3.8 for Lp(a) and 31.2 ± 3.2 mmmol MDA/mg protein, respectively. The fatty acid composition of Lp(a) and LDL is shown in Table 1. A comparison of the antioxidant composition showed that Lp(a) contained less alpha-tocopherol (1.8 ± 0.28 nmol/mg) than LDL (2.4 ± 0.31 nmol/mg).

After 48 hrs of incubation, we observed a decrease in cell growth (results not shown) in cells exposed to oxidized Lp(a). Hence, oxidized Lp(a) causes a hindrance in cell growth similar to such caused by oxidized LDL or PMA [18].

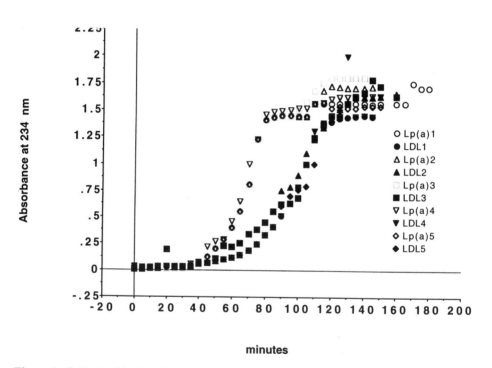

Figure 1. *Oxidation kinetics of Lp(a) and LDL by measuring the increase of absorbance at 234 nm. 200 ug of lipoprotein in oxygen saturated PBS (pH 7.4), 10 uM CuCl₂ final concentration.*

TABLE 1. Fatty acid composition of Lp(a) and LDL.

Fatty Acid	Lp(a)	LDL
16:0	25.4 ± 2.8	24.3 ± 3.1
16:1	1.8 ± 0.4	2.0 ± 0.8
18:0	6.8 ± 0.9	7.2 ± 1.2
18:1	17.6 ± 2.1	19.6 ± 2.6
18:2	43.1 ± 3.7	41.1 ± 3.8
20:4	5.3 ± 1.2	5.8 ± 1.1

Percent of total fatty acids (mean ± SD) from five samples.

Figure 2. Shows the effect of native and oxidized Lp(a) and LDL along with PMA on the expression of surface antigen MAC-1 on U937 cells. Oxidized Lp(a) induced expression of the adhesion molecule MAC-1 which was comparable to oxidized LDL.

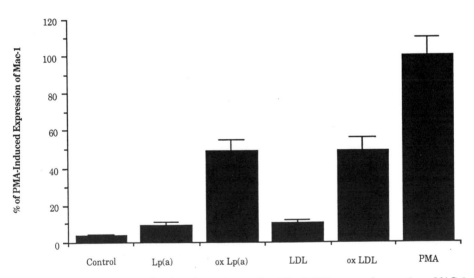

Figure 2. *Effect of native, oxidized Lp(a) native and oxidized LDL on surface antigen MAC-1 expression on U937 cells. Mean % relative cell number to PMA expression (± SD).*

Figure 3. Shows the effect of native and oxidized Lp(a) and LDL along with PMA on the adhesion of U937 cells to cultured human umbilical vein endothelial cells. Oxidized Lp(a) induced adhesion of U937 cells to endothelial cells which was comparable to that of oxidized LDL.

DISCUSSION

Results from the oxidation lag phase and from the oxidation rates of Lp(a) and LDL from the same individuals demonstrated that Lp(a) was less resistant to oxidative modification than the corresponding LDL. Analysis of the fatty acid composition showed that LDL and Lp(a) had very similar fatty acid compositions [14]. In addition, comparison of the antioxidant composition showed that Lp(a) contained less a-tocopherol than LDL raising the speculation that the higher susceptibility of Lp(a) to oxidation may be due to the lower content of antioxidants [14, 15].

The present investigation showed that oxidized Lp(a) could induce adhesion molecules as surface antigens on U937 monocytic cells and that the effect was

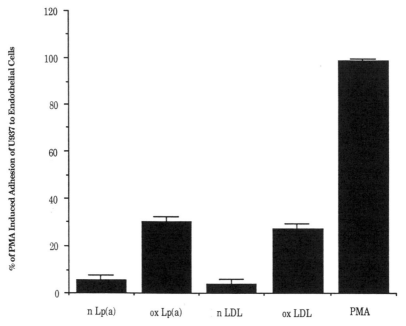

Figure 3. *Effect of native and oxidized Lp(a), LDL on adhesion of U937 cells to endothelial cells. Data are presented as precent (mean) of PMA-induced adhesion in three experiments (± SD).*

comparable to that of oxidized LDL. It showed a strong increase in the adhesion of U937 cells to endothelial cells after incubation with oxidized Lp(a) as compared to native Lp(a). The nature of the mechanism by which oxidized Lp(a) as well as oxidized LDL induces expression of surface antigens on U937 and adhesion of activated U937 to endothelial cells remains unknown. Oxidized Lp(a) particles like oxidized LDL are chemoattractants for monocytes and they may trap cells in the arterial wall. In addition, oxidized lipoproteins like free radicals may trigger the initiation of cascades of autocrine and paracrine events leading to the secretion of cytokines and growth factors (i.e., transforming growth factor) which have been shown to inhibit clot lysis and enhance cell migration and proliferation, all typical characteristics of angiogenic events.

ACKNOWLEDGEMENT

We are indebted to the American Heart Disease Prevention Foundation, INC., and to the World Health Organization for financial aid. We thank David Apanay for his excellent technical assistance.

REFERENCES:

1. Reidy, M.A., J. Fingerle, and V. Lindner, *Factors controlling the development of arterial lesions after injury.* Circulation, 1992. **86**(Suppl. III): p. 43.
2. Lerman, A. and J.C. Burnett, Jr., *Intact and altered endothelium in regulation of vacsometion.* Circulation, 1992. **86**(Suppl III): p. 12.
3. Schwartz, R.S., D.R. Holmes, and E.J. Topol, *The restenosis paradigm revisited: an alternative proposal for cellular mechanism.* J. Am. Coll. Cardiol., 1992. **20**: p. 1284.
4. Vlodavsky, L., *et al.*, *Extracellular matrix-resident basic fibroblast growth factor: implication for the control of angiogenesis.* J. Cell. Biochem, 1991. **47**: p. 167.
5. Dahlen, G.H., *et al.*, *Association of levels of lipoprotein Lp(a), plasma lipids, and other lipoproteins with coronary artery disease documented by angiography.* Circulation, 1986. **74**: p. 758.
6. Hearn, J.A., *et al.*, *Predictive value of lipoprotein (a) and other serum lipoproteins in the angiographic diagnosis of coronary artery disease.* Am. J. Cardiol., 1990. **66**: p. 1176.
7. Hearn, J.A., *et al.*, *Usefulness of serum lipoprotein (a) as a predictor of restensis after percutaneous coronary angioplasty.* Am. J. Cardiol., 1992. **69**: p. 736.
8. Loscalzo, J., *Lipoprotein (a): a unique risk factor for atherothrombotic disease.* Arteriosclerosis, 1990. **10**: p. 672.
9. Loscalzo, J., *et al.*, *Lipoprotein (a), fibrin binding, and plasminogen activation.* Arteriosclerosis, 1990. **10**: p. 240.
10. Utermann, G., *The mysteries of lipoprotein (a).* Science, 1989. **246**: p. 904.
11. Scanu, A.M., *Lp(a): a link between thrombosis and atherosclerosis.* Ear. J. Epidem., 1992. **8**: p. 76.
12. Hajjar, K.A., *et al.*, *Lipoprotein (a) modulation of endothelial cell surface fibrinolysis and its potential role in atherosclerosis.* Nature, 1989. **339**: p. 303.
13. Scanu, A.M. and G.M. Fless, *Lipoprotein (a): heterogeneity and biological relevance.* J. Clin. Invest., 1990. **85**: p. 1709.
14. Sattler, W., *et al.*, *Oxidation of lipoprotein Lp(a). A comparison with low-density lipoproteins.* Biochem. Biophys. Acta., 1991. **1081**: p. 65.
15. Naruszewicz, M., E. Selinger, and J. Davignon, *Oxidative modification of lipoprotein (a) and the effect of b-carotein.* Metabolism, 1992. **41**: p. 1215.
16. Parthasarathy, S., D. Steinberg, and J.L. Witztum, *The role of oxidized low-density lipoproteins in the pathogenesis of atherosclerosis.* Annual. Rev. Med., 1992. **43**: p. 219.
17. Esterbauer, H., *et al.*, *Continuous monitoring of in vitro oxidation of human low density lipoproteins.* Free Rad. Res. Commun., 1989. **6**: p. 67.
18. Frostegard, J., *et al.*, *Oxidized low density lipoprotein induces differentiation and adhesion of human monocytes and the monocytic cell line U937.* Proc. Natl. Acad. Sci. U.S.A., 1990. **47**: p. 904.

ALTERED PROLIFERATION
OF RETINAL MICROVASCULAR CELLS IN RESPONSE TO
NON-ENZYMATIC GLYCOSYLATED MATRIX PROTEINS

Theodosia A. Kalfa,[1] Mary E. Gerritsen,[2] and Effie C. Tsilibary[1]

[1]Department of Laboratory Medicine and Pathology,
 University of Minnesota Medical School
[2]Miles Laboratories
 West Haven, CT

INTRODUCTION

Diabetic retinopathy is characterized by microaneyrisms derived from vessels which have selectively lost their pericytes and demonstrate focal proliferation of endothelial cells. The proliferation of retinal endothelial cells and the pericytic degeneration followed by replacement with pericytic "ghosts" is an almost invariable and early feature of diabetic retinopathy [1].

Two general pathophysiologic mechanisms have been proposed to link hyperglycemia with the observed pathologic and functional damage in tissues susceptible to diabetic complications: non-enzymatic glycosylation of long-lived proteins[2] and the polyol pathway[3]. The hypothesis that diabetic retinopathy is caused through abnormal polyol-inositol metabolism has been investigated in numerous studies involving diabetic and galactosemic dogs[4,5], rats [6] and diabetic humans[7] using aldose reductase inhibitors and the results have been contradictory. Moreover other studies have related incidence and severity of diabetic retinopathy to collagen-linked fluorescence that has been attributed to AGE products (after-products following non-enzymatic glycosylation of proteins)[8].

We selected therefore to study further the effect of nonenzymatic glycosylation of matrix proteins on the proliferation of retinal microvascular cells.

Our initial approach was to examine the proliferative effect of isolated, diabetically modified, matrix molecules as a substrate for culturing retinal microvascular cells.

MATERIALS AND METHODS

Non-enzymatic glycosylation of BM Macromolecules

Type IV collagen or laminin, major components of basement membranes (BM), were

Angiogenesis: Molecular Biology, Clinical Aspects
Edited by M.E. Maragoudakis *et al*, Plenum Press, New York 1994

isolated from EHS tumor grown subcutaneously in mice, according to protocols previously described[9,10]. For modifying the proteins with nonenzymatic glycosylation the following procedure was followed: the proteins were dialyzed extensively against PBS, pH 7.4 including 10mM EDTA, 0.02% NaN_3, 50µg/ml PMSF and NEM, at 4°C. In the case of type IV collagen the above buffer contained additionally 0.5M NaCl. The used buffer was designed to limit proteolytic degradation and polymerization. The protein solutions were centrifuged at 20,000 rpm for 20 min at 4°C in a Beckman L8-M ultracentrifuge to clear large aggregates and they were incubated in the absence of sugar or in the presence of either 50 mM or 500 mM glucose for 72 hrs at 29°C. Under these conditions approximately 1 mole and 10 moles of glucose were incorporated per mole of macromolecule, respectively (Table 1)[11]. Following extensive dialysis against PBS, samples were used to coat 96 well Immulon #1 tissue culture plates.

Coating Plates

Type IV collagen was used in the concentrations of 25 µg/ml and 100 µg/ml while laminin was used in the concentrations of 100 µg/ml and 200 µg/ml for the Retinal Pericytes and Retinal Endothelial Cells, respectively. These concentrations were selected according to initial experiments designed to determine conditions for optimal adhesion of each different cell type to type IV or laminin, respectively. 50 µl of each protein solution was added per well and allowed to dry overnight at 29°C. Quadruplicates or sets of six wells were used for each control as well as glycosylated protein. When this procedure was complete, the plates were used immediately or they stored at 4°C for at most 2 weeks before use.

Cells

Bovine Retinal Pericytes (BRP) and *Bovine Retinal Endothelial Cells (BRE)* were isolated from bovine retinas[12]. Cells were cultured in 75-cm^2 flasks (Falcon, Oxnard, CA). For the BRE cells the flasks were coated with 5 ml/flask of 100 µg/ml fibronectin (Sigma, St. Louis, MO) and after a 30 min incubation at room temperature the residual matrix solution was removed. The media used were RPMI 1640 containing 10% calf serum for the BRP and DMEM (4500 mg/l glucose) containing 20% fetal calf serum for the BRE cells, supplemented with 50 IU/ml Penicillin and 50 µg/ml Streptomycin. Sera, media and antibiotics were purchased from Sigma, St. Louis, MO. Cells between passages 5 and 13 were used for the experiments.

The cells were released from the tissue culture flasks for passaging and use in experiments by washing twice with Versene (sterile PBS + 0.5 mM EDTA) (GIBCO BRL, Grand Island, NY) followed by treatment with trypsin-EDTA for 1 min. The concentrationof the trypsin solution used was 0.05% trypsin containing 0.53 mM EDTA for the BRP and 0.015% trypsin containing 0.15 mM EDTA for the BRE cells.

Table 1. GLUCOSE INCORPORATION (in moles glucose per mole of each macromolecule)[11]

	type IV collagen	laminin
50 mM glucose	1.35	0.90
500 mM glucose	12.02	11.04

Cell Proliferation Assays

The protein-coated plates were sterilized before use with UV irradiation for 2 hrs. The irradiation time was determined experimentally to have a minimal effect on the cell-adhesion related properties of the proteins (type IV collagen and laminin) while it prevented bacterial contamination. Subsequently the wells were blocked with HBSS (Sigma, St. Louis, MO) containing 10% BSA + 50 IU/ml Penicillin and 50 μg/ml Streptomycin., 200 μl/well, for 2 hrs at 37°C.

BRP were in several instances synchronized for the experiment by serum deprivation and 1.2 mM hydroxyurea, according to method previously used for various cells [13] included retinal pericytes [14]. More specifically, the culture medium was removed and after thorough washes with Versene, the cells were incubated in RPMI-1640 supplemented with 1% CS for 48 h. Then, the serum-deficient medium was replaced by complete growth medium containing 10% CS. 6 hrs later, concentrated hydroxyurea stock solution was added to reach a final concentration of 1.2 mM. After 12 h, the cells were washed twice with RPMI 1640 and used in the proliferation assay. The BRE cells were not used after synchronization as they were extremely sensitive to the procedure.

To perform the assay the cells following releasing with trypsin were washed twice in the corresponding medium by centrifugation at 1000 rpm x 5 min and they were resuspended in culture medium. Cells were then seeded onto sterile 96-well Immulon #1 plates (5,000 BRP/well and 20,000 BRE cells/well) which had been previously coated with either control or glycosylated type IV collagen or laminin (as described previously). The usual number of plates used in each experiment was six to eight, reflecting the number of time-points followed. Cells were allowed to adhere for 5-8 hrs (BRP) or 20-24 hrs (BRE cells), at 37°C and non-adherent or non-viable cells were removed by washing 2 times with 100 μl/well Versene. The growth medium was replaced in all but one plates and the cells were allowed to grow at 37°C. The plate which did not get fed was considered as the time zero of the experiment and the number at time zero was subtracted from each further time interval. The following time-points were examined every 24 hrs afterward. For each time point the cells were released from their respective wells after gentle wash with 100 μl/well Versene twice and treatment with 100 μl/well trypsin-EDTA for 10 min, at 37°C. The number of cells/well was determined using an automated Coulter counter, by transfering the trypsinized cells in vials filled with 8 ml Isoton II (Fisher Scientific, Pittsburgh, PA). Values were expressed as the mean and standard deviation of 4 or 6 replicates, depending on the experiment, and the results were confirmed by at least three repeats of each experiment.

RESULTS

The glycation method used resulted in chemical modification mainly of the Amadori type in glycosylated proteins. We could not detect AGE; products therefore if present, these should occur only in minimal amounts. Presumably then, the effects on the growth rate on the two retinal cell types we examined should be the result of primarily Amadori-modified type IV collagen and laminin.

The growth curves of the BRP and the BRE cells cultured on control and glucose modified basement membrane macromolecules were studied. The numbers of cells initially adhered on the substrate molecules, at time 0 (not significantly different between

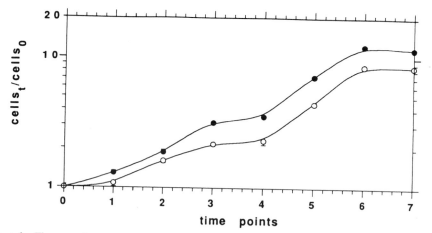

Figure 1. The growth curve of the retinal pericytes on the glycated by 50 mM glucose type IV collagen (—o—) was significantly lower from the control (—●—) already on the first day after adhesion (p<0.0005 by one tailed t-test). Each point represents the mean ±sd of cell growth in six replicates of a representative experiment.

control and glycated substrates, at this time), were used for the calculation of the proliferation curves shown on figures 1-4.

The retinal pericytes on the glycated type IV collagen (50 mM glucose) had significantly lower proliferation rate than on the control protein, even after time point 1 (24 hrs after the adhesion) (Figure 1). The difference in the proliferation rate continued to increase until the 5th day after time zero when the cell numbers were 7.19 and 4.44 the initial number of the cells adhered on the control and glycated type IV collagen,

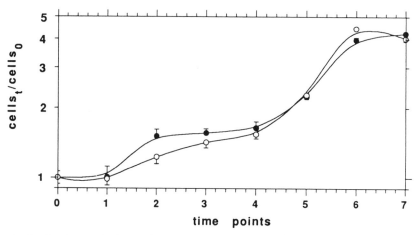

Figure 2. On the glycated laminin (50 mM glucose:—o—), BRP growth rate was significant lower from the control (—●—) 48 hrs after adhesion (p<0.0005 by one tailed t-test) and with lower significance (p<0.005, p<0.005) for the third and fourth day, respectively. There was no signifficant difference after the fifth day. Each point represents the mean ±sd of cell growth in six replicates of a representative experiment.

114

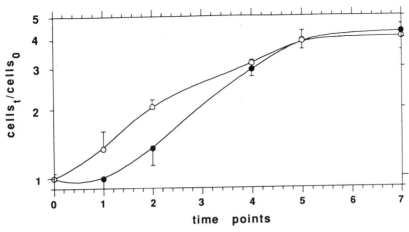

Figure 3. BRE cells displayed higher growth rate on modified (500mM glucose: —o—) than on control type IV collagen (—•—) the first 48 hrs after adhesion. (p<0.05 on time point 1, p<0.005 on time point 2, by one tailed t-test). Each point represents the mean ±sd of cell growth in quadruplicates of a representative experiment.

respectively. This corresponds to a growth rate on the glycated type IV collagen equal to 62% of the control. The difference although in smaller percentage rate, remained statistically significant until the 7th time point (the latest time-point of the experiment).

On laminin the retinal pericytes were proliferating much slower than on type IV collagen (Figure 2). 48 hrs after adhesion the cell number was 1.52 and 1.23 the initial number of the cells adhered on the control and glycated (50 mM glucose) laminin, respectively. This difference equivalent to 20% less growth on glycated laminin at this time point was gradually overcome by the fourth day. 7 days after time zero the cell number has been increased only 4.2-4.0 times respectively.

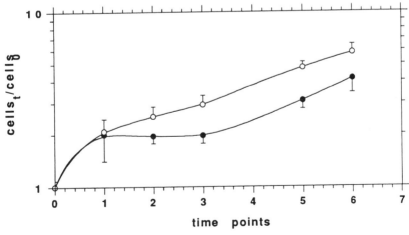

Figure 4. BRE cells were starting to proliferate faster on glucose (50 mM) modified laminin (—o—) than on control (incubated with no sugar) laminin (—•—) 48 hrs after adhesion. They kept a consistently higher growth rate till the end of the experiment (p<0.005 by one tailed t-test). Each point represents the mean ±sd of cell growth in six replicates of a representative experiment.

The growth curves of the retinal pericytes on type IV collagen or laminin modified by 500 mM glucose were similar to those obtained from seeding on macromolecules modified by 50 mM glucose. Non-synchronized retinal pericytes used in proliferation assays gave similar results (data not shown).

The retinal endothelial cells had a faster growth rate on modified (500 mM glucose) type IV collagen only for the three days after seeding with the greatest observed difference on the second day: 150% of the control, i.e. 2.04 against 1.34 the initial number of the cells adhered on modified and control substrate respectively (Figure 3). After the fourth day the rate of growth was similar on control or glycosylated type IV collagen.

On laminin modified by 50 mM glucose the faster growth rate of BRE was consistent from the 2nd till the 6th day after adhesion and reached the maximum difference (154%) on the fifth day (Figure 4). Laminin modified by 500 mM glucose, as a substrate for retinal endothelial cells, had erratic and inconsistent results and was therefore not examined further.

DISCUSSION

We present in this report evidence that non-enzymatic glycosylation of type IV collagen and laminin, two major components of the retinal microvascular matrix has an effect on the growth rate of retinal endothelial cells, which sit on the retinal extracellular matrix and retinal pericytes which are ensheathed in this matrix. In diabetic retinopathy, there is an early disappearence of retinal pericytes. As the disease progresses, microaneurisms form with focal areas of neovascularization and proliferation of the endothelial cells. These findings are unique for diabetic retinopathy[1,15]. Because non-enzymatic glycosylation of basement membrane components, in particular type IV collagen has been observed at least in the kidneys of diabetic rats and diabetic human subjects[16-18], we addressed the question whether this chemical modification has an effect on the growth rate of retinal microvascular cells. Under our experimental conditions approximately 1-10 moles of glucose were incorporated per mole protein[11]. We seeded freshly released bovine retinal pericytes (BRP) and retinal endothelial cells (BRE) onto solid phase-immobilized, control or glycosylated type IV collagen and laminin. Media were kept constant in all instances: BRP were cultured in RPMI containing 10 mM glucose and BRE were culured in DMEM which contained 20 mM glucose. Each medium was tested and found to be optimal for the maintenance of these cell types in culture.

When BRP were seeded onto glycosylated type IV collagen (tIV), there was approximately 40% decrease of their growth rate when compared to BRP grown on control tIV. This difference was maintained throughout the time course of the experiment. When BRP were grown on glycosylated laminin, there was an initial decrease of approximately 20% which was overcome within 4 days. Therefore both glycosylated proteins had an effect of suppressing the growth of BRP, albeit in response to type IV collagen the effect was more persistent. It is possible that BRP *in situ* use primarily tIV to interact with their surrounding basement mebrane matrix. When tIV undergoes diabetic modifications due to increased plasma glucose concentrations, interactions between BRP and glycosylated tIV become altered and result in phenotypic changes such as growth suppression. In preliminary experiments in which EHS-derived, whole basement membrane matrix was glycosylated and used for seeding BRP we observed a substantial decrease of their growth which persisted thoughout the time course of the experiment (data not shown).

In the case of BRE cells, a substantial increase of proliferation was observed on both glycosylated tIV and laminin. This effect was more enhanced and persistent in response to glycosylated laminin. The proliferative effect of tIV was gradually overcome by these cells. It is possible that BRE cells *in situ* use primarily laminin to interact with the underlying basement membrane, but this hypothesis awaits for further examination.

It is noteworthy that non-enzymatic glycosylation *per se* of each of the basement membrane-components that we examined could to some extent simulate the effect of diabetes on the retinal microvascular cells. More specifically, *in situ* there is a decreased growth of BRP evidenced by their disappearence in the early stages of diabetic retinopathy accompanied by an increased proliferation rate of BRE, as the formation of neovessels indicates. We conclude that abnormally increased levels of glucose in the plasma which are commonly observed in diabetes result in glycosylation of the retinal microvascular matrix. This in turn, results, at least in part, in altered recognition of this matrix by the retinal cells and ensuing differences of the cellular phenotype, which alter the growth rate of these cells. According to this hypothesis, the different proliferative response of each retinal cell type to the glycosylated substrate should reflect differences of the cell surface and/or signaling mechanisms which transmit information from the cell surface to the cell. The obtained evidence therefore indicates the paramount importance of glycemic control in diabetes mellitus, as the most efficient means to prevent non-enzymatic glycosylation of the retinal and possibly other extracellular matrices which may be involved in the development of diabetic compications.

Acknowledgments This work was supported by the following grants: JDF 190361, JDF 132043, ADA, AHA Minnesota Affiliate, NIH R01-DK43574, NIH R29-DK39216 to E.C.T.

REFERENCES

1. D.G. Cogan, D. Toussaint, and T. Kuwaraba, Retinal Vascular Patterns: Diabetic Retinopathy, *Arch Opthalmol.* 66:366-378 (1961).
2. M. Brownlee, A. Cerami, and H. Vlassara, Advanced products of nonenzymatic glycosylation and the pathogenesis of diabetic vascular disease, *Diabetes Metab Rev.* 4:437-51 (1988).
3. D.A. Greene, S.A. Lattimer, and A.A. Sima, Sorbitol, phosphoinositides, and sodium-potassium-ATPase in the pathogenesis of diabetic complications, *N Engl J Med.* 316:599-606 (1987).
4. T.S. Kern and R.L. Engerman, Retinal polyol and myo-inositol in galactosemic dogs given an aldose- reductase inhibitor, *Invest Ophthalmol Vis Sci.* 32:3175-7 (1991).
5. T.S. Kern and R.L. Engerman, Aldose reductase inhibition fails to prevent retinopathy in diabetic and galactosemic dogs, *Diabetes.* 42:820-825 (1993).
6. W.J. Robison, M. Nagata, N. Laver, T.C. Hohman, and J.H. Kinoshita, Diabetic-like retinopathy in rats prevented with an aldose reductase inhibitor, *Invest Ophthalmol Vis Sci.* 30:2285-92 (1989).
7. Sorbinil Retinopathy Trial Research Group, A randomized trial of sorbinil, an aldose reductase inhibitor, in diabetic retinopathy, *Arch Ophthalmol.* 108:1234-44 (1990).
8. V.M. Monnier, V. Vishwanath, K.E. Frank, C.A. Elmets, P. Dauchot, and R.R. Kohn, Relation between complications of type I diabetes mellitus and collagen-linked fluorescence, *N Engl J Med.* 314:403-8 (1986).
9. H.K. Kleinman, M.L. McGarvey, L.A. Liotta, P.G. Robey, K. Tryggvason, and G.R. Martin, Isolation and characterization of type IV procollagen, laminin and heparan sulfate proteoglycan from the EHS sarcoma, *Biochemistry.* 21:6188-6193 (1982).

10. R. Timpl, H. Rhode, P. Gehron-Robey, S.I. Rennard, J.-M. Foidart, and G.R. Martin, Laminin: a glycoprotein from basement membranes, *J. Biol. Chem.* 254:9933-9937 (1979).
11. C.S. Haitoglou, E.C. Tsilibary, M. Brownlee, and A.S. Charonis, Altered cellular interactions between endothelial cells and nonenzymatically glucosylated laminin/type IV collagen, *J Biol Chem.* 267:12404-7 (1992).
12. A. Capetandes and M.E. Gerritsen, Simplified methods for consistent and selective culture of bovine retinal endothelial cells and pericytes, *Invest Ophthalmol Vis Sci.* 31:1738-44 (1990).
13. W.B. Jakoby and I.H. Pastan."Cell Culture,"Academic Press, INC, San Diego, CA (1979).
14. W. Li, S. Shen, M. Khatami, and J.H. Rockey, Stimulation of retinal capillary pericyte protein and collagen synthesis in culture by high-glucose concentration, *Diabetes.* 33:785-9 (1984).
15. N. Ashton, Pathogenesis of diabetic retinopathy, *in* "Diabetic Retinopathy,"H.L. Little, R.L. Jack, A. Patz, and P.H. Forsham, Thieme-Stratton Inc, New York (1983).
16. M.P. Cohen, E. Urdanivia, M. Surma, and V.Y. Wu, Increased glycosylation of glomerular basement membrane collagen in diabetes, *Biochem Biophys Res Commun.* 95:765-9 (1980).
17. M.P. Cohen, E. Urdanivia, and V.Y. Wu, Nonenzymatic glycosylation of glomerular basement membrane, *Renal Physiol.* 4:90-5 (1981).
18. A.J. Perejda and J. Uitto, Nonenzymatic glycosylation of collagen and other proteins: relationship to development of diabetic complications, *Coll Relat Res.* 2:81-8 (1982).

A NOVEL LAMININ DOMAIN INVOLVED IN ADHESION OF ENDOTHELIAL CELLS

Aristidis Charonis, Nitsa Koliakos, George Koliakos, Gregg Fields, Lorrel Regger, Carrie Lynch, Mark Spreeman, Anne Hunter and Howard Higson

Department of Lab. Med. Pathology
University of Minnesota Medical School
Minneapolis, MN 55455, U.S.A.

Endothelial cells interact with various cell types and with extracellular matrices; both types of interactions are important for their functional integrity. Most endothelial cells are underlined by a basement membrane, which is a very specialized form of extracellular matrix. It contains several macromolecules present exclusively in it and absent from any other extracellular matrix compartment. These macromolecules are: type IV collagen, laminin, entactin/nidogen, perlekan(a form of heparan sulfate proteoglycan), chondroitin sulfate proteoglycan. Most of them exist in various isoforms which interact in complicated ways to form the heteropolymeric and heterogeneous structure which is defined as basement membrane[1].

Endothelial cells interact with the extracellular matrix using several types of cell surface macromolecules. So far, two families of these macromolecules have been studied in some detail: integrins and proteoglycans. These interactions are crucial in determining the polarity and the differentiated phenotype of endothelial cells. Cell surface proteoglycans exhibit diversity in the chemical structure(type of disaccharide, degree of sulfation) of their side chains; it has been observed that they may contain stretches of highly sulfated heparin-like domains[2].

The purpose of our studies is to gain a better understanding at the molecular level of the interactions between endothelial cells and a major basement membrane glycoprotein, laminin. Laminin is a large (850kDa) macromolecule that consists of three polypeptide chains originally described from a tumor-derived laminin type as B1, B2 and A. At the electron microscopic level, it appears as an assymetric cross with three short arms each exibiting two (or three) globular domains and one long arm with a large terminal globular domain[3]. In the past, we have used heparin as a model system and examined its interactions with laminin, in order to understand interactions between laminin and endothelial cell surface associated proteoglycans. Electron microscopic studies using the technique of rotary shadowing have revealed three major heparin-binding sites on laminin: on the globule of the long arm, on the outer globule of a short arm and on one of the inner globules of a short arm. Elution of heparin bound to a laminin affinity column with linear salt gradient produced three peaks at 0.15M, 0.17M and 0.20M NaCl.Electron microscopic data in the presence of increasing salt concentrations suggested that the strongest binding was at the globule of the long arm and the weakest at the inner globule of the short arm[4].

In order to map in a more precise way the putative heparin-binding site on the outer globule of the short arm, we have synthesized several peptides from this domain, using as a selection criterion their charge density and hydropathy index. In solid phase direct and competition binding experiments, we have observed that one peptide, derived from the outer globular domain of the B1 chain of laminin, named peptide AC15, had the ability to bind very specifically to heparin. This peptide represented amino acids 202-218 of the B1 chain and had the following sequence: RIQNLLKITNLRIKFVK [4].

Further studies were initiated in order to examine the possible role of this laminin peptide in adhesion, spreading and migration of endothelial cells. For our studies we have used two types of endothelial cells: one derived from a large vessel, bovine aorta, and the other derived from a microvascular bed, human dermis. In most of the studies we have used as control peptides other synthetic peptides derived from the laminin amino acid sequence that had similar length, hydropathy index and negative charge density. The data were practically identical for both types of endothelial cells used.

In order to study the role of peptide AC15 in promoting cell adhesion, we performed direct and competition binding assays. For the direct assays, increasing amounts of peptides were first adsorbed on the bottom of plastic wells. In separate experiments using radioactively labeled peptides we had identified the exact amount for each peptide that is retained in the well during the course of the experiment(data not shown). After coating, wells were washed with buffer and blocked with BSA in order to prevent nonspecific

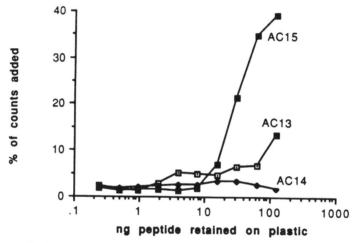

Figure 1. Direct adhesion of metabolically labeled endothelial cells on plastic wells coated with increasing amounts of peptide AC15 and control peptides AC13 and AC14.

sticking. Endothelial cells at 70% confluency were metabolically labeled by overnight exposure to radioactive methionine, detached from the culture dish by using trypsin/EDTA and washed in culture medium. After checking for their viability, they were introduced in the peptide-coated wells and allowed to adhere for 60 min. At the end of the incubation period, wells were washed three times and the radioactivity associated with each well was counted. The results shown in figure 1, demonstrate that peptide AC15 was able to promote cell adhesion at levels much higher than control peptides; we observed a dose dependent increase in adhesion starting from very low coating levels. Competition experiments were performed by following a similar protocol. In this case, the same amount of laminin(0.005 mg) was coated on all plastic wells. Metabolically labeled endothelial cells were allowed to adhere in the presence of increasing concentrations of competing peptides. We observed that increasing concentrations of peptide AC15 were able to inhibit endothelial cell adhesion to laminin, whereas no inhibition was observed when control peptides were used(data not shown).

We then studied the role of peptide AC15 in promoting spreading of endothelial cells. For these experiments, laminin was used as a positive control and BSA as a negative control.

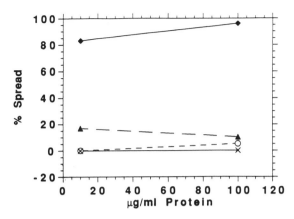

Figure 2. Percent of spreading of endothelial cells adhering to plastic wells coated at two different concentrations with laminin(closed diamonds), BSA(crosses), peptide AC15(closed triangles) and control peptide AC13(open circles).

Plastic wells were coated with two different concentrations of laminin, BSA and peptides and endothelial cells(without previous labeling) were detached as described above and allowed to adhere for 60 min. At the end of the experiment, cells were fixed and stained and the extent of spreading was quantitated by two investigators independently. The data shown in figure 2, demonstrate that only laminin was able to promote cell spreading; despite the fact that peptide AC15 was supporting adhesion at levels comparable to laminin, it failed to promote cell spreading. We explored the possibility that lack of spreading was due to lack of conformation of peptide AC15. For this purpose, we coupled peptide AC15 to ovalbumin and coated the complex on plasic wells. This complex again, failed to support cell spreading(data not shown).

We finally studied the possible role of peptide AC15 in endothelial cell motility. For these experiments, we performed competition assays in Boyden chambers. Filters were floated on

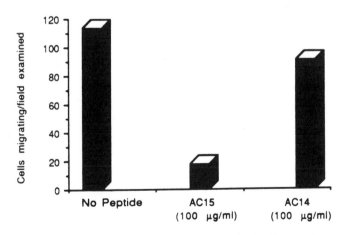

Figure 3. Migration of endothelial cells into laminin-coated filters in the absence of any peptide or in the presence of peptide AC15 and control peptide AC14.

a laminin solution of 0.12mg/ml for 16 hrs and then the migration of endothelial cells into the laminin coated filter was examined when the upper chamber solution contained along with the endothelial cells no peptide, peptide AC15, or a control peptide. As shown in figure 3, the presence of peptide AC15 was able to inhibit cell migration through the filter; in contrast, under control conditions, laminin coated on the filter was able to promote cell migration as expected. These results indicate that peptide AC15 may be involved in endothelial cell motility.

Peptide AC15 is derived from the outer globule of one of the short arms of laminin, formed by the B1 chain. Elastase digestion can produce various laminin fragments, among them fragment E4, which contains this portion of laminin[5]. Therefore, experiments have been initiated, where the adhesion of endothelial cells to fragment E4 was studied. Serially diluted fragment E4 or BSA were coated on plastic wells and the adhesion of metabolically labeled endothelial cells was quantitated, as described above. The results shown in figure 4 suggest that fragment E4 can promote specific adhesion of endothelial cells. However, this adhesion was by two orders of magnitude less than the adhesion observed for the intact laminin molecule(data not shown). Many factors can contribute to this difference: It is possible that fragment E4 is altered by the digestion with elastase and that the area where peptide AC15 is located could be affected. Alternatively, peptide AC15 may be present in laminin in a "cryptic" conformation that makes it less active in the native molecule; in this case, peptide AC15 could exhibit higher activity only in cases where laminin is being degraded.

In conclusion, our studies suggest that peptide AC15 is a laminin domain which can bind to heparin and promotes the adhesion and motility of endothelial cells. More work needs to be done to define its role in the native laminin molecule and to identify the endithelial cell surface molecules that possibly interact with this laminin domain.

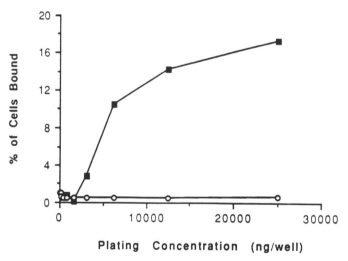

Figure 4. Adhesion of metabolically labeled endothelial cells on plastic wells coated with BSA(open circles) or laminin fragment E4(closed squares).

ACKNOWLEDGEMENTS

This work was supported by a grant from the American Heart Association to A. Charonis.

REFERENCES

1. A.S.Charonis and E.C.Tsilibary, Assembly of Basement Membrane Proteins , in "Organization and Assembly of Plant and Animal Extracellular Matrix", R.P.Mecham, ed., Academic Press, New York (1990).
2. H.B.Nader, C.P.Dietrich, V.Buonassisi, and P.Colburn, Heparin sequences in the heparan sulfate chains of an endothelial cell proteoglycan, Proc.Natl.Acad.Sci.USA 84:3565 (1987).
3. G.R.Martin and R.Timpl, Laminin and other basement membrane components, Annu.Rev.Cell Biol. 3:57 (1987).
4. K.K.Koliakos, G.G.Koliakos, E.C.Tsilibary, L.T.Furcht and A.S.Charonis, Mapping of three major heparin-binding sites on laminin and identification of a novel heparin-binding site on the B1 chain, J.Biol. Chem. 264:17971 (1989).
5. J.C.Schittny and P.D.Yurchenco, Terminal short arm domains of basement membrane laminin are critical to its self-assembly, J.Cell Biol. 110:825 (1990).

REGULATION OF ANGIOGENESIS: THE ROLE OF PROTEIN KINASE, NITRIC OXIDE, THROMBIN AND BASEMENT MEMBRANE SYNTHESIS

M.E. Maragoudakis, N. E. Tsopanoglou, G. C. Haralabopoulos, E. Sakkoula, E. Pipili-Synetos and E. Missirlis

Department of Pharmacology
University of Patras Medical School
261 10 Rio Patras, Greece

INTRODUCTION

Angiogenesis plays a key role in the pathology of many disease states (Folkman and Shing, 1992). Practically every medical specialty deals with disease states where a derangement of angiogenesis is evident. In addition, angiogenesis is a prominent event in many physiological processes such as the establishment of placenta, the changes in the mammary gland during lactation and in the endometrium during ovulation (Auerbach et al., 1991). Angiogenesis is also important in wound healing and repair processes (Knighton et al., 1990).

There are fundamental differences between physiological and pathological angiogenesis (Table I). Physiological angiogenesis such as that occuring in ovulation and wound healing can last for a relatively short time (days or weeks) and then return to a quiescent state. Physiological angiogenesis is tightly regulated both temporally and spatially. The formation of new capillaries is rapid yet controlled and transient, returning to physiological state levels.

Table I. Differences between pathological and physiological angiogenesis.

Physiological Repair	Processes	Pathological angiogenesis
Ovulation	Wound healing	Heamangiomas
Uterus	Peptic ulcer	Occular neo-varscularization
Placenta	Myocardial infarction	arthritis, psoriasis, cancer
Duration: Days or Weeks		Years
Regulation: Stringent regulation, self-limited, Returns to quiescent state		Uncontrolled

On the contrary in situations such as diabetic retinopathy, chronic inflammation etc. angiogenesis persists for months or years in an uncontrolled fashion and contributes to the pathology of the disease. The new blood vessels formed under these pathological conditions may be defective (Konerding et al., 1992), thus they leak causing haemorrhages and occlusions as for example in diabetic retinopathy and in solid tumors.

The lack of angiogenesis in most normal tissues is probably the result of interaction of a complex multifunctional regulatory system, which maintains a balance between stimulation and suppression of angiogenesis (Figure 1).

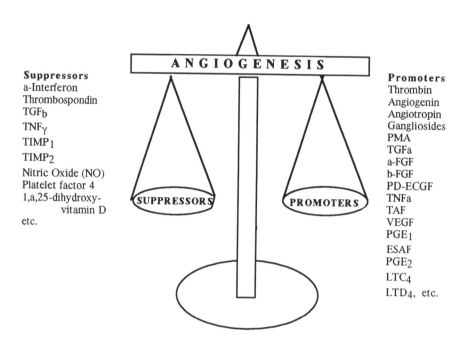

Suppressors
a-Interferon
Thrombospondin
TGFb
TNFγ
TIMP1
TIMP2
Nitric Oxide (NO)
Platelet factor 4
1,a,25-dihydroxy-
     vitamin D
etc.

Promoters
Thrombin
Angiogenin
Angiotropin
Gangliosides
PMA
TGFa
a-FGF
b-FGF
PD-ECGF
TNFa
TAF
VEGF
PGE1
ESAF
PGE2
LTC4
LTD4, etc.

Figure 1. Angiogenesis under physiological conditions is an extensively slow process under tight regulatory control maintained through a balance of multiple endogenous angiosuppressors and angiopromoters.

The ability, to initiate the angiogenic cascade, however, is ever present in all tissues. It is likely that excess of any of the many endogenous promoters or lack of suppressors of angiogenesis can set it off at a moment's notice, as it happens in the activation of the cascade of blood coagulation.

Understanding the detailed mechanisms and the interaction of the multiple factors involved in the regulation of angiogenesis is of paramount importance in locating the site of derangement of this process in disease states. It will also help in the development of promoters or inhibitors of angiogenesis for clinical application.

Angiogenesis and Basement Membrane Synthesis

The existing methodology for assessing angiogenesis *in vivo* is based on the morphological evaluation of new blood vessel formation in model systems. This is cumbersome expensive, time consuming and gives only semiquantitative results (Gullino, 1981). We have proposed that the rate of basement membrane synthesis can serve as a convenient and sensitive biochemical index of angiogenesis, thus providing a method for assessing angiogenesis quantitatively (Maragoudakis et al., 1988a). A major component of basement membrane is type IV collagen which can be measured as collagenous protein solubilized by clostridial collagenase.

We have evaluated a number of promoters and inhibitors of angiogenesis by this methodology and compared the results with the most rigorous tests of morphological evaluation ie. the Harris-Hooker method (Harris-Hooker et al., 1983) and by computer assisted image analysis. A very good agreement of the results was obtained in all cases (Maragoudakis et al., 1992). Based on these results we are using the rate of collagenous protein synthesis and an index of angiogenesis and confirm our findings by morphological evaluation.

Basement membranes are not only structural elements of the blood vessels, but also their individual components appear to have profound effect on the phenotypic differentiation of endothelial cells (Madri et al., 1991). Cultured human umbilical vein endothelial cells (HUVECS) plated on matrigel form tube-like structures (Grant et al., 1989). We have found that although matrigel contains laminin and type IV collagen in excess, the tube formation by HUVECS is absolutely dependent on new collagenous protein synthesis by these cells. When inhibitors of basement membrane collagen biosynthesis were present, the alignment of HUVECS to form tube-like structures was not evident (Haralabopoulos et al., unpublished observations).

This requirement for new collagen synthesis in the formation of new blood vessels was also established *in vivo*. We have found that in the chick chorioallantoic membrane system (CAM) inhibition of basement membrane collagen biosynthesis prevents angiogenesis (Maragoudakis et al., 1988). The same results were obtained in animal models, where it was shown that inhibitors of basement membrane collagen biosynthesis are effective antitumor agents, probably as a result of inhibition of angiogenesis (Maragoudakis et al., 1990). Therefore, we are using basement membrane biosynthesis as a target for developing inhibitors of angiogenesis with antitumor properties (Maragoudakis et al., 1993).

Regulation of Angiogenesis by Protein Kinases

The multiplicity of endogenous factors that have been reported thus far to either promote or suppress angiogenesis in the various *in vitro* and *in vivo* models of angiogenesis is reminiscent in many ways of the situation that existed in the blood clotting cascade before the interrelation and interaction of the various clotting factors was elucidated. Similarly in angiogenesis we are faced with a complex cascade of events leading from the activation of the normally quiescent endothelial cell to the formation of blood vessels with many factors involved in the process. This cascade, similarly to blood coagulation, is under tight regulatory control. This regulation involves the interaction of endothelial cells with other cell types such as pericytes, the interaction with extracellular matrix proteins and the involvement of endogenous promoters or suppressors of angiogenesis through surface receptors each connected to a different transduction system.

Such a complex system requires that overall controls of the muliple transduction mechanisms must exist so that a sensitive and effective regulation of neovascularization or disappearance of blood capillaries is maintained under physiological or pathological conditions.

We have considered the overall regulation of angiogenesis through two major signalling pathways namely protein kinase C (PKC) and the cyclic-AMP-dependent protein kinase A (PKA). These pathways are known to mediate the effects of growth factors, hormones, neurotransmitters and oncogenes on cell proliferation and cellular responses (Powis, 1991).

We have established in the CAM system that activators of PKC such as 4-β-phorbol-12-myristate-13-acetate (4β-PMA), 1,2-dioctanoyl-sn-glycerol (OAG) caused a marked increase in angiogenesis. On the contrary the specific inhibitor of PKC Ro-318220 and the commonly used inhibitor of PKC 1-(5-isoquinoline-sulfonyl)2-methyl piperazine (H7) suppressed angiogenesis (Tsopanoglou, Pipili-Synetos and Maragoudakis, 1993). Both basal and phorbol ester-stimulated angiogenesis was suppressed by Ro-318220 to levels below that of controls. The results are dose-dependent and specific. The inactive analog of phorbol ester 4-a-phorbol-12-myristate-13-acetate (4-a-PMA) and the 1,2-dioleoyl-sn-glycerol (diolein), which either do not activate or can not reach PKC did not promote angiogenesis.

Contrary to PKC activators, the activation of PKA caused a suppression of angiogenesis (Tsopanoglou, Haralabopoulos and Maragoudakis, 1994, in press). Forskolin, which activates adenylate cyclase and elevates the intracellular levels of c-AMP and the Sp-diastereoisomer of adenosine-cyclic-3,5'monophosphothioate (Sp-c-AMPS), a cell permeable analog of c-AMP, which mimics endogenous c-AMP caused a suppression of collagenous protein biosynthesis and angiogenesis. These results were obtained in the CAM system using collagenous protein synthesis as an index of angiogenesis and morphological evaluation of blood vessel formation using computer assisted image analysis.

The aforementioned opposite modulation of angiogenesis by activators of PKC and elevated c-AMP levels was further confirmed by the suppression of 4-β-PMA-stimulated angiogenesis by either forskolin or Sp-c-AMPS. On the contrary the Rp-diastereoisomer of adenosine cyclic-3',5'-monophosphotioate (Rp-c-AMPS), which antagonises endogenous c-AMP biochemical actions, had no effect on angiogenesis and did not suppress 4-β-PMA-stimulated angiogenesis.

Similar results were obtained in the HUVEC tube formation assay system. In this system the PKC inhibitor Ro318220 caused a dose-dependent inhibition of tube formation and 4-β-PMA reversed this effect. Also forskolin and Sp-c-AMPS caused inhibition of tube formation (Tsopanoglou, Haralabopoulos and Maragoudakis, 1994, in press).

We conclude from these studies that activation of PKC and PKA have opposing effects in the modulation of angiogenesis. This seems to be a general cross-talk phenomenon between the two major signalling pathways, which has been documented in many other physiological processes (Houslay, 1991).

The Involvement of Leucotrienes in the Regulation of Angiogenesis

Leucotrienes have profound effects on a number of physiological systems (Samuelson et al., 1987). Many inflammatory diseases such as asthma, arthritis and psoriasis are linked with elevated synthesis of leucotrienes by neutrophils, mast cells and macrophages. We have found (Maragoudakis et al., 1992) that LTC_4 and LTD_4 are potent promoters of angiogenesis while LTB_4 is without effect in the CAM system.

The Role of Thrombin in Angiogenesis

It has been well established that both growth and metastasis of solid tumors are angiogenesis dependent (Folkman, 1985). On the other hand the relationship between hypercoagulability and cancer has been recognized for more than 125 years (Rickles and Edwards, 1983). First in 1872 Trouseau observed that spontaneous coagulation is common in cancerous patients and this observation has been verified by the accumulation of clinical and pathological data. The molecular mechanism for this association between blood coagulation and cancer growth and metastasis is not yet completely understood.

We have established that thrombin is a powerful promoter of angiogenesis. This newly reported action of thrombin may play a key role in the metastatic spread of cancer and in wound healing based on its ability to induce angiogenesis (Tsopanoglou, Pipili-Synetos and Maragoudakis, 1993).

Using human α-thrombin we have shown that there is a dose-depended (up to 1.0 iu thrombin/disc) increase of angiogenesis in the CAM system.

Similarly γ-thrombin has an angiogenesis-promoting effect in the CAM. γ-Thrombin, however, has the catalytic but not the anion binding exocite of α-thrombin, which is required for clotting activity. Both α- and γ-thrombin cause a maximum increase of angiogenesis in the CAM (up to 80% above control levels). We conclude from these type of experiments that thrombin promotes angiogenesis by a mechanism independent of fibrin formation.

This effect of thrombin on angiogenesis is specific as shown by the fact that hirudin, which binds both the catalytic and the anion binding exocites of thrombin, abolishes the effects of both α- and γ-thrombin on angiogenesis.

P-PACK-thrombin, which lacks the catalytic but retains the anion binding exocite, has no effect by itself on angiogenesis, but when combined with α- or γ-thrombin, it abolishes their effect. This indicates that both catalytic and anion binding exocites of thrombin are essential for its angiogenic action.

Heparin which binds thrombin and accelerates the inactivation of thrombin by antithrombin III, also abolishes the effect of α- and γ-thrombin on angiogenesis (Tsopanoglou, Pipili-Synetos and Maragoudakis, 1993). The involvement of thrombin in angiogenesis could be mediated by a number of mechanisms. Thrombin could promote angiogenesis by destroying thrombospondin (TSP). Thrombospondin is a thrombin sensitive protein (TSP), which is known to inhibit angiogenesis (Good et al., 1990). In addition many cell types and mediators may be involved in the promotion of angiogenesis by thrombin. Factors that are essential for vascular proliferation, such as PAF, PA, metalloproteases may be generated as a result of thrombin action.

Table II. Effects of PKC inhibitors and PKA activators on thrombin-promoted angiogenesis.

Treatment	n	Collagenous protein biosynthesis (CPM/mg protein x10^{-3})		
		Control	Treatment	
α-Thrombin (1 IU/disc)	80	16.8±0.8	29.9±1.4	***
α-Thrombin (1 IU/disc) + Ro 318220 (10 μg/disc)	13	13.9±1.9	9.4±0.9	***
α-Thrombin (1 IU/disc) + Forskolin (500 ng/disc)	24	15.9±1.7	14.8±1.1	ns

Control contains the vehicle only. Experimental details are described in Tsopanoglou, Haralabopoulos and Maragoudakis (1994).

The promotion of angiogenesis by thrombin is reversed by both PKC inhibitors such as Ro318220 and PKA activators such as forskolin (Table II). These results indicate that the actions of thrombin on angiogenesis are mediated through the PKC and PKA signalling pathways.

Other investigators have shown that fibrin and fibrin degradation products are angiogenic (Knighton et al., 1982). We have shown that fibrin formation is not a prerequisite for the promotion of angiogenesis by thrombin. We have investigated the relative contribution of thrombin itself and fibrin in stimulating angiogenesis.

When fibrin is formed by increasing amounts of fibrinogen while keeping thrombin constant at 0.1 IU, which by itself is not angiogenic in the CAM system, there is a dose dependent increase in angiogenesis. Also when the maximum dose of thrombin (1.0 IU) for angiogenic activity was used and fibrinogen was added, there was a further increase in angiogenesis caused by fibrin formation (Table III). These results suggest that the promotion of angiogenesis during the blood clotting cascade is independently stimulated by fibrin and thrombin itself.

Table III. Relation contribution of fibrin and thrombin itself in promoting angiogenesis in the CAM

Treatment	n	Collagenous protein biosynthesis (CPM/mg protein x10-3)		
		Control	Treatment	
Thrombin (0.1 IU/disc)	6	6.04±0.8	7.0±0.8	ns
Thrombin (0.3 IU/disc)	6	8.10±1.2	10.1±2.0	*
Thrombin (1 IU/disc)	6	7.20±0.3	12.1±1.3	***
Fibrinogen (500 µg/disc)	14	13.50±1.6	15.4±2.1	ns
Thrombin (0.1 IU/disc) + Fibrinogen (500 µg/disc)	6	5.40±0.6	8.2±0.4	***
Thrombin (0.3 IU/disc) + Fibrinogen (500µg/disc)	6	6.10±0.3	9.7±1.6	***
Thrombin (1 IU/disc) + Fibrinogen (500 µg/disc)	6	5.60±0.7	9.5±0.5	***

Controls contain the vehicle only. Experimental details are described in Tsopanoglou, Haralabopoulos and Maragoudakis (1994).

Nitric Oxide (NO) Involvement in the Regulation of Angiogenesis

Nitric oxide (NO) is an endogenous vasoactive molecule with multiple functions. Table IV describes some of the functional roles of this molecule and the list is rapidly expanding.

We have investigated the role of NO in regulating angiogenesis (Pipili-Synetos, Sakoula and Maragoudakis, 1993) and have found that sodium nitroprusside (SNP), which spontaneously generates NO, caused a marked and dose-dependent suppression of angiogenesis. On the other hand inhibitors of NO synthase such as N^G-monomethyl-L-arginine (L-NMMA) or N^G-nitro-L-arginine methylester (L-NAME) stimulated angiogenesis in the CAM.

Table IV. Physiological actions of nitric oxide (NO)

It helps to maintain the blood pressure
It helps to kill foreign invaders in the immune response
It is a major biochemical mediator of penile erections
It is probably a major biochemical component of long terms memory
It plays a key role in modulating angiogenesis

Arginine, which is the endogenous substrate for NO generation by NO synthase, caused an inhibition of angiogenesis by itself. When arginine was combined with the competitive inhibitors of NO synthase (L-NMMA and L-NAME) it abolished their angiogenesis promoting effects. NO formation, therefore, appears to be an important endogenous regulator of angiogenesis since L-NMMA and L-NAME, which prevent NO generation, stimulate the angiogenic cascade. Agents which cause an increase in the availability of NO such as SNP, L-arginine and superoxide dismutase (SOD), caused a suppression of unstimulated angiogenesis to an extend of about 34%. The effect of agents that increase the availability of NO in basal angiogenesis is far more pronounced under conditions where angiogenesis is stimulated. For example, when angiogenesis is promoted by thrombin or PMA, the presence of SNP suppressed the thrombin or PMA-stimulated angiogenesis below the basal levels. Similarly (SOD, which as a free radical scavenger increases the availability of endogenous NO, completely reverses the angiogenesis-promoting effect of thrombin and PMA (Pipili-Synetos et al., 1994). These results suggest that NO may suppress angiogenesis irrespective of the stimulus and as such it may serve as the main break, which keeps angiogenesis under normal physiological conditions in the quiescent state.

Regulation of angiogenesis and the development of angiosuppressors or angiopromoters

The aforementioned studies on the regulation of angiogenesis aim at the understanding of the mechanisms of derangement of angiogenesis in disease states. In addition it may provide the basis for developing promoters or inhibitors of angiogenesis for clinical applications.

Angiosuppressors have the potential for clinical application in chronic inflammation, ocular diseases, psoriasis, haemangiomas, solid tumors etc. (Folkman, 1992; Gullino, 1981). On the other hand promoters of angiogenesis are of potential use in situations such as non-healing wounds or fractures in diabetic or old people, also in irradiated tissue, burns, peptic ulcers, etc. (Gullino, 1992).

Indeed the first successful treatment of an angiogenic disease has been reported. A child with pulmonary heamangioma, a usually fatal disease, was successfully treated with interferon-alpha-2a, an inhibitor of angiogenesis (White et al., 1989). Other angiogenesis inhibitors are in clinical trials as antitumor agents.

Clinical application of promoters of angiogenesis is probably forthcoming in the near future. Many reports point to the fact that in ischemic injury endogenous angiogenic agents are released, including b-FGF, which stimulates angiogenesis and growth of preexisting collaterals. This release is often inadequate to prevent clinical manifestations of ischemic disease. This was shown by Yanagisawa-Miwa et al. (1992) in a canine experimental myocardial infarct model where by intracoronary injection of b-FGF they could salvage the infarcted myocardium by stimulating angiogenesis.

The healing of peptic ulcers was also shown to be accelerated by acid-stable b-FGF. In fact there is compelling evidence that the conventional antiulcer drugs, i.e. the acid neutralizing agents and the H_2-blockers, may be acting by protecting the endogenously released b-FGF from acid degradation (Folkman et al., 1991).

New leads for developing angiogenesis-promoting agents based on our findings with thrombin and PKC activators are promising. While these exciting possibilities exist for clinical applications of promoters of angiogenesis, there is also concern for the potential dangers of systemic application of these agents for fear of activating angiogenesis in latent tumors, which may flare up with such a treatment.

The clinical application of angiosuppressors does not raise this type of concern. However, progress in the development of potent, non-toxic and active *in vivo* inhibitors of angiogenesis has been extremely slow. The approach towards this goal has been empirical, because of our lack of understanding of the key steps and the control mechanisms of angiogenesis. The multiplicity of the endogenous angiogenic factors reported thus far makes it unlikely that inhibition of one of them eg. b-FGF could be an effective approach. The tissue is likely to overproduce another angiogenic factor to compensate for the missing one.

A more realistic approach would be to inhibit a single essential and preferably a rate-limiting step in the angiogenesis cascade. We have identified such a step in the biosynthesis of BM of the newly formed vessels. Like angiogenesis, BM synthesis is an extremely slow process under normal physiological conditions. Therefore, inhibiting BM synthesis and angiogenesis for a relatively short time in the adult is not likely to have serious side effects. This was shown thus far for three angiogenesis inhibitors in animal models. 8,9-dihydroxy-7-methyl-benzo(b)quinolizinium bromide (GPA 1734) inhibits BM synthesis and angiogenesis and suppresses tumor growth without any cytotoxic effects to tumor cells or toxic effects to animals (Maragoudakis, Sarmonika and Panoutsacopoulou 1988, Missirlis, Karakiulakis and Maragoudakis, 1990). Similarly tricyclodecan-9-yl-xanthate (D609) inhibits BM synthesis angiogenesis and tumor growth (Maragoudakis et al., 1990). A newly described agent, titanocene dichloride, has the same actions (Bastaki et al., 1994).

We conclude from these studies that BM plays a pivotal role in angiogenesis. BM is not only a structural essential element of all blood vessels but it also plays the role of a local hormone for the activated endothelial cell and the expression of the angiogenic phenotype. As such, BM biosynthesis is an ideal target for developing suppressors of angiogenesis.

ACKNOWLEDGEMENTS

This work was supported from grants from the Greek National Drug Organization and the Greek Ministry of Science and Technology. We thank Anna Marmara for typing the manuscript.

REFERENCES

Auerbach, R., Auerbach, W., and Polakowski, I., 1991, Assays for angiogenesis: A review, *Pharmac. Therap.* 51:1-11.

Bastaki, M., Missirlis, E., Klouras, N., Karakiulakis, G., and Maragoudakis, M., 1994, *Eur. J. Pharmacol.* (in press).

Folkman, J., 1985, Towards an understanding of angiogenesis: Seach and discovery, *Persp. in Biol. and Med.* 29:10-35.

Folkman, J., Szabo, S., Storroff, M., McNeil, P., Li, W., and Shing, Y., 1991, Duodenal ulcer: Discovery of a new mechanism and development of angiogenic therapy that accelerates healing, *Ann. of Surg.* 214:414-426.

Folkman, J., and Shing, Y., 1992, Angiogenesis: a mini review, *J. Biol. Chem.* 267:10931-10933.

Good, D.J., Polverini, P.J., Rastinejad, F., LeBeau, M.M., Lemons, R.S., Frazier,

W.A., and Bouck, N.P., 1990, A tumor suppressor-dependent inhibitor of angiogenesis is immunoligically and functionally indistinguishable from a fragment of thrombospondin., *Proc. Natl. Acad. Sci.* USA 87:6624-6628.

Grant, D.S., Tashiro, K.I., Segui-Real, B., Yamada, Y., Martin, G.R., and Kleinman, H., 1989, Two different laminin domains mediate the differentiation of human endothelial cells into capillary-like structures in vitro, *Cell* 58:933-943.

Gullino, P.M., 1981, Angiogenesis and Neoplasia, *New Engl. J. Med.* 305:884-885.

Gullino, P.M., 1992, On promoters of angiogenesis and therapeutic potential, in "Angiogenesis in Health and Diseases", M.E. Maragoudakis, P. Gullino and P.I. Lelkes, eds., Plenum Press, pp. 287-294.

Harris-Hooker, S.A., Gajdusek, C.M., Wight, T.N., and Schwartz, S.M., 1983, Neovascular response induced by cultured aortic endothelial cells, *J. Cell Physiol.* 114:302-310.

Houslay, M.D., 1991, Crosstalk: a pivotal role of protein kinase C in modulating relationships between signal transduction pathways, *Eur. J. Biochem.* 195:9-27.

Knighton, D.R., Hunt, T.K., Thakral, K.K. and Goodson, W.H., 1982, Role of platelets and fibrin in the healing sequence: An in vivo study of angiogenesis and collagen synthesis, *Annals of Surgery* 196:379-388.

Knighton, D.R., Phillips, G.D., and Fiegel, V.D., 1990, Wound healing angiogenesis: Indirect stimulation of basic fibroblast growth factor, *J. Trauma* 30 (suppl. 12):S134-S144.

Konerding, M.A., van Ackern, C., Steinberg, F., Streffer, C., 1992, The development of the tumor vascular system: 2D and 3D approaches to network formation in Human xenografted tumors, in: "Angiogenesis in Health and Diseases", M.E. Maragoudakis, P. Gullino and P.I. Lelkes, eds., Plenum Press, pp. 173.

Madri, J.A., Bell, L., Marx, M., Merwin, J.R., Basson, C., and Prinz, C, 1991, Effects of soluble factors and extracellular matrix components on vascular cell behavior in vitro and in vivo: Models of de-endothelialization and repair, *J. Cellul. Biochem.* 45:123-130.

Maragoudakis, M.E., Panoutsacopoulou, M. and Sarmonika M., 1988, Rate of basement membrane synthesis as an index of angiogenesis, *Tissue and Cell* 20:531-539.

Maragoudakis, M.E., Tsopanoglou, N.E., Bastaki, M. and Haralabopoulos, G., 1992, Evaluation of promoters and inhibitors of angiogenesis using basement membrane biosynthesis as an index, in: "Angiogenesis in Health and Diseases", M.E. Maragoudakis, P. Gullino and P.I. Lelkes, eds., Plenum Press, pp. 275.

Maragoudakis, M.E., Sarmonika M., and Panoutsacopoulou, M., 1988, Inhibition of basement membrane synthesis prevents angiogenesis, *J. Pharm. Exp. Therap.* 244:729-733.

Maragoudakis, M.E., Missirlis, E., Sarmonika M., Panoutsacopoulou, M., and Karakiulakis, G., 1990, Basement membrane biosynthesis as a target to tumor therapy, *J. Pharm. Exp. Therap.* 253:753-757.

Maragoudakis, M.E., Missirlis, E., Karakiulakis, G.D., Sarmonika, M., Bastaki, M., and Tsopanoglou, N., 1993, Basement membrane biosynthesis as a target for developing inhibitors of angiogenesis with antitumor properties. *Kidney Int.* 43: 147-150.

Maragoudakis, M.E. Angiogenesis. In Annual of Cardiac Surgery, M. Yacoub & J. Pepper (eds) , Current Science Ltd. London, pp. 13-19 (1993).

Missirlis, E., Karakiulakis, G., and Maragoudakis, M.E., 1990, Antitumor effects of GPA1734 in rat Walker 256 carcinoma, *Invest. New Drugs* 8:145-147.

Pipili-Synetos, E., Sakoula, E., & Maragoudakis, M.E., 1993, Nitric oxide is involved in the regulation of angiogenesis. *Br. J. Pharmacol.* 108:855-857.

Pipili-Synetos, E., Sakoula, E., Haralabopoulos, G., Andricopoulou, P., Peristeris, P. and Maragoudakis, M.E., 1994, Nitric oxide is an endogenous antiangiogenic mediator, *Br. J. Pharmacol.* (in press).

Powis, G., 1991, Signalling targets for anticancer drug development, *TIPS Reviews* 12:188-194.

Rickles, F.R., and Edwards, R.L., 1983, Activation of blood coagulation in cancer: Trousseau's syndrome revisited, *Blood* 64:14-31.

Samuelson, B., Dahlen, S.E., Lindgren, J.A., Rouzel, C.A., and Serhan, C.N., 1987, Leukotriene and lipoxins: structure, biosynthesis and biological effects, *Science* 237:1171-1176.

Tsopanoglou, N.E., Pipili-Synetos, E. and Maragoudakis, M.E., 1993, Protein Kinase C involvement in the regulation of angiogenesis. *J. Vascular Research* (in press).

Tsopanoglou, N.E., Pipili-Synetos, E., and Maragoudakis, M.E., 1993, Thrombin promotes angiogenesis by a mechanism independent of fibrin formation. *Am. J. Physiology* (in press).

Tsopanoglou, N.E., Haralabopoulos, G., and Maragoudakis, M.E., 1994, Opposing effects on modulation of angiogenesis by protein kinase C and cyclic-AMP-mediated pathways, *J. Vascular Res.* (in press).

White, C.W., Sondheimer, H.M., Crouch, E.C., Wilson, H., and Fan, L.L., 1989, Treatment of pulmonary hemangimatosis with recombinant interferon alpha-2a, *N. Engl. J. Med.* 320:1197-1200.

Yamagisawa-Miwa, Uchida, Y., Nakamura, F., Tomaru, T., Kido, H., Kamijo, T., Sugimoto, T., Kaji, K., Utsuyama, M., Kurashima, C. and Ito, H., 1992, Salvage of infarcted myocardium by angiogenic action of basic fibroblast growth factor, *Science* 257:1401-1404.

BASEMENT MEMBRANE LAMININ-DERIVED PEPTIDE SIKVAV PROMOTES ANGIOGENESIS AND TUMOR GROWTH

Hynda K. Kleinman, Derrick S. Grant, and Maura C. Kibbey

National Institute for Dental Research
National Institutes of Health
Bethesda, MD 20892 USA

INTRODUCTION

The basement membrane is a thin extracellular matrix which underlies endothelial cells in vessels and forms a barrier to the passage of macromolecules and cells (Martin et al, 1988). Basement membranes also provide structural support and are very biologically active (Kleinman et al, 1987; Beck et al, 1990). The major and constant components of basement membranes include laminin, collagen IV, entactin, heparan sulfate proteoglycan and various growth factors (Martin et al, 1988; Vukicevic et al, 1992). These components interact with each other to form a highly elastic and organized structure.

Basement membranes are very biologically active with the response dependent on the specific cell type (Kleinman et al, 1993). Such activity has been demonstrated using either specific antibodies, isolated matrices such as the lens, or a sarcoma tumor extract reconstituted in vitro. The most direct results have been obtained using Matrigel, an extract of the Engelbreth-Holm-Swarm (EHS) tumor which can be coated on culture dishes (Kleinman et al, 1986). This extract contains the major basement components and various growth factors including EGF, bFGF, TGFß, PDGF, and insulin-like growth factor 1 (Vukicevic et al, 1992). This extract is a liquid at 4°c and gels at 37°c. It can be used as a culture substratum or cells can be resuspended in the cold liquid before the gel is formed. In vitro, Matrigel promotes epithelial and endothelial cell differentiation and a greatly reduced proliferative rate is often observed with normal cells (Kubota et al, 1988). Cells differentiate in a highly specific manner. For example, salivary gland cells form gland-like structures (Kibbey et al, 1992b) while oviduct cells form tubes with secretory villi directed toward the lumen (Joshi, 1991). In addition, sea urchin micromeres differentiate and form spicules (Benson and Chuppa, 1990) and the entire life cycle of the avian malaria parasite Plasmodium is completed on Matrigel (Warburg and Miller, 1992). Furthermore, endothelial cells form capillary-like structures with a lumen (Kubota et al, 1988). Matrigel has also been used in vivo to facilitate tumor growth (Kibbey et al, 1992a), assay for angiogenesis (Passaniti et al, 1993), increase neural graft survival (Haber et al, 1988), and stimulate epithelialization of intestinal defects (Thompson, 1990).

BIOLOGICAL ACTIVITIES OF LAMININ

Laminin, a major glycoprotein in basement membranes, is one of several very active species in this matrix. Laminin, as isolated from the EHS tumor, has been found to promote cell adhesion, migration, growth, differentiation, neurite outgrowth, tumor growth and metastases, and increased activity of tyrosine hydroxylase and collagenase IV (Kleinman et al, 1993b). Laminin is composed of three chains designated A(M_r = 400,000), B1(M_r = 210,000), and B2(M_r = 200,000) (Figure 1).

A

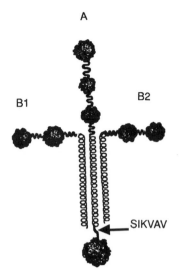

B1   B2

SIKVAV

Figure 1. Schematic model of laminin. The SIKVAV (ser-ile-lys-val-ala-val) sequence is located on the long arm of the A chain as designated.

Several homologues of laminin have been described which are also biologically active. When cells adhere to laminin, they generally undergo a distinct and cell type specific morphogenesis unlike fibronectin-mediated adhesion where cell spreading is most commonly observed. On laminin, for example, Sertoli cells become more columnar (Suarez-Quian et al, 1985) whereas Schwann cells become more elongated (McGarvey et al, 1984). Laminin is of particular interest because of its ability to promote neurite outgrowth with many neuronal cells and to facilitate nerve regeneration in vivo. Laminin has also been shown to promote the malignant phenotype. Laminin-adherent tumor cells in vitro are more malignant when injected in vivo (Terranova et al, 1984; Jun et al, 1993). The level of laminin 32/67Kd receptors correlates with malignancy (Wewer al, 1983) and intravenous coinjection of laminin with melanoma cells results in increased numbers of colonies on the surface of the lungs (Barsky et al, 1984). Furthermore, antibodies to laminin decrease lung colonization. Laminin also increases the activity of collagenase IV, an enzyme responsible in part for tumor spread (Turpeeniemi-Hujanen et al, 1986). These studies demonstrate important functions for laminin in development and in disease. It is likely that specific clinical formulations of laminin may be used in facilitating tissue repair and in inhibiting certain disease processes.

Various active domains of laminin have been described at the synthetic peptide level (Beck et al, 1990; Kleinman et al, 1993b). These laminin-derived synthetic peptides have different biological functions. The most studied peptides include YIGSR (residues 929-933 on the B1 chain), RGD (residues 1118-1128 on the A chain), and SIKVAV (residues 2099-2105 on the A chain). All three peptides promote cell adhesion but are cell type specific. For example, RGD is most active with endothelial cells (Grant et al, 1989) whereas SIKVAV is most active with neuronal cells (Tashiro et al, 1989). Interestingly YIGSR inhibits angiogenesis in vitro and in vivo (Sakamoto et al, 1990) whereas SIKVAV has the opposite activity (Kibbey et al, 1992a; Grant et al, 1992). It is likely that all of the active sites on laminin are not available (active) simultaneously due to both conformational changes and steric blocking by interacting macromolecules.

The SIKVAV peptide from the laminin A chain is unusually active with a variety of cells (Table I). This peptide sticks very well to plastic and promotes the adhesion of many cells. Cells migrate well to this peptide in Boyden chamber assays and appear more migratory on surfaces coated with the peptide (Tashiro et al, 1989l; Grant et al, 1992). This peptide has been found to promote cell growth but it is probably not the main growth promoting site on laminin

Table 1 Biological activities of SIKVAV

adhesion
migration
growth
neurite outgrowth
angiogenic in vitro and in vivo
experimental tumor metastases
subcutaneous tumor growth
plasminogen activator activation
collagenase IV activity

been localized to the upper cross region (Panayotou et al, 1989). SIKVAV has been found to increase plasminogen activation some 20-fold and this results in increased collagenase IV activity (Stark et al, 1990). Changes in the proteases may account in part for the increased angiogenic and tumorigenic activities of this peptide (Grant et al, 1992; Kanemoto et al, 1990; Schnaper et al, 1993).

LAMININ-DERIVED SIKVAV CONTAINING PEPTIDE PROMOTES ANGIOGENESIS

The SIKVAV peptide promotes angiogenic activity in vitro. When human umbilical vein endothelial cells are plated on a basement membrane Matrigel substratum, they attach, migrate, and align into capillary-like structures which have a lumen (Kubota et al, 1988). In the presence of the SIKVAV peptide, this morphological differentiation is altered (Grant et al, 1992). The capillary-like structures invade the Matrigel and the cells at the intersections of the tubes form "sprouts" which appear to be generating additional tubes. The invasion into the Matrigel is likely facilitated by increased protease levels since some material from the matrix substratum is released into the medium. Furthermore, zymography demonstrated an increase in the 72 Kd collagenase IV activity. Proteases have recently been shown to be important in the Matrigel-induced formation of capillary-like structures (Schnaper et al, 1993) and likely are important in vivo. Such data demonstrate that the SIKVAV peptide has important effects on endothelial cells related to angiogenic behavior.

Several in vivo assays have been used to assess the angiogenic activity of SIKVAV in vivo (Table 2). A new assay involves subcutaneously-injected Matrigel as a vehicle for delivery of angiogenic agents (Passaniti et al 1992; Kibbey et al, 1992a). This assay has the advantage of not requiring surgical procedures. A known angiogenic factor, basic fibroblast growth factor, demonstrated vessel formation in the implant by three days and the number of vessels was easily quantitated with an Optomax Image Analyzer (Passaniti et al, 1992). The unsupplemented Matrigel implant showed very little cellular infiltrate. When the laminin-derived SIKVAV containing peptide was tested in this model, angiogenesis was observed in a dose-dependent manner (Kibbey et al, 1992a). This peptide was also active in the chick yolk sac and chorioallantoic membrane assays (Grant et al, 1992). In the sponge angiogenesis assay, SIKVAV was also very active in a dose-dependent manner in promoting vessel ingrowth. Control peptides with the same amino acids but in a scrambled order showed no activity in any of the in vitro or in vivo assays. In all of these assays, the SIKVAV containing peptides were as active as known angiogenic compounds such as bFGF and tumor necrosis factor. These in vivo data support the in vitro result and suggest that a portion of laminin may be important in development and in the normal vascular response to injury.

Table 2. Angiogenesis assays where SIKVAV is active

in vitro Matrigel tube assay
subcutaneous Matrigel assay in mice
subcutaneous polyvinyl sponge implant in mice
chick chorioallantoic membrane
chick yolk sack membrane

SIKVAV PROMOTES TUMOR GROWTH DUE TO ANGIOGENIC ACTIVITY

The SIKVAV peptide has been found to increase experimental metastases and tumor growth (Table 3) when coinjected intravenously with B16F10 melanoma cells. An increase of up to five-fold in the number of colonies on the surface of the lungs was observed and the effect was dose-dependent (Kanemoto et al, 1990). The effect appeared to be directly on the tumor cells since if the peptide was injected one hour before the tumor cells, no increase in lung colony formation was observed (Sweeney et al, 1991). If the peptide was injected from one to eight hours after the tumor cells, the number of lung colonies was greatly increased. As expected, the peptide had no effect on cellular arrest in the lungs. When coinjected subcutaneously with Matrigel, the size of the

Table 3. Effect of SIKVAV on experimental metastases and subcutaneous tumor growth

SIKVAV amt(ng)	Number of lung colonies(ranges)[a]	Tumor Volume (mm^3)[b]
0	22(15-35)	$4,900 \pm 1,000$
1	42(20-85)	$12,000 \pm 3,000$

a Data from Kanemoto et al, 1990
b Data from Kibbey et al, 1992a

tumors was much greater than the Matrigel plus tumor cells alone (no peptide) (Table 3). Since we had found that the peptide when mixed with Matrigel in the subcutaneous assay could promote angiogenesis, we determined the number of vessels in the tumors grown in the presence or absence of the peptide and with or without Matrigel. When melanoma cells are mixed with Matrigel and injected subcutaneously not only is tumor size increased but the number of vessels per area is also

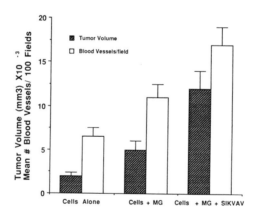

Figure 2. Correlation of tumor volume with the vessels per field in subcutaneous tumors grown under different conditions. B16F10 melanoma cells (100,000 cells) were injected either alone, with 0.5 ml Matigel, or 0.5 ml Matrigel plus 1 mg of 13 mer SIKVAV. After 3 weeks, tumors were harvested after measurement with a caliper. Blood vessels were quantitated on histologic sections.

increased (Fig 1). These data demonstrate that tumor cells in the presence of Matrigel increase vessel formation and thus angiogenesis likely contributes to the growth promoting effect of Matrigel on tumor growth (Kibbey et al, 1992a. The number of vessels per field was also further increased when SIKVAV was premixed with the melanoma cells and Matrigel. These data demonstrate that increased vascularization of tumors results in a significant increase in tumor size.

SUMMARY

The basement membrane glycoprotein laminin contains a potent angiogenic site. The SIKVAV-laminin derived peptide is active both in vitro and in vivo in promoting angiogenesis and can give a significant growth advantage to tumor cells. Antiangiogenic strategies directed against SIKVAV should reduce tumor growth and metastases.

References

Barsky, L.H., Rao, C.N., Williams, J.E., and Liotta, L.A. (1984) Laminin molecular domains which alter metastasis in a murine model. J. Clin. Invest. 74:843-848.

Beck, K., Hunter, I., and Engel, J. (1990) Structure and function of laminin: Anatomy of a multidomain protein. FASEB J. 4:148-160.

Benson, S., and Chuppa, S. (1990) Differentiation in vitro of sea urchin micromeres on extracellular matrix in the absence of serum. J. Exp. Zool. 256:222-226.

Grant, D.S., Kinsella, J.L., Fridman, R., Auerbach, R., Piasecki, B. A., Yamada, Y., Zain, M., and Kleinman, H.K. (1992) Interaction of endothelial cells with a laminin A chain peptide (SIKVAV) in vitro and induction of angiogenic behavior in vivo. J. Cellul. Physiol. 153:614-625.

Grant, D.S., Tashiro, K.-I., Segui-Real, B., Yamada, Y., Martin, G.R., and Kleinman, H.K. (1989) Two different laminin domains mediate the differentiation of human endothelial cells into capillary-like structures in vivo. Cell 58:933-934.

Haber, S., Finklestein, S.D., Benowitz, L.I., Sladek, J.R., and Collier, T.J. (1988) Matrigel enhances survival and integration of grafted dopamine neurons into striatum. In: "Progress in Brain Research" (D.M. Gash and J.R. Sladek, Jr., Eds.), Vol. 78, pp. 427-433. Elsevier, New York.

Joshi, M.S. (1991) Growth and differentiation of the cultured secretory cells of the cow oviduct on reconstituted basement membrane. J. Exp. Zool. 260:229-238.

Jun, S.H., Thompson, E.W., Goltardis, M., Torri, J., Yamamura, K., Kibbey, M.C., Kim, W.H., and Kleinman, H.K. (1993) Laminin adhesion selected primary human colon cancer cells are more tumorigenic than the parental and non-adherent cells. Int. J. Cancer, Submitted.

Kanemoto, T., Reich, R., Greatorex, D., Adler, S.H., Yamada, Y., and Kleinman, H.K. (1990) Identification of an amino acid sequence from the laminin A chain which stimulates metastases formation and collagenase IV production. Proc. Natl. Acad. Sci. USA 87:2279-2283.

Kibbey, M.C., Grant, D.S., and Kleinman, H.K. (1992a) Role of the SIKVAV site of laminin in promotion of angiogenesis and tumor growth: An in vivo Matrigel model. J. Natl. Cancer Inst. 84:1633-1638.

Kibbey, M.C., Royce, L.S., Dym, M.S., Baum, B.J., and Kleinman, H.K. (1992b) Glandular-like morphogenesis of the human submandibular tumor cell line A253 on basement membrane components. Exp. Cell Res. 198:343-357.

Kleinman, H.K., Graf, J., Iwamoto, Y., Kitten, G.T., Ogle, R.D., Sasaki, M., Yamada, Y., Martin, G.R., and Luckenbill-Edds, L. (1987) Role of basement membrane, In: Differentiation in Molecular and Cellular Aspects of Basement Membranes, Ed: R. Timpl and D.G. Rohrbach, pp. 309-326.

Kleinman, H.K., McGarvey, M.L., Hassell, J.R., Star, V.L., Cannon, F.B., Laurie, G.W., and Martin, G.R. (1986) Basement membrane complexes with biological activity. Biochemistry 25:312-318.

Kleinman, H.K., Weeks, B.S., Schnaper, H.W., Kibbey, M.C., Yamamura, K., and Grant, D.S. (1993) The laminins: A family of basement membrane glycoproteins important in cell differentiation and tumor metastases. Vitamins and Hormones 47:161-186.

Kubota, Y., Kleinman, H.K., Martin, G.R., and Lawley, T.J. (1988) Role of laminin and basement membrane in the morphological differentiation of human endothelial cells into capillary-like structures. J. Cell. Biol. 107:1589-1598.

Martin, G.R., Timpl, R., and Kuhn, K., (1988) Basement membrane proteins: Molecular structure and
function. 39:1-50.

McGarvey, M.L., Baron van Evercooren, A., Kleinman, H.K., and DuBois-Dalcq, M. (1986) Synthesis and effects of basement membrane components in cultured rat Schwann cells. Dev. Biol. 105:18-28.

Panayotou, G., End, P., Aumailley, M., Timpl, R., and Engel, J. (1989) Domains of laminin with growth-factor activity. Cell 56:93-101.

Passanti. A., Taylor, R.M., Pili, R., Guo, Y, Long, P.V., Haney, J.A., Pauly, R.R., Grant, D.S., and Martin, G.R. (1992). A new quantative method for assessing angiogenesis and antiangiogenesis using reconstructed basement membrane, heparin and FGF. Lab. Invest. 67:519-528.

Tashiro, K., Sephel, G.C., Weeks, B., Sasaki, M., Martin, G.R., Kleinman, H.K., and Yamada, Y. (1989) A synthetic peptide containing IKVAV sequence in the A chain of laminin mediates cell attachment, migration, and neurite outgrowth. J. Biol. Chem. 264:16174-16182.

Sakamoto, N., Iwahana, M., Tanaka, N.G., and Osada, Y. (1991) Inhibition of angiogenesis and tumor growth by a synthetic laminin peptide CDPGYIGSR-NH$_2$ Cancer Res. 51, 903-906.

Schnaper, W.H., Grant, D.S., Stetler-Stevenson, W.G., Fridman, R., O'Orazi, G., Bird, B.E., Hoythya, M., Fuerst, T.R., French, D.L., Quigley, J.P., and Kleinman, H.K. (1993) Type IV collagenase activity promotes endothelial cell formation into capillary-like structures on basement membrane in vitro. J. Cell. Physiol., in press.

Stack, S., Gray, R.D., and Pizzo, S.V. (1990) Modulation of plasminogen activation and type IV collagenase activity by a synthetic peptide derived from the laminin A Chain. Biochemistry 30:2073-2077.

Suarez-Quian, C.A., Hadley, M.A., and Dym, M. (1985) Effects of substrate on the shape of Sertoli cells in vitro. Ann. NY Acad. Sci. 438:417-434.

Sweeney, T.M., Kibbey, M.C., Zain, M., Fridman, R., and Kleinman, H.K. (1991) Basement membrane and the laminin peptide containing SIKVAV promotes tumor growth and metastases. Cancer Metast. Rev. 10:245-254.

Thompson, J.S. (1990) Basement membrane components stimulate epithelialization of intestinal defects in vivo. Cell Tissue Kinet. 23:443-451.

Terranova, V.P., Williams, J.E., Liotta, L.A., and Martin, G.R. (1984) Modulation of the metastatic activity of melanoma cells by laminin and fibronectin. Science 226:982-985.

Turpeeniemi-Hujanen, T., Thorgeisson, U.P., Rao, C.N., and Liotta, L.A. (1986) Laminin increases the release of type IV collagenase. J. Biol. Chem. 261:1883-1889.

Vukicevic, S., Kleinman, H.K., Luyten, F.P., Roberts, A.B., Roche, N.S., and Reddi, A.H. (1992) Identification of multiple growth factors in basement membrane matrigel suggests caution in interpretations of cellular activity related to extracellular matrix components. Exp. Cell Res. 202:1-8.

Warburg, A. and Mille, L.H. (1992) Sporogonic development of a malaria parasite in vitro. Science 255:448-450.

Wewer, U.M., Liotta, L.A., Jaye, M., Ricca, G.A., Drohan, W.N., Slaysmith, A.O., Rao, C.N., Wirth, P., Coligan, J.E., Albrechtsen, R., Mudry, M., and Sobel, M.E. (1986) Altered levels of laminin receptor mRNA in various human carcinoma cells that have different abilities to bind laminin. Proc. Natl. Acad. Sci. U.S.A. 83:7137-7141.

RELEASE OF GANGLIOSIDES IN THE MICROENVIRONMENT

AND ANGIOGENIC RESPONSE IN VIVO

Pietro M. Gullino

Department of Biomedical Sciences
University of Torino
Via Santena, 7
10126 Torino, ITALY

This is a summary of results obtained in experiments stimulated by the following observation: melanosarcoma cells injected in the center of a rabbit cornea migrate toward the periphery and as soon as they reach a distance of 1.5–2.0 mm from the limbus a large number of microvessels develop to colonize the neoplastic cell population. Thereafter, rapid growth of a solid tumor ensues. The objective of our experiments was to analyse the mechanism of this event on the assumption that by blocking angiogenesis the formation of a solid tumor could be avoided or impaired. The experimental approach was based on the following hypothesis: In the adult organism angiogenesis is a multifactorial event: the analysis of tissue composition at the onset of an angiogenic response may suggest which events have determinant infuence on the process.

The rabbit cornea was utilized as the experimental model for the following reasons: (a) it is avascular i.e. the number of vessels in the background is zero and the tissue is free of plasma and erithocytes therefore it closely reflects the composition of the tissue environment. (b) it is transparent i.e. new vessels are visible with a stereo–microscope and any inflammatory event that may complicate the experiment is rapidly detectable by opacity. (c) it is bilateral i.e. experimental treatment and control are possible in the same animal. (d) it is an immunopriviledged site and (e) the size is sufficient to obtain samples useful for analysis of tissue composition.

Prostaglandin E1 (PGE1) and basic fibroblastic growth factor (bFGF) were utilized as angiogenesis triggers for 3 main reasons: (a) PGE1 is angiogenic in the rabbit cornea and gives a well reproducible response. (b) bFGF was utilized as "control" because it is a molecule quite different from PGE1 but showing similar biological properties as angiogenesis inducer. (c) the delivery of both molecules using the slow-release pellet procedure[1] offers a good reliability in predicting the expected response. This last condition is particularly important, because the assessment of tissue composition was performed on corneal fragments removed at the end of day 3 after angiogenic stimulation, i.e. before new vessels were visible and budding from the limbal network was just beginning. Thus, cornea composition was evaluated before blood was carried within the corneal parenchima.

First we observed that the cornea became richer in sialic acid. The increment was 2-3 fold that of the contralateral cornea treated with the slow-release pellet lacking the angiogenic stimulus, and occurred regardless of the angiogenesis trigger.[2] Among the tissue molecules bearing sialic acid, we concentrated on gangliosides particularly because we had observed that in tumor bearing animals continuous injections of gangliosides favoured metastasis[3] and several highly metastatic neoplastic cell populations are known to shed gangliosides.[4]

At the end of day 3 after angiogenic stimulation we found in the cornea a quantity of GM3, GM2, GD3 gangliosides about twice that of the contralateral unstimulated cornea carrying the slow-release pellet alone. Besides an increment in the total content of these gangliosides there was also a change in their relative proportion: GM3 increased about 30%, GM2 doubled and GD3 augmented 2.7 fold.[5] Experiments were performed to assess how pertinent these changes were to the angiogenic response. Two approaches were followed: an _in vitro_ analysis of the ganglioside influence on microvascular endothelium, utilizing growth and motility as reference parameters, and the _in vivo_ effect of ganglioside enrichment of the cornea on tissue neovascularization.

Bovine adrenal capillary endothelium (BACE) was transferred from standard medium to DMEM + 0.2% calf serum. This subsistence medium ensured good survival of these cells for 72h. If gangliosides GM1 or GD1b or GT1b were added (6.25 µM final concentration) the cell number about doubled over the 72h period. If bFGF (0.6µM) was added to the subsistence medium, the survival of microvascular endothelium was prolonged beyond the 72h period. However, within the same period of time the number of cells tribled if gangliosides were added as above. In these experiments a second observation was made: If sialic acid was removed from the ganglioside molecule, the growth increment was not observed. Addition of sialic acid alone to the culture medium was also ineffective.[5,6]

The same experiment was repeated in comparing chemiotactic activity on BACE induced by GM1 or GD1b or GT1b, added alone or in mixture. The presence of gangliosides in the medium enhanced cellular migration. At the final concentration of 50µg/ml the increment in motility increased in parallel with the number of sialic acid molecules present in the added ganglioside i.e. the number of cells migrated was about twice for GM1, thrice for GD1b and four fold for GT1b. The mixture of the 3 gangliosides in equal proportions did not improve migration above the expected average number of cells observed for each ganglioside. As found for growth, removal of sialic acid from the ganglioside molecule nullified the increment of migration while sialic acid alone was ineffective on improving motility.[6,7]

The effects on microvascular endothelium due to gangliosides mixed in different proportions was evaluated by comparing the cell number after 72h incubation. GM3 + GM2 + GD3 mixtures added to cultures were prepared either with the 3 compounds in the same proportion of the normal cornea or in the proportion of the angiogenesis stimulated cornea. The conclusion was that as the GM3:GD3 ratios decreased, the endothelial cell growth increased. This was seen for total ganglioside concentration in the medium between 1.8 to 3.6 uM;[8] toxic effects occurred beyond these levels.

On the whole the experiments on growth and migration of microvascular endothelium in vitro indicated that both events were improved by the increase of total gangliosides content of the medium but prevalence of GM3 over GD3 in the added mixtures had depressing effects on both growth and migration.

In the next set of experiments the validity of this conclusion was tested in vivo. A pellet containing 10 ng PGE1 was implanted in each cornea of the same rabbit. At this dose angiogenesis was not induced. Local increment of tissue gangliosides was achieved by implanting close to the PGE1 pellet, a second pellet containing either GM1 or GT1b in doses from 0 to 200 ng. Addition of the second pellet enhanced the number of angiogenesis positive implants from zero to about 50%. When the ganglioside-bearing pellet was removed, the newly formed vessels disappeared; when the pellet was replaced, the vessels reappeared. This behaviour was observed when the angiogenesis trigger was either PGE1 or bFGF. Thus, local enrichment of tissue gangliosides promoted angiogenesis when the dose of the angiogenesis trigger was too low to induce neovascularization.

The angiostatic effect of GM3 as compared to GD3 was tested as follows: Each cornea of the same rabbit received one PGE1 pellet of 50ng. Then a second pellet was implanted bearing 50ng of GD3 in the right cornea and 50ng GM3 in the left cornea. Ten days later 75% of the right corneas showed angiogenesis as contrasted to only 12% of the left corneas. The experiment was repeated using a PGE1 dose of 250ng/pellet inducing angiogenesis in 77% of implants. The addition of GM3 in an adjacent pellet reduced the number of positive implants in a dose dependent manner from 77 to 25%.[5,8]

In conclusion: As an approach to understanding the events that occur during angiogenesis we utilized the rabbit cornea as the experimental model. The changes in tissue composition were evaluated before the cornea was colonized by the newly formed capillaries. At this time the total content of tissue gangliosides increased and the relative proportion among them changed with a relative reduction of the GM3:GD3 ratios. Repetition of the same situation under experimental conditions revealed that increment of total tissue gangliosides in the culture medium favoured motility and growth of microvascular endothelium. In vivo enrichment of tissue gangliosides enhanced the number of angiogenic responses in corneas where the dose of the angiogenesis trigger was unable to do so by itself. Prevalence of GM3 had angiostatic effect as compared to other gangliosides like GD3 or GM1. Gangliosides were not angiogenic but they could modulate the angiogenic response.

These observations are pertinent to the study of tumor growth and metastasis formation. It has been observed that (a) neoplastic cells progressing toward neoplastic transformation acquire angiogenic capacity,[9,10] (b) the interstitial fluid of solid tumors sampled in vivo is very rich in prostaglandins type E, which is strongly angiogenic,[11] (c) during neoplastic transformation of dermal fibroblasts there is accumulation of basic fibroblastic growth factor, a known angiogenic molecule,[12](d) the metastasizing capacity of neoplastic cell populations is positively correlated with cell surface sialylation[4] and (e) in vivo, neoplastic cells shed gangliosides in the microenvironment therefore total quantity and relative proportions change, i.e. the GM3:GD3 ratio is 19:1 in normal melanocytes but 1:15 in melanosarcoma cells.[13] A change in the GM3:GD3 ratios occurring in a microenvironment enriched by gangliosides shed by the cell population may trigger neovascularization. Thus, angiogenesis may be induced by an "angiogenetic environment" more than an "angiogenesis factor".

This hypothesis could satisfy at least two observations: (a) in the adult organism angiogenesis occurs under a variety of conditions ranging from neoplasia to diabetic retinopathy, psoriasis, rheumatoid arthritis, wound healing etc. and (b) angiogenic molecules presently known are in such a large number and so different in chemical structure and biological properties as to indicate that the angiogenic response is a multifactorial event. Angiogenic molecules, like PGE1 or bFGF, normally present in the microenvironment at doses insufficient to trigger angiogenesis, may become effective when appropriate changes of the microenvironment occur. Gangliosides, per se not angiogenic, are molecules with determinant influence in producing these changes, and modulate the angiogenic response.

REFERENCES

1. R. Langer and J. Folkman, Polymers for the sustained release of proteins and other macromolecules, Nature (London) 263:797 (1976).
2. G. Alessandri, K.S. Raju, and P.M. Gullino, Interaction of gangliosides with fibronectin in the mobilization of capillary endothelium,Invasion and Metastasis 6:145 (1986).
3. G. Alessandri, S. Flippeschi, P. Sinibaldi, F. Mornet, P. Passera, F. Spreafico, A.P.M. Cappa, and P.M. Gullino PM, Influence of gangliosides on primary and metastatic neoplastic growth in human murine cells, Cancer Res. 47:4243 (1987).
4. G. Yogeeswaran, and P.L. Salk, Metastatic potential is positively correlated with cell surface sialylation of cultured murine tumor cell lines, Science (Wash DC) 212:1514 (1981).

5. M. Ziche, G. Alessandri, and P.M. Gullino, Gangliosides promote the angiogenic response, Lab. Invest. 61:629 (1989).

6. G. Alessandri, G. De Cristan, M. Ziche, A.P.M. Cappa, and P.M. Gullino, Growth and motility of microvascular endothelium are modulated by the relative concentration of gangliosides in the medium, J. Cell Physiol. 151:23 (1992).

7. G.De Cristan, L. Morbidelli, G. Alessandri, M. Ziche, A.P.M. Cappa, and P.M. Gullino, Synergism between gangliosides and basic fibroblastic growth factor in favouring survival, growth and motility of capillary endothelium, J. Cell Physiol. 144:505 (1990).

8. M. Ziche, L. Morbidelli, G. Alessandri, P.M. Gullino, Angiogenesis can be stimulated or repressed in vivo by a change in GM3:GD3 ganglioside ratio, Lab. Invest. 67:711 (1992).

9. M. Ziche, and P.M. Gullino, Angiogenesis and neoplastic progression in vitro, J. Nat. Cancer Inst. 69:483 (1982).

10. S.S. Brem, H. Jensen, and P.M. Gullino, Angiogenesis as a marker of preneoplastic lesions of the human breast, Cancer 41:239 (1978).

11. M. Ziche, J. Jones, and P.M. Gullino, Role of prostaglandin E1 and copper in angiogenesis, J. Nat. Cancer Inst. 69:475 (1982).

12. J. Kandel, E. Bossy-Wetzel, I.F. Radvany, M. Klagbrun, and J.Folkman, Neovascularization is associated with a switch to the export of bFGF in the multi-step development of fibrosarcoma, Cell 66: 1095 (1991)

13. M.H. Ravindranath, T. Tsuchida, and R.F. Irie, Diversity of ganglioside expression in human melanoma, in: "Gangliosides and Cancer," H.F. Oettgen, ed., VCF, p.81, New York (1989).

ANGIOGENESIS IN VITRO: CYTOKINE INTERACTIONS AND BALANCED EXTRACELLULAR PROTEOLYSIS

Michael S. Pepper, Jean-Dominique Vassalli,
Lelio Orci and Roberto Montesano

Department of Morphology,
University Medical Center,
1, Rue Michel-Servet
1211 Geneva 4
Switzerland

INTRODUCTION

The blood vascular system is composed of a series of vessels of varying structural complexity. All blood vessels are lined by a monolayer of quiescent endothelial cells, which provide a structural and functional barrier between circulating blood and the surrounding tissues. In their simplest form, as represented by capillaries, blood vessels are surrounded by a basement membrane composed of type IV collagen, laminin, proteoglycans and other glycoproteins. A second cell type, namely the pericyte, is often associated with the capillary wall.

During development, all blood vessels begin as simple endothelial-lined capillaries (Evans, 1909), to which other vessel wall cells such as smooth muscle cells and fibroblasts are progressively added to form vessels of greater structural complexity. Two processes have been implicated in the formation of new capillaries: vasculogenesis, the in situ differentiation of mesodermal precursors into endothelial cells, and angiogenesis, the formation of new vessels by a process of sprouting from preexisting vessels (Risau et al., 1988; Pardenaud et al., 1989). While vasculogenesis appears to be limited to the early embryonic period, angiogenesis is known to occur throughout life. In addition to its role during development, angiogenesis occurs in physiological settings such as corpus luteum formation and wound healing. Angiogenesis may however be detrimental to the organism. This occurs in pathological settings such as proliferative retinopathy as a complication of diabetes. Angiogenesis is also necessary for the continued growth of solid tumors, and through newly formed vessels contributes to the hematogenous spread of metastasis (Folkman and Klagsbrun, 1987).

Angiogenesis: Molecular Biology, Clinical Aspects
Edited by M.E. Maragoudakis *et al*, Plenum Press, New York 1994

The series of morphogenetic events which result in the formation of new capillary blood vessels has been well described (Folkman and Klagsbrun, 1987). Angiogenesis begins with localized breakdown of the basement membrane of the parent vessel. Endothelial cells then migrate into the surrounding matrix within which they form a capillary sprout. Sprout elongation occurs as a result of further migration and of endothelial cell proliferation proximal to the migrating front. Fusion with the tip of another maturing sprout produces a capillary loop. A functional capillary results once a lumen has been formed, and maturation is completed by reconstitution of the basement membrane. The formation of new capillaries is regulated by a large number of polypeptide and non-polypeptide factors (Folkman and Klagsbrun, 1987; Klagsbrun and D'Amore, 1991). In attempting to classify their activities, we have suggested that the multiple cell functions which occur during angiogenesis should be considered as belonging either to a phase of activation which includes initiation and progression, or to a phase of resolution which includes inhibition of the activation phase and vessel maturation (Pepper et al., 1993b). While a great deal is known about those factors which induce the activation phase, very little is known about the factors involved in the phase of resolution.

It has been proposed that a number of separable although interrelated elements of the angiogenic process may be facilitated by extracellular proteolytic activity. These can be divided into two categories. The first includes those processes in which protease activity is necessary to overcome the mechanical barriers imposed by the surrounding extracellular matrix (Pepper and Montesano, 1990). These include the initial degradation of the investing basement membrane, cell migration, and the formation of a lumen, which by definition requires the formation of an extracellular space within a three-dimensional matrix. The second category includes those processes which modulate cytokine activity either by direct proteolytic activation of latent cytokines such as transforming growth factor-ß (TGF-ß), or indirectly by releasing matrix-bound cytokines such as basic fibroblast growth factor (bFGF) and thereby increasing their bioavailability (Flaumenhaft and Rifkin, 1992).

The in vitro studies which we will describe in this review were aimed at exploring 1) whether extracellular proteolysis is modulated during endothelial cell migration; 2) whether interactions exist between angiogenesis-modulating cytokines; 3) the role of balanced extracellular proteolysis in capillary morphogenesis.

ENDOTHELIAL CELL PROTEOLYTIC PROPERTIES ARE MODULATED DURING MIGRATION

One of the systems of extracellular proteolysis which has been most extensively studied in the context of cell migration and invasion is the plasminogen activator (PA)-plasmin system (Moscatelli and Rifkin, 1988; Vassalli et al., 1991). The central component of this system is plasmin, a protease of tryptic specificity capable of degrading certain matrix components (such as laminin and proteoglycans) and also of activating metalloprotease zymogens (which degrade collagens and certain other glycoproteins). Plasmin is generated from its inactive precursor plasminogen, by the activity of two PAs, urokinase-type PA and tissue-type PA (u-PA and t-PA). u-PA has been implicated in processes of cell migration and tissue remodelling, while t-PA is believed to be involved mainly in intravascular thrombolysis. u-PA activity can be

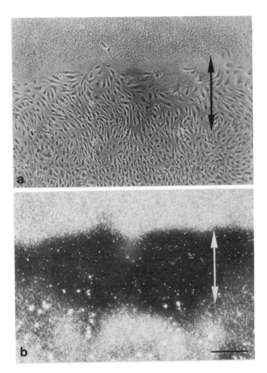

Figure 1. u-PA activity is induced in migrating endothelial cells. Wounded monolayers of microvascular endothelial cells were overlaid 24 hours after wounding with a mixture containing casein, agar and plasminogen. (a) Phase contrast view of the wound edge; arrows indicate the limits of the zone of caseinolysis. (b) Dark field illumination of the same region: an area of caseinolysis is seen as a dark band between bands of unlysed casein above and below (limits of the region of caseinolysis indicated by the white arrows). Bar = 250μm. (From Pepper et al., J. Cell Biol. 105: 2535-2541 (1987), with copyright permission from Rockefeller University Press.)

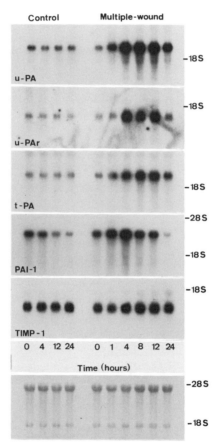

Figure 2. Induction of u-PA, u-PAr, t-PA and PAI-1 mRNAs in multiple-wounded monolayers of microvascular endothelial cells. Hybridization of Northern blots containing total cellular RNA extracted from non-wounded (Control) and multiple-wounded monolayers revealed a marked increase u-PA, u-PAr, t-PA and PAI-1 mRNAs in multiple-wounded cultures, with striking differences in the kinetics of induction. A similar increase in TIMP-1 mRNA was not observed. (From Pepper et al., J. Cell Biol. 122: in press (1993), and Pepper et al., J. Cell. Physiol. 153: 129-139 (1992), with copyright permission from Rockefeller University Press and Wiley Liss Inc.)

localized to the cell surface through binding to a specific high-affinity receptor (u-PAr), and u-PA and t-PA are subject to inhibition by specific physiological PA inhibitors (PAIs), PAI-1 and PAI-2.

In our attempts to determine whether the proteolytic properties of endothelial cells are modulated during migration, we have focussed on the PA system. The migration assay we have used is performed in two dimensions as follows: a region of a confluent monolayer is mechanically scraped away, also referred to as "mechanical wounding", following which cells lining the wound edge spontaneously migrate in to cover the denuded region; although no additional exogenous stimulus is required, the migratory response is dependent on endogenous bFGF (Sato and Rifkin, 1988).

Using a substrate overlay technique, in which wounded monolayers are overlaid with a thin layer of agar containing plasminogen and casein, we observed the appearance of zones of casein lysis over migrating cells (Figure 1). Casein is a substrate for plasmin, and the fact that its degradation was dependent on the presence of plasminogen in the overlay, indicated the presence of PA activity associated with the migrating cells. A kinetic analysis revealed a temporal correlation between cell migration and PA production. Thus, when monolayers were overlaid 8 days after wounding, at which stage the wound had closed, lytic activity was no longer observed. By co-incorporating specific inhibitors of either u-PA or t-PA into the overlay, we found that the cell-associated PA on migrating cells was u-PA (Pepper et al., 1987).

To provide sufficient material for zymographic and mRNA analysis, the number of migrating cells in the tissue culture dishes was increased by multiple-wounding a confluent monolayer; under these conditions, caseinolysis was also increased over wound edge cells. Zymography revealed an increase in cell-associated u-PA activity in multiple-wounded cultures, confirming our results obtained with the overlay technique. Zymography also revealed an increase in t-PA, bound to PAI-1, in the culture supernatant of multiple-wounded cultures. t-PA is secreted into the culture medium, which explains why it is not detected in the overlay technique which primarily detects cell-associated u-PA activity (Pepper et al., 1993a).

To determine whether the increase in cell-associated PA activity in multiple-wounded monolayers was due to receptor-bound u-PA, monolayers were acid-treated. Acid treatment removed most of the cell-associated u-PA activity which could be recovered in the acid wash. To verify that acid treatment had unmasked previously occupied u-PA binding sites, acid-treated monolayers were incubated with human u-PA. Human u-PA bound efficiently to acid-treated monolayers, and this could be inhibited by preincubating the cells with a peptide corresponding to the receptor binding region in the NH_2-terminus of mouse u-PA. These results demonstrate that the increase in PA activity on migrating endothelial cells is due to receptor-bound u-PA (Pepper et al., 1993a).

In experiments designed to assess the effect of wounding on protease and inhibitor mRNA expression, the width of the wounds in multiple-wounded cultures was chosen to allow for complete closure after 24 hours. By Northern analysis, we observed an increase in both u-PA and u-PAr mRNA levels, and a subsequent decrease in u-PAr mRNA which was tightly linked to the time of wound closure. t-PA mRNA was also increased (Figure 2). In situ hybridization studies were performed to determine the localization of the mRNA increase. The increase in both u-PA and u-PAr mRNAs was localized primarily to migrating cells at the edge of wounded monolayers. To determine whether the increase in u-PAr mRNA could be translated into an increase in receptor expression, wounded monolayers were incubated with ^{125}I-labelled human u-PA. This revealed an increase in the binding of radioactive u-PA to migrating cells. The specificity of this binding was determined by preincubating the monolayers with a peptide corresponding to the receptor-binding region of mouse u-PA. Taken together, these results demonstrate that the induction of PA activity on migrating endothelial cells is due to increased production of u-PA, which binds to u-PAr whose expression is upregulated on the same cells (Pepper et al., 1993a).

What about the other side of the proteolytic coin, that is to say what happens to production of PAI-1 during wound-induced migration ? By reverse zymography, we observed an increase in PAI-1 activity in multiple-wounded cultures. This was confirmed by Northern analysis in which an ephemeral increase in PAI-1 mRNA was observed with a subsequent decrease which was independent of wound closure (Figure 2). This appeared to be relatively specific for PAI-1 in microvascular endothelial cells, since a tissue inhibitor of metalloproteases (TIMP-1) was not increased under the same conditions (Figure 2). In situ hybridization studies also revealed that the increase in PAI-1 mRNA was localized to migrating cells (Figure 3) (Pepper et al., 1992b).

What are the mechanisms responsible for the increase in u-PA, u-PAr and PAI-1 expression in migrating endothelial cells ? Wounding induces a number of alterations in cell functions in cells lining the wound edge, including cell division and cell migration. We have observed that inhibition of endothelial cell division neither inhibits the induction of urokinase activity at the wound edge nor affects the increase in u-PA, u-PAr or PAI-1 mRNA levels in multiple-wounded cultures. We have demonstrated that endothelial cells migrating into the wound are moving to a state of low density. In low density cultures, in which endothelial cells also proliferate and migrate, u-PA, u-PAr and PAI-1 mRNAs were all increased; however, this increase could not be prevented by inhibition of endothelial cell proliferation. Taken together, these results indicate that the observed changes in expression of different components of the PA system are likely to be related to cell migration and the associated reduction in cell density rather than to proliferation (Pepper et al., 1987, 1992b, 1993a).

Sato and Rifkin (1988) have reported that wound-induced endothelial cell migration is decreased by addition of anti-bFGF antibodies to cultures after wounding. We found that under similar conditions, these antibodies completely inhibited the increase in u-PA activity associated with migrating cells (Pepper et al., 1993a). In addition, the increase in u-PA, u-PAr and PAI-1 mRNAs in multiple-wounded monolayers was markedly inhibited in the presence of the antibodies (Pepper et al., 1992b, 1993a). These findings are consistent with the observation that bFGF increases u-PA (Moscatelli et al., 1986; Montesano et al., 1986; Pepper et al., 1990), u-PAr (Mignatti et al., 1991; Pepper et al., 1993a) and PAI-1 (Saksela et al., 1988; Pepper et al., 1990) expression in endothelial cells. That bFGF was released into the medium as a consequence of wounding was suggested by the finding that the increase in u-PA and u-PAr mRNAs was greater when culture medium was not changed after wounding, and by the observation that the efficiency of the anti-bFGF antibodies was increased when medium was removed immediately after wounding and replaced with fresh medium containing the antibodies (Pepper et al., 1993a). These results suggest two possible mechanisms of bFGF release. First, bFGF is released from dead or damaged cells as a consequence of wounding (McNeil et al., 1989; Gadjusek and Carbon, 1989; Muthukrishnan et al., 1991), and second, bFGF is released from migrating cells (McNeil et al., 1989). Although none of our observations allow us to exclude either possibility, our results clearly suggest that endogenous bFGF released as a consequence of wounding is required for the increase in u-PA, u-PAr and PAI-1 expression in migrating endothelial cells.

Although we have focussed on u-PA, u-PAr and PAI-1 expression, we have also observed that t-PA is increased in endothelial cell monolayers in response to multiple-wounding. The mechanisms for this increase are likely to differ from those discussed above, since t-PA induction in BME cells in response to bFGF is minimal (Pepper et al., 1990; 1991a). In addition, t-PA mRNA is decreased in low density BME cultures (M. S. Pepper, unpublished observation), in contrast to what we observed for u-PA, u-PAr and PAI-1 mRNAs.

What is the functional significance of the increase in u-PAr expression on migrating endothelial cells ? Although it has consistently been observed that u-PA is absent from resting endothelial cells in vivo, this enzyme is induced in endothelial cells during neovascularization of ovarian follicles, corpus luteum and maternal decidua (Bacharach et al., 1992), and also during inflammation (Grøndahl-Hansen et al., 1989), situations in which endothelial cell migration is also induced. The first and currently most apparent function is therefore that of increasing the efficiency of extracellular proteolysis and localizing it to the immediate pericellular environment. However, additional catalytically-independent functions have been attributed to the u-PA/u-PAr interaction. These include mitogenesis (Rabbani et al., 1990, 1992), chemotaxis (Gudewicz and Gilboa, 1987; Del Rosso et al., 1990) and differentiation (Nusrat and Chapman, 1991). Similar non-proteolytic functions including chemotaxis (Fibbi et al., 1988) and chemokinesis (Odekon et al., 1992) have been observed in endothelial cells. Our findings on the co-induction of u-PA and its receptor provide a molecular basis for both proteolytic and putative non-proteolytic functions of the receptor in migrating endothelial cells. They also point to the possible existence of an autocrine loop in which u-PA, through interaction with its receptor, increases its own synthesis and possibly that of other components of the PA system in migrating cells.

Finally, we also demonstrate an increase in PAI-1 activity and mRNA in multiple-wounded endothelial cell cultures. However, the kinetics of the PAI-1 increase differed from those observed for u-PA and its receptor in that the PAI-1 increase was ephemeral, with a subsequent decrease which was independent of wound closure. We suggest that expression of protease inhibitors at very early stages of cell migration may provide a mechanism which confines extracellular matrix degradation to the immediate pericellular environment. PAI-1 bound to fibrin may protect blood clots from premature dissolution. This is particularly important in angiogenesis, which frequently occurs in situations of fibrin deposition such as wound healing.

To summarize: we have demonstrated that the increase in PA activity on migrating endothelial cells is due to receptor-bound u-PA, which in turn can be accounted for by an increase in both u-PA and u-PAr expression by these cells. The increase in receptor-bound u-PA is accompanied by an increase in expression of the physiological PA inhibitor PAI-1 by the same cells, although with different kinetics. Furthermore, the increase in u-PA, u-PAr, and PAI-1 expression in migrating cells can all be inhibited by antibodies to bFGF. The demonstration that the u-PAr is increased during cell migration strengthens the hypothesis that cell-associated protease activity is an important element in the cohort of functions expressed by migrating and invading cells. This activity is likely to be appropriately limited in time and space by the concomitant production of PAI-1.

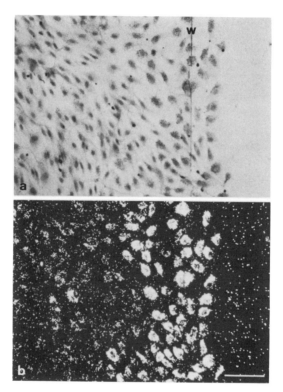

Figure 3. The PAI-1 increase in wounded microvascular endothelial cell monolayers is localized to migrating cells. (a) Monolayer 4 hours after wounding counterstained with methylene blue. The original wound edge (w) is visible. (b) Dark field view of the same cells. In situ hybridization using an antisense [^{3}H]-labelled PAI-1 RNA probe reveals an increase in PAI-1 mRNA expression in migrating cells. Bar = 100μm. (From Pepper et al., J. Cell. Physiol. 153: 129-139 (1992), with copyright permission from Wiley Liss Inc.).

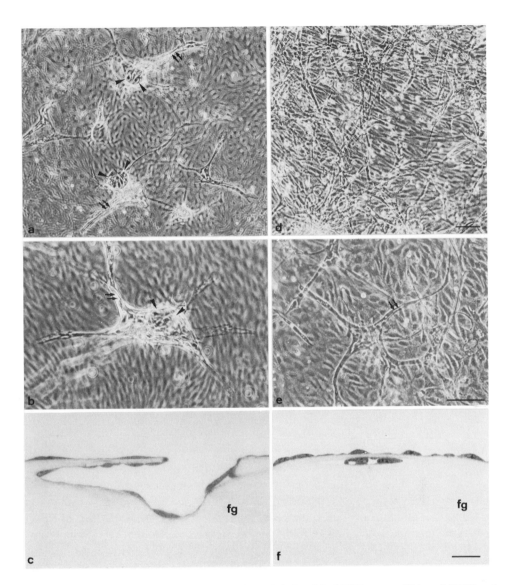

Figure 4. Morphological analysis of invading cell cords induced by bFGF or co-addition of bFGF and TGF-ß1. bFGF (30ng/ml) was added without (**a-c**) or with 500pg/ml TGF-ß1 (**d-f**) to confluent monolayers of microvascular endothelial cells on fibrin gels. The resulting capillary-like tubular structures were viewed by phase-contrast microscopy (**a, b, d** and **e**) and were further assessed by examination of semi-thin sections (**c** and **f**). bFGF induced endothelial cells cells to invade from a circular opening in the surface monolayer (arrow-heads in **a** and **b**), to form well organized cell cords with a clearly visible refringent lumen (arrows in **a** and **b**), which tapered down progressively in the distal part of the cords. These observations were confirmed by semi-thin sectioning, in which the proximal part of the cords was often seen to be cavernous (**c**). When 500pg/ml TGF-ß1 was co-added with bFGF, the total additive length of the invading cell cords was increased (compare **a** and **d**). Clearly distinguishable lumina were present beneath the surface monolayer (white refringent line indicated by the arrows in **e**), although lumen size was decreased to a more physiological size when compared to cultures treated with bFGF alone (compare **c** and **f**). Bars: **a** and **d**, 100μm; **b** and **e**, 100μm; **c** and **f**, 20μm. (From Pepper et al., Exp. Cell Res. 204: 356-363 (1993), with copyright permission from Academic Press Inc.)

157

INTERACTIONS BETWEEN ANGIOGENESIS-MODULATING CYTOKINES

In attempting to classify the activities of angiogenesis-modulating cytokines, it is useful to consider angiogenesis as occurring in two phases: a phase of activation and a phase of resolution. The phase of activation encompasses initiation and progression, and includes: a) basement membrane degradation; b) cell migration and extracellular matrix invasion; c) endothelial cell proliferation; and d) capillary lumen formation. The phase of resolution encompasses termination and vessel maturation, and includes: a) inhibition of endothelial cell proliferation; b) cessation of cell migration; and c) basement membrane reconstitution (Pepper et al., 1993b). With respect to angiogenesis-modulating cytokines, these can be divided into those which induce the activation phase through their direct interaction with endothelial cells themselves, namely bFGF (Klagsbrun and D'Amore, 1991) and vascular endothelial growth factor (VEGF) (Ferrara et al., 1992a), and those which act indirectly by inducing the production of direct-acting angiogenic cytokines by inflammatory cells. This latter group consists principally of TGF-ß and tumor necrosis factor-α, both of which are potent inflammatory cell chemoattractants (Klagsbrun and D'Amore, 1991). While much is known about those factors which induce the activation phase, very little is known about the factors involved in the phase of resolution. A possible candidate is TGF-ß (Flaumenhaft et al., 1992), for its direct effect on endothelial cells, at least in vitro, is to inhibit endothelial cell proliferation and migration and also to reduce extracellular proteolysis, all of which are components of the activation phase. TGF-ß has also been reported to promote the organization of single endothelial cells embedded in three-dimensional collagen gels into tube-like structures (Madri et al., 1988; Merwin et al., 1990), a model which could be representative of capillary maturation.

Very little is known about interactions between angiogenesis-modulating cytokines. It is highly likely however that endothelial cells are rarely (if ever) exposed to a single cytokine during physiological and pathological processes. In order to explore potential interactions between angiogenesis-modulating cytokines, we have taken advantage of an in vitro model of angiogenesis (Montesano and Orci, 1985), which consists of cultivating endothelial cells on the surface of a three-dimensional collagen or fibrin gel. In control cultures, the cells form a monolayer on the surface of the gel. When the monolayer is treated with an angiogenic factor such as bFGF (Montesano et al., 1986) or VEGF (Pepper et al., 1992a), the cells invade the underlying gel; by phase contrast microscopy, by focussing beneath the surface monolayer, branching and anastomosing cell cords can be seen within the gel (Figure 4a, b, d, e). In cross section, the presence of tube-like structures resembling capillaries can be observed beneath the surface monolayer (Figure 4c, f). Invasion can be quantitated by measuring the total additive length of all cells which have penetrated into the underlying gel.

Using this three-dimensional model, we have assessed the effect of simultaneous addition of bFGF and VEGF on the in vitro angiogenic response. When added separately at equimolar concentrations, bFGF was about twice as potent as VEGF (Figure 5). However, when bFGF and VEGF were co-added, the resulting effect on invasion was far greater than additive and occurred with greater rapidity than the response to either cytokine alone (Pepper et al., 1992a) (Figure 5). If the synergism

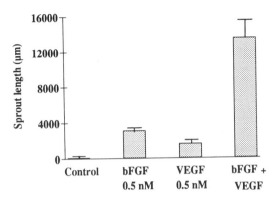

Figure 5. Synergistic effect of bFGF and VEGF on in vitro angiogenesis. Endothelial cell invasion was quantitated by measuring the total additive length of all cell cords which had penetrated beneath the surface monolayer. At equimolar concentrations (0.5nM) bFGF was about twice as potent as VEGF. Co-addition of the two cytokines induced an invasive response which was greater that additive. (From Pepper et al., Biochem. Biophys. Res. Commun. 189: 824-831 (1992), with copyright permission from Academic Press Inc.)

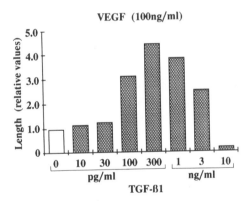

Figure 6. Quantitative analysis of the effect of TGF-ß1 on VEGF-induced collagen gel invasion. Confluent monolayers of microvascular endothelial cells were co-treated for 4 days with VEGF (100 ng/ml) and TGF-ß1 or either cytokine alone, and the total length of all invading cell cords determined as described in Materials and Methods. Median values from at least 3 experiments are shown relative to controls, and were calculated from the mean length in μm of three randomly selected fields per experiment of cultures co-treated with VEGF and TGF-ß1/(cultures treated with VEGF only + cultures treated with TGF-ß1 only) in the same experiment. (From Pepper et al., Exp. Cell Res. 204: 356-363 (1993), with copyright permission from Academic Press Inc.)

which we have observed in vitro also exists in vivo, angiogenesis would be more prominent in tumors which release more than one angiogenic factor. This observation may justify anti-angiogenesis strategies based on the neutralization of a single angiogenic factor.

In attempting to understand the mechanisms responsible for this synergistic effect, we tested conventional parameters such as proliferation, two-dimensional migration and PA-mediated proteolysis. In none of these situations was the effect of simultaneous addition of bFGF and VEGF greater than additive (Pepper, 1992); we are therefore unable to demonstrate a direct correlation between any one of these parameters and the synergism we observed in collagen gels. Our observations nonetheless highlight the importance of a three-dimensional environment: had we relied exclusively on traditional two-dimensional assays of proliferation, migration or proteolysis, the synergism between bFGF and VEGF would not have been detected.

TGF-ß is an angiogenesis-modulating cytokine that has variously been described as angiogenic or anti-angiogenic. In vivo, TGF-ß is a potent inducer of angiogenesis (Roberts et al., 1985; Yang and Moses, 1990), whose effect is believed to be mediated by secretory products of TGF-ß-recruited connective tissue and inflammatory cells (Wahl et al., 1987; Wisemann et al., 1988; Yang and Moses, 1990; Phillips et al., 1992). In vitro however, TGF-ß inhibits a number of essential components of the angiogenic process. These include endothelial cell proliferation (Baird and Durkin, 1986; Fràter-Schröder et al., 1986; Müller et al., 1987; Antonelli-Orlidge et al., 1989), migration (Heimark et al., 1986; Müller et al., 1987; Sato and Rifkin, 1989) and extracellular proteolytic activity (Saksela et al., 1987; Pepper et al., 1990; 1991b). Results from three-dimensional in vitro assays demonstrate that the response to TGF-ß varies depending on the assay used. Thus, TGF-ß inhibits endothelial cell invasion of three-dimensional collagen (Müller et al., 1987) or fibrin (Pepper et al., 1991c) gels, as well as the explanted amnion (Mignatti et al., 1989). In addition, TGF-ß inhibits bFGF-induced capillary lumen formation within three-dimensional fibrin gels (Pepper et al., 1990). These results support the notion that TGF-ß is a direct-acting inhibitor of extracellular matrix invasion and tube formation. However, it has also been reported that TGF-ß promotes organization of endothelial cells into tube-like structures (Madri et al., 1988; Merwin et al., 1990). These apparently conflicting results may be reconciled by considering that TGF-ß might have different functions on vessel formation at different stages of the angiogenic process. Thus when acting directly on endothelial cells, it may inhibit invasion and vessel formation, and once sprout formation has occurred, TGF-ß may be necessary for the inhibition of further endothelial cell replication and migration, and induce vessel organization and functional maturation.

An additional possibility is that the direct effect of TGF-ß on endothelial cell function is concentration dependent, particularly since this cytokine has been described as a bifunctional regulator in a variety of other biological processes (Nathan and Sporn, 1991). The effect of a wide range of concentrations of TGF-ß1 on bFGF- or VEGF-induced angiogenesis was assessed in our three-dimensional in vitro model. We found that in the presence of TGF-ß1, bFGF- or VEGF-induced invasion was increased at 200-500pg/ml TGF-ß1 and decreased at 5-10ng/ml TGF-ß1 (Figure 6). The inhibitory effect at relatively high concentrations is in accord with previous studies in which endothelial cell invasion of three-dimensional collagen gels (Müller et al., 1987) or the explanted amnion (Mignatti et al., 1989) were inhibited by TGF-ß1 at 1-10ng/ml. Taken together, these results clearly demonstrate that the effect of TGF-ß1 on bFGF- or VEGF-induced in vitro angiogenesis is concentration-dependent. The mechanisms responsible for this biphasic effect are not known. One hypothesis is based on alterations in the net balance of extracellular proteolysis (see below). However, we also have evidence to suggest that integrin expression is differentially affected at these different concentration of TGF-ß1 (M. S. Pepper, unpublished observation). The relative contribution of these parameters, namely proteases and integrins, is currently under investigation

To summarize the effects of TGF-ß on the angiogenic response, it could be stated that the direct effect of TGF-ß on endothelial cells not only varies at different stages of the angiogenic process, but is also concentration dependent. Thus in addition to its indirect angiogenic effect, TGF-ß could both promote and inhibit angiogenesis when acting directly on endothelial cells.

The identification of a cytokine which inhibits angiogenesis by acting directly on endothelial cells, and which lacks paradoxical in vivo and in vitro effects which are often difficult to explain, would facilitate and possibly simplify our current understanding of the physiological factors which regulate angiogenesis. It has recently been observed that leukemia inhibitory factor (LIF) is a potent inhibitor of aortic endothelial cell proliferation in vitro (Ferrara et al., 1992b). Although LIF was initially purified and cloned using a bioassay based on its ability to induce monocyte differentiation, it is one of a growing number of cytokines which are characterized by pleiotropy and functional redundancy (reviewed by Hilton, 1992).

Using our three-dimensional in vitro model, we have found that LIF is a potent inhibitor of angiogenesis and that it lacks the concentration-dependent stimulatory effect characteristic of TGF-ß1. The inhibitory effect was observed on both microvascular and aortic endothelial cells and occurred irrespective of the angiogenic stimulus, which included bFGF, VEGF or the synergistic effect of the two factors in combination. The inhibitory effect in three-dimensional collagen gels could be correlated with inhibition of proliferation and migration in conventional two-dimensional assays in aortic endothelial cells. In addition, LIF decreased the proteolytic potential of aortic and microvascular endothelial cells by increasing their expression of plasminogen activator inhibitor-1. These results demonstrate that LIF is a potent inhibitor of in vitro angiogenesis, and that this effect correlates with a decrease in endothelial cell proliferation, migration and extracellular proteolysis (M. S. Pepper, N. Ferrara, L. Orci, R. Montesano, manuscript submitted).

To summarize: using a three-dimensional model of in vitro angiogenesis, we demonstrate that important interactions exist between different cytokines in the in vitro angiogenic response. Synergism was observed between bFGF and VEGF, TGF-ß1 had a biphasic effect on bFGF- or VEGF-induced invasion, and LIF inhibited bFGF- or VEGF-induced angiogenesis. We suggest that the temporally-coordinated and concentration-dependent activity of a number of cytokines is necessary for the control of different elements of the angiogenic process in specific and appropriate settings in vivo.

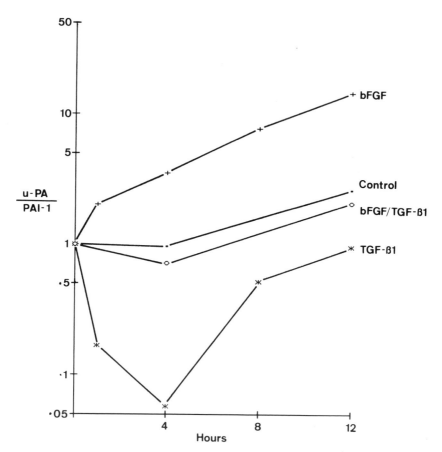

Figure 7. bFGF and TGF-ß1 modulate the potential proteolytic balance as reflected by the ratio of u-PA:PAI-1 mRNAs. The values for each mRNA, and hence the ratio, were arbitrarily taken to be 1 at time = 0. Following exposure to bFGF, an increase in potential proteolytic activity (u-PA:PAI-1 mRNA ratio) is observed. In contrast, TGF-ß1 decreases the u-PA:PAI-1 mRNA ratio, which is lowest after 4 hours. In control cultures and in cultures exposed to both agents simultaneously, a small increase in u-PA:PAI-1 mRNA ratio is observed after 12 hours. (From Pepper et al., J. Cell Biol. 111: 743-755 (1990), with copyright permission from Rockefeller University Press.)

PROTEOLYTIC BALANCE AND CAPILLARY MORPHOGENESIS

That the balance between proteases and protease inhibitors might be important for normal capillary morphogenesis was first demonstrated using the model of three-dimensional in vitro angiogenesis described above. In these experiments however, fibrin was substituted for collagen, the reasoning being that angiogenesis often occurs in a fibrin-rich matrix, for example during wound healing. In striking contrast to what we had observed with collagen gels, we noted that upon addition of the angiogenic stimulus, the fibrin substrate was progressively lysed (Montesano et al., 1987). The absence of a three-dimensional matrix scaffold therefore precluded invasion and the formation of capillary-like tubes. However, inhibition of fibrinolysis by addition of the serine protease inhibitor Trasylol was permissive for the formation of tube-like capillaries in the underlying gel (Montesano et al., 1987). These findings demonstrated that inhibition of excessive proteolysis was necessary for the preservation of an intact matrix scaffold into which endothelial cells could be induced to migrate to form capillary-like tubular structures.

This study highlighted the notion that although increased protease activity is clearly associated with the invasive phenotype, protease inhibitors are likely to play an equally important, albeit permissive role during angiogenesis, by preserving extracellular matrix integrity. Additional indirect support for this hypothesis came from the following observations. While studying the effects of bFGF and VEGF on the increase in extracellular proteolysis in endothelial cells, we found that this was mainly due to an increase in u-PA expression. However we also noted that bFGF and VEGF increased expression of PAI-1, a physiological PA inhibitor, although the amplitude and the kinetics of the PAI-1 mRNA increase were markedly different from those of u-PA mRNA (Pepper et al., 1990; 1991a). PAI-1 was also increased in endothelial cells migrating in two dimensions, with the kinetics of the PAI-1 mRNA increase being identical to those observed following addition of bFGF or VEGF to confluent quiescent monolayers of endothelial cells (Pepper et al., 1992b - see the first part of this review).

In attempting to interpret these observations, we used the ratio of u-PA to PAI-1 mRNA as a reflection of the potential proteolytic activity of the cells at each time point. In controls there was a small increase in the u-PA:PAI-1 mRNA ratio and hence in extracellular proteolysis with time in culture (Figure 7). In response to bFGF or VEGF, the ratio was markedly increased above controls (Pepper et al., 1990; 1991a) (Figure 7). These results show that although u-PA and PAI-1 are both increased by bFGF or VEGF, the net balance of extracellular proteolysis as represented by the u-PA:PAI-1 mRNA ratio is always positive in response to these cytokines. Having previously observed that protease inhibitors play an important permissive role in the protection of the matrix against excessive proteolysis, our working hypothesis is that the transitory induction of PAI-1 by migrating endothelial cells serves to protect the matrix during the initial phases of migration.

What happens in the presence of greater concentrations of PAI-1 ? In attempting to answer this question, we used a cytokine which greatly increases PAI-1 production by endothelial cells, namely TGF-ß1 (Saksela et al., 1987; Pepper et al., 1990; 1991b). In addition to increasing PAI-1, we also observed that TGF-ß1 increased u-PA expression in microvascular endothelial cells. However, when using the u-PA:PAI-1 mRNA ratio, the net response to TGF-ß1 was always anti-proteolytic (Figure

7). Of particular interest was the finding that when bFGF and TGF-ß1 were added simultaneously, levels of proteolysis, as represented by the u-PA:PAI-1 mRNA ratio, mimicked levels seen in controls (Pepper et al., 1990) (Figure 7). The u-PA:PAI-1 ratio represents total extracellular proteolysis, which we would like to suggest includes both cell-associated protease activity and PAI-1 deposition into the extracellular matrix. Knowing that TGF-ß1 was capable of modulating bFGF-induced proteolysis, we next assessed its effect on bFGF-induced capillary-like tube formation in vitro. Experiments aimed at addressing this problem were performed in fibrin rather that collagen gels in order to assay more specifically for the PA system. In the presence of bFGF alone, cells formed tube-like structures with a widely patent lumen (Figure 4a-c). In contrast, when TGF-ß1 was co-added with bFGF, lumen diameter was markedly reduced (Pepper et al., 1990; 1993b) (Figure 4d-f). Although this was true both at 500pg/ml and 5ng/ml TGF-ß1, doses which increased and decreased bFGF-induced invasion respectively (see the second part of this review), the presence of a lumen was less frequently observed at 5ng/ml than at 500pg/ml (M. S. Pepper, unpublished observation). Furthermore, lumen size at 500pg/ml TGF-ß1 was reduced to a size which was physiologically more relevant (Pepper et al., 1993b) (Figure 4d-f). Since the creation of a hollow space (i.e. the lumen) within the fibrin gel is dependent on fibrinolysis, these qualitative findings suggest that the antiproteolytic effect of TGF-ß1, resulting from a large increase in PAI-1, is responsible, at least in part, for the reduction in lumen size.

What about the biphasic effect of TGF-ß1 on bFGF- or VEGF-induced invasion which was described in the second part of this review; could this be explained by alterations in extracellular proteolytic activity ? Since the magnitude of PAI-1 induction is proportional to TGF-ß1 concentration in the range used in these studies (10pg/ml-10ng/ml) (M. S. Pepper, unpublished observation), our working hypothesis is that at potentiating doses of TGF-ß1 (200-500pg/ml), PAI-1 is present at an optimal concentration which preserves matrix integrity without reducing protease activity below a critical level required for invasion. This increases cell-matrix contacts, which in turn results in an increase in migration. However at higher concentrations of TGF-ß1 (5-10ng/ml), the further increase in PAI-1 production reduces proteolytic activity below the critical level required for invasion, and hence invasion is decreased.

What happens when there is too much proteolysis ? Is this situation compatible with normal capillary morphogenesis ? It had previously shown that mice expressing the polyoma virus middle T oncogene (mT) develop endothelial cell tumors which manifest as hemangiomas (Bautch et al., 1987; Williams et al., 1988). We have developed an in vitro correlate of hemangioma formation by embedding mT-expressing endothelial cells into three-dimensional fibrin gels (Montesano et al., 1990). In contrast to normal endothelial cells which form a network of capillary-like tubes (Figure 8a), mT-expressing endothelial form endothelial-lined cysts which bear a striking resemblance to the cavernous hemangiomas which were seen in vivo (Figure 8b, c).

Could the morphogenetic behaviour of the mT-expressing cells be associated with a perturbation in proteolytic activity ? Zymography and northern analysis revealed that u-PA expression was greater in mT-expressing endothelial cells than in normal endothelial cells. Furthermore, this effect appeared to be specific for endothelial cells, since the increase in u-PA mRNA was not seen in three non-endothelial cell types expressing mT. In addition, by reverse zymography we were unable to detect PAI-1 activity in the cell extracts or culture supernatants of mT-expressing endothelial cells, which was in contrast to the large amounts of PAI-1 present in three endothelial cell

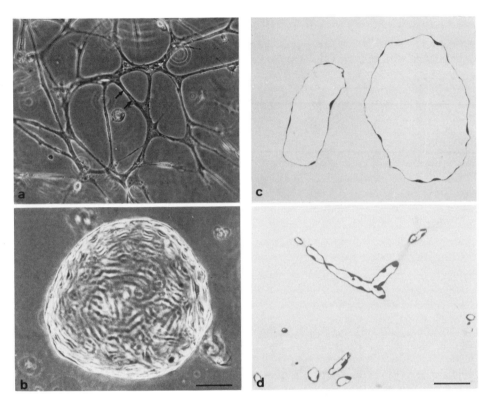

Figure 8. Morphogenetic behaviour of normal mouse brain endothelial cells (BECs) and mT-expressing endothelioma cells grown within three-dimensional fibrin gels. (a) Phase contrast view of a network of branching and anastomosing cords formed by primary mouse BECs; fine slit-like lumina are indicated by the small arrows. (b) Spherical cyst formed by mT-expressing endothelial cells. In this picture, the focus is approximately on the equatorial plane of the cyst. The endothelial cells lining the floor and the roof of the cavity appear blurred in the center of the cyst. (c) Semi-thin section of mT-expressing endothelial cells showing the formation of hemangioma-like sacs lined by a continuous monolayer of flattened endothelial cells. (d) Semi-thin section of mT-expressing endothelial cells grown in the presence of Trasylol, an inhibitor of serine proteases; the cells have formed branching tubules resembling capillary blood vessels. (c) and (d) are printed at the same magnification. Bars: (a), (b) = 100μm; (c), (d) = 250μm. (From Montesano et al., Cell 62: 435-445 (1990), with copyright permission from Cell Press.)

types not expressing mT. Levels of PAI-1 mRNA were also markedly lower in mT-expressing endothelial cells when compared to normal endothelial cells (Montesano et al., 1990).

Knowing that mT-expressing endothelial cells produce more u-PA than their normal counterparts, and very little if any PAI-1 at all, we asked what would happen if we attempted to reduce the excessive proteolysis by adding protease inhibitors to the culture system. By adding Trasylol, an inhibitor of serine proteases, at the time of embedding, we found that the mT-expressing endothelial cells instead of forming cysts now formed a branching network of capillary-like tubes (Montesano et al., 1990) (Figure 8d). These results demonstrate that excessive proteolytic activity is incompatible with normal capillary morphogenesis, but that by reducing this activity by the addition of protease inhibitors, one can restore normal morphogenetic properties to the endothelial cells.

To summarize: our results demonstrate that although endothelial cell migration and invasion are associated with increased extracellular proteolytic activity, protease inhibitors play an equally important albeit permissive role during angiogenesis by preserving matrix integrity. Furthermore, at optimal concentrations, protease inhibitors may potentiate invasion and regulate lumen formation to a physiologically relevant size. However, when protease activity is reduced below a critical level, invasion and lumen formation are inhibited. Finally, excessive proteolytic activity is not compatible with normal capillary morphogenesis, since it results in the formation of endothelial-lined cysts. Normal morphogenetic properties can be restored to endothelial cells by reducing this excessive proteolytic activity.

SUMMARY AND PERSPECTIVES

Using a number of experimental models, we have, we believe, been able to accurately recapitulate many important components of the angiogenic process. Two major principles have emerged from the in vitro studies described in this review.

The first supports the notion that a precise protease-antiprotease equilibrium allows for localized pericellular matrix degradation during cell migration, while at the same time protecting the extracellular matrix against inappropriate destruction. Excessive proteolytic activity is incompatible with normal capillary morphogenesis since this results in the formation of aberrant vascular structures. In the absence of sufficient proteolysis on the other hand, invasion and lumen formation are inhibited.

The second principle arises from the observation that important interactions exist between different angiogenesis-modulating cytokines. Thus the activity of these cytokines should not be considered in isolation, for as we have shown, this activity is contextual, and depends on the presence and concentration of other cytokines in the pericellular environment of the responding endothelial cell.

The ultimate hope of a biomedical research project such as has been elaborated in this review, is to be able to contribute to the treatment of certain diseases in man. Modulation of the angiogenic process has been proposed as an alternative/adjunct to current therapeutic strategies in several angiogenesis-related or -dependent diseases. Foremost amongst these is the inhibition of solid tumor growth and the spread of metastasis. Inhibition of angiogenesis is also of potential benefit in treating ocular neovascularization, arthritis, psoriasis and hemangiomas. On the other hand, stimulation of angiogenesis may be of benefit in wound healing and fracture repair, in the treatment of peptic ulcers, in the healing of ischaemic ulcer, and in stimulating collateral vessel growth in myocardial infarction.

Can the principles which we have elaborated be applied to the clinical setting ? The answer is almost certainly yes, although our responses to this question must for the moment remain speculative. With respect to the role of balanced extracellular proteolysis, tilting the balance in the direction of anti-proteolysis may decrease endothelial cell invasion and hence new vessel formation, which may be useful in settings where inhibition of angiogenesis is desired. Based on our previous in vitro observations on hemangioma formation, anti-proteolysis could also be envisaged as an adjunct to current modes of hemangioma treatment. With respect to the contextual activity of angiogenesis-modulating cytokines, it is reasonable to suggest that in situations where stimulation of angiogenesis is desired, the benefit derived from co-addition of two cytokines whose interaction is synergistic would be greater than that derived from the addition of one of these cytokines alone. On the other hand, if the synergism which we have observed in vitro also applies in vivo, angiogenesis would be far more prominent in solid tumors which release more than one angiogenic factor. This observation may justify anti-angiogenesis strategies based on neutralization of a single factor, which may greatly decrease the angiogenic response by suppressing the synergistic effect.

What of the future ? It is clear that is has now become critical to test our hypotheses in vivo, particularly if we wish to apply our findings to the clinical setting. A major effort is therefore currently underway to test these hypotheses in appropriate animal models. The in vitro models which we have described can also be exploited in the search for new agents, either pharmacologic or physiologic, which either stimulate or inhibit angiogenesis. A number of potentially important findings are emerging from this approach. First, Fotsts et al. (1993) have recently demonstrated that genistein, which is found in high concentration in the urine of individuals consuming a diet rich in soya, is a potent inhibitor of angiogenesis in our three-dimensional in vitro model. Second, using the same in vitro model, we have recently identified a putative angiogenic factor secreted by Swiss 3T3 fibroblasts, whose physico-chemical properties differ from those of previously described angiogenic polypeptides (Montesano et al., 1993).

ACKNOWLEDGEMENTS

We are grateful to our colleagues both in Geneva and elsewhere, who have contributed so generously to these studies: Dominique Belin, Napoleone Ferrara, Theodore Fotsis, Werner Risau, André-Pascal Sappino, Lothar Schweigerer and Erwin Wagner. This work was supported by grants form the Swiss National Science Foundation, and grants in aid from the Sir Jules Thorn Charitable Overseas Trust and the Juvenile Diabetes Foundation.

REFERENCES

Antonelli-Orlidge A., Saunders K.B., Smith S.R. and D'Amore P.A. (1989): An activated form of transforming growth factor ß is produced by cocultures of endothelial cells and pericytes. Proc. Natl. Acad. Sci. USA 86: 4544-4548.

Bacharach E., Itin A., and Keshet E. (1992): In vivo patterns of expression of urokinase and its inhibitor PAI-1 suggest a concerted role in regulating physiological angiogenesis. Proc. Natl. Acad. Sci. USA 89: 10686-10690.

Baird A. and Durkin T. (1986): Inhibition of endothelial cell proliferation by type-ß transforming growth factor: interactions with acidic and basic fibroblast growth factors. Biochem. Biophys. Res. Commun. 138: 476-482.

Bautch V.L., Toda S., Hassell J.A. and Hanahan D. (1987): Endothelial tumors develop in transgenic mice carrying polyoma middle T oncogene. Cell 51: 529-538.

Del Rosso M., Fibbi G., Dini G., Grappone C., Pucci M., Caldini R., Magnelli L., Fimiani M., Lotti T. and Panconesi E. (1990): Role of specific membrane receptors in urokinase-dependent migration of human keratinocytes. J. Invest. Dermatol. 94: 310-316.

Evans H.M. (1909): On the development of the aortae, cardinal and umbilical veins, and the other blood vessels of vertebrate embryos from capillaries. Anat. Rec. 3: 498-519.

Ferrara N., Jakeman L., Houck, K. and Leung D.W. (1992a): Molecular and biological properties of the vascular endothelial growth factor family of proteins. Endocrine Rev. 13: 18-32.

Ferrara N, Winer J. and Henzel W.J. (1992b): Pituitary follicular cells secrete an inhibitor of aortic endothelial cell growth: identification as leukemia inhibitory factor. Proc. Natl. Acad. Sci. USA 89: 698-702.

Fibbi G., Ziche M., Morbidelli L., Magnelli L. and Del Rosso M. (1988): Interaction of urokinase with specific receptors stimulates mobilization of bovine adrenal capillary endothelial cells. Exp. Cell Res. 179: 385-395.

Flaumenhaft R and Rifkin D.B. (1992): The extracellular regulation of growth factor action. Mol. Biol. Cell 3: 1057-1065.

Folkman J. and Klagsbrun M. (1987): Angiogenic factors. Science 235: 442-447.

Fotsis T., Pepper M. Aldercreutz H., Fleischmann G., Hase T., Montesano R. and Schweigerer L. (1993): Genistein, a dietary-derived inhibitor of in vitro angiogenesis. Proc. Natl. Acad. Sci. USA 90: 2690-2694.

Fràter-Schröder M., Müller G., Birchmeier W. and Böhlen P. (1986): Transforming growth factor-beta inhibits endothelial cell proliferation. Biochem. Biophys. Res. Commun. 137: 295-302.

Gajdusek C.M. and Carbon S. (1989): Injury-induced release of basic fibroblast growth factor from bovine aortic endothelium. J. Cell. Physiol. 139: 570-579.

Grøndahl-Hansen J., Kirkeby L.T., Ralfkiaer E., Kristensen P., Lund L.R. and Danø K. (1989): Urokinase-type plasminogen activator in endothelial cells during acute inflammation of the appendix. Am. J. Pathol. 135: 631-636.

Gudewicz P.W. and Gilboa N. (1987): Human urokinase-type plasminogen activator stimulates chemotaxis of human neutrophils. Biochem. Biophys. Res. Commun. 147: 1176-1181.

Heimark R.L., Twardzik D.R. and Schwartz S.M. (1986): Inhibition of endothelial regeneration by type-beta transforming growth factor from platelets. Science 233: 1078-1080.

Hilton D.J. (1992): LIF: lots of interesting functions. Trends Biochem. Sci. 17: 72-76.

Klagsbrun M. and D'Amore P:A. (1991): Regulators of angiogenesis. Ann. Rev. Physiol. 53: 217-239.

Nathan C. and Sporn M. (1991): Cytokines in context. J. Cell Biol. 113: 981-986.

Madri J.A., Pratt B.M. and Tucker A.M. (1988): Phenotypic modulation of endothelial cells by transforming growth factor-ß depends on the composition and organization of the extracellular matrix. J. Cell Biol. 106: 1357-1384.

McNeil P.L., Muthukrishnan L., Warder E. and D'Amore P.A. (1989): Growth factors are released by mechanically wounded endothelial cells. J. Cell Biol. 109: 811-822.

Merwin J.R., Anderson J.M., Kocher O., van Itallie C.M. and Madri J.A. (1990): Transforming growth factor beta1 modulates extracellular matrix organization and cell-cell junctional complex formation during in vitro angiogenesis. J. Cell. Physiol. 142: 117-128.

Mignatti P., Mazzieri R. and Rifkin D.B. (1991): Expression of urokinase receptor in vascular endothelial cells is stimulated by basic fibroblast growth factor. J. Cell Biol. 113: 1193-1202.

Mignatti P., Tsuboi R., Robbins E. and Rifkin D.B. (1989): In vitro angiogenesis on the human amniotic membrane: requirement for basic fibroblast growth factor-induced proteinases. J. Cell Biol. 108: 671-682.

Montesano R. and Orci L. (1985): Tumor-promoting phorbol esters induce angiogenesis in vitro. Cell 42: 469-477.

Montesano R., Pepper M.S., Vassalli J.-D. and Orci L. (1987): Phorbol ester induces cultured endothelial cells to invade a fibrin matrix in the presence of fibrinolytic inhibitors. J. Cell. Physiol. 132: 460-466.

Montesano R., Pepper M.S. and Orci L. (1993): Paracrine induction of angiogenesis in vitro by Swiss 3T3 fibroblasts. J. Cell Sci., in press.

Montesano R., Vassalli J.-D., Baird A., Guillemin R. and Orci L. (1986): Basic fibroblast growth factor induces angiogenesis in vitro. Proc. Natl. Acad. Sci. USA 83: 7297-7301.

Moscatelli D., Presta M. and Rifkin D.B. (1986): Purification of a factor from human placenta that stimulates capillary endothelial cell protease production, DNA synthesis and migration. Proc. Natl. Acad. Sci. USA 83: 2091-2095.

Moscatelli D. and Rifkin D.B. (1988): Membrane and matrix localization of proteases: a common theme in tumor invasion and angiogenesis. Biochim. Biophys. Acta 948: 67-85.

Müller G., Behrens J., Nussbaumer U., Böhlen P. and Birchmeier W. (1987): Inhibitory action of transforming growth factor ß on endothelial cells. Proc. Natl. Acad. Sci. USA 84: 5600-5604.

Muthukrishnan L., Warder E. and McNeil P.L. (1991): Basic fibroblast growth factor is efficiently released from a cytosolic storage site through plasma membrane disruptions of endothelial cells. J. Cell. Physiol. 148: 1-16.

Nusrat A.R. and Chapman H.A. (1991): An autocrine role for urokinase in phorbol ester-mediated differentiation of myeloid cell lines. J. Clin. Invest. 87: 1091-1097.

Odekon L.E., Sato Y. and Rifkin D.B. (1992): Urokinase-type plasminogen activator mediates basic fibroblast growth factor-induced bovine endothelial cell migration independent of its proteolytic activity. J. Cell. Physiol. 150: 258-263.

Pardenaud L., Yassine F., Dieterlen-Lièvre F. (1989): Relationship between vasculogenesis, angiogenesis and haemopoiesis during avian ontogeny. Development 105: 437-485.

Pepper M.S. (1992): Angiogenesis in vitro: interactions between angiogenesis-modulating cytokines and the role of balanced extracellular proteolysis. M.D. thesis, Faculty of Medicine, University of Geneva. 120p.

Pepper M.S., Belin D., Montesano R., Orci L. and Vassalli J.-D. (1990): Transforming growth factor beta 1 modulates basic fibroblast growth factor-induced proteolytic and angiogenic properties of endothelial cells in vitro. J. Cell Biol. 111: 743-755.

Pepper M.S., Ferrara N., Orci L. and Montesano R. (1991a): Vascular endothelial growth factor (VEGF) induces plasminogen activators and plasminogen activator inhibitor-1 in microvascular endothelial cells. Biochem. Biophys. Res. Commun. 181: 902-906.

Pepper M.S., Ferrara N., Orci L. and Montesano R. (1992a): Potent synergism between vascular endothelial growth factor and basic fibroblast growth factor in the induction of angiogenesis in vitro. Biochem. Biophys. Res. Commun. 189: 824-831.

Pepper M.S. and Montesano R. (1991): Proteolytic balance and capillary morphogenesis. Cell Diff. Dev. 32: 319-328.

Pepper M.S., Montesano R., Orci L. and Vassalli J.-D. (1991b): Plasminogen activator inhibitor-1 is induced in microvascular endothelial cells by a chondrocyte-derived transforming growth factor-beta. Biochem. Biophys. Res. Commun. 176: 633-638.

Pepper M.S., Sappino A.-P., Montesano R., Orci L. and Vassalli J.-D. (1992b): Plasminogen activator inhibitor-1 is induced in migrating endothelial cells. J. Cell. Physiol. 153: 129-139.

Pepper M.S., Sappino A.-P., Stöcklin R., Montesano R., Orci L. and Vassalli J.-D. (1993a): Upregulation of urokinase receptor expression on migrating endothelial cells. J. Cell Biol., in press.

Pepper M.S., Vassalli J.-D., Montesano R. and Orci L. (1987): Urokinase-type plasminogen activator is induced in migrating capillary endothelial cells. J. Cell Biol. 105, 2535-2541.

Pepper M.S., Vassalli J.-D., Montesano R. and Orci L. (1991c): Chondrocytes inhibit endothelial sprout formation in vitro: evidence for involvement of a transforming growth factor-beta. J. Cell. Physiol. 146: 170-179.

Pepper M.S., Vassalli J.-D., Orci L. and Montesano R. (1993b): Biphasic effect of transforming growth factor-ß1 on in vitro angiogenesis. Exp. Cell Res. 204: 356-363.

Phillips G.D., Whitehead R.A. and Knighton D.R. (1992): Inhibition by methylprednisolone acetate suggests an indirect mechanism for TGF-ß induced angiogenesis. Growth Factors 6: 77-84.

Rabbani S.A., Desjardins J., Bell A.W., Banville D., Mazar A., Henkin J., and Goltzman D. (1990): An amino-terminal fragment of urokinase isolated from a prostate cancer cell line (PC-3) is mitogenic for osteoblast-like cells. Biochem. Biophys. Res. Commun. 173: 1058-1064.

Rabbani S.A., Mazar A., Bernier S., Haq M., Bolivar I., Henkin J., and Goltzman D. (1992): Structural requirements for the growth factor activity of the amino-terminal domain of urokinase. J. Biol. Chem. 267: 14151-14156.

Risau W., Sariola H., Zerwes H.-G., Sasse J., Ekblom P., Kemler R. and Doetschman T. (1988): Vasculogenesis and angiogenesis in embryonic stem cell derived embryoid bodies. Development 102: 471-478.

Roberts A.B., Sporn M.B., Assoian R.K., Smith J.M., Roche N.S., Wakefield L.A., Heine U.I., Liotta L.A., Falanga V., Kehrl J.H. and Fauci A.S. (1986): Transforming growth factor ß: Rapid induction of fibrosis and angiogenesis in vivo and stimulation of collagen formation in vitro. Proc. Natl. Acad. Sci. USA 83: 4167-4171.

Saksela O., Moscatelli D. and Rifkin D.B. (1987): The opposing effects of basic fibroblast growth factor and transforming growth factor beta on the regulation of plasminogen activator in capillary endothelial cells. J. Cell Biol. 105: 957-963.

Sato Y. and Rifkin D.B. (1988): Autocrine activities of basic fibroblast growth factor: regulation of endothelial cell movement, plasminogen activator synthesis and DNA synthesis. J. Cell Biol. 107: 1199-1205.

Sato Y. and Rifkin D.B. (1989): Inhibition of endothelial cell movement by pericytes and smooth muscle cells: activation of a latent transforming growth factor-ß1-like molecule by plasmin during co-culture. J. Cell Biol. 109: 309-315.

Vassalli J.-D., Sappino A.-P. and Belin D. (1991): The plasminogen activator/plasmin system. J. Clin. Invest. 88: 1067-1072.

Wahl S.M., Hunt D.A., Wakefield L.M., McCartney-Francis N., Wahl L.M., Roberts A.B. and Sporn M.B. (1987): Transforming growth factor beta (TGF-beta) induces monocyte chemotaxis and growth factor production. Proc. Natl. Acad. Sci. USA 84: 5788-5792.

Weisman D.M., Polverini P.J., Kamp D.W. and Leibovich S.J. (1988): Transforming growth factor-beta (TGFß) is chemotactic for human monocytes and induces their expression of angiogenic activity. Biochem. Biophys. Res. Commun. 157: 793-800.

Williams R.L., Courtneidge S.A. and Wagner E.F. (1988): Enbryonic lethalities and endothelial tumors in chimeric mice expressing polyoma virus middle T oncogene. Cell 52: 121-131.

Yang E.Y. and Moses H.L. (1990): Transforming growth factor ß1-induced changes in cell migration, proliferation and angiogenesis in the chicken chorioallantoic membrane. J. Cell Biol. 111: 731-741.

ENDOTHELIAL PLASMINOGEN ACTIVATORS AND MATRIX METALLOPROTEINASES IN FIBRINOLYSIS AND LOCAL PROTEOLYSIS

Victor W.M. van Hinsbergh, Roeland Hanemaaijer and Pieter Koolwijk

Gaubius Laboratory IVVO-TNO
P.O. Box 430
2300 AK Leiden
The Netherlands

INTRODUCTION

Fibrin is a temporary matrix, which is formed after wounding of a blood vessel and when plasma leaks from blood vessels forming a fibrous exudate, often seen in areas of inflammation and in tumors.[1] The fibrin matrix acts as a barrier preventing further blood loss, and provides a scaffolding in which new microvessels can infiltrate during wound healing. A proper timing of the outgrowth of microvessels, angiogenesis, as well as the subsequent (partial) disappearance of these vessels is essential to ensure adequate wound healing and to prevent the formation of scar tissue. It is generally believed that plasminogen activators play an important role in the migration of leukocytes and endothelial cells, and in the dissolution of the fibrin matrix.[2,3] Plasminogen activators are serine proteases, which enzymatically convert the zymogen plasminogen into the active protease plasmin, the prime protease that degrades fibrin.

The production of plasminogen activators by endothelial cells not only contributes to the proteolytic events related to the formation of microvessels in a wound, but also plays a crucial role in the prevention of thrombosis. If fibrin becomes deposited within the lumen of a blood vessel, cessation of the blood flow may occur accompanied by ischemia and eventually death of the distal tissues. The endothelium contributes considerably to the maintenance of blood fluidity by exposing anticoagulant molecules, by providing factors that interfere with platelet aggregation, and by its ability to stimulate fibrinolysis. Lysis of intravascularly generated fibrin must occur rapidly. However, it should be limited to a local area, because a general elevation of fibrinolysis upon wounding would result in rebleeding. Hence, endothelial cells apply plasminogen activators for initial events in angiogenesis, for the degradation of the temporary fibrin matrix during wound healing, and for the immediate dissolution of a fibrinous thrombus originating within a blood vessel[4] (Figure 1). They are able to execute all these processes adequately by orchestrating in time and space the production of both types of plasminogen activators as well as that of specific inhibitors and cellular receptors for plasminogen activators and plasminogen.

Angiogenesis: Molecular Biology, Clinical Aspects
Edited by M.E. Maragoudakis *et al*, Plenum Press, New York 1994

171

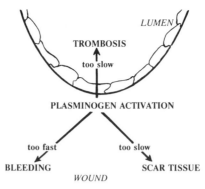

Figure 1. Plasminogen activation by the endothelium is controlled in time and space to meet various functions: prevention of thrombosis without bleeding and angiogenesis, which is necessary for adequate wound repair (from: Van Hinsbergh, *Ann. N. Y. Acad. Sci.* 667:151 (1992); with permission).

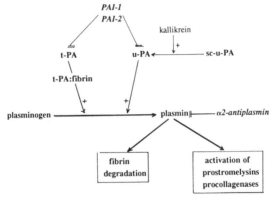

Figure 2. Schematic representation of the plasminogen activation system. +: activation; -: inhibition. Inhibitors are indicated in italics. PA: plasminogen activator; t-PA: tissue-type PA; u-PA: urokinase-type PA; sc-u-PA: single-chain u-PA; PAI: PA inhibitor.

COMPONENTS OF THE FIBRINOLYTIC SYSTEM

Figure 2 summarizes the proteases and the inhibitors involved in fibrinolysis. Fibrin degradation and probably also activation of several matrix metalloproteinases can be accomplished by the serine protease plasmin, which is formed from its zymogen plasminogen by plasminogen activators (PAs). Two types of plasminogen activators are presently known: tissue-type plasminogen activator (t-PA) and urokinase-type plasminogen activator (u-PA).[5,6] The genes of plasminogen, t-PA and u-PA have been characterized. All three proteins are synthesized as a single polypeptide chain, which is converted by proteolytic cleavage to a molecule with two polypeptide chains connected by a disulphide bond. The carboxy-terminal parts of the molecules (the so called B-chain) contain the proteolytically active site, whereas the amino-terminal parts of the molecule (the A-chain) are build up of domains that determine the interaction of the proteases with matrix proteins and cellular receptors. The cleavage of plasminogen and single-chain u-PA is necessary to disclose the proteolytic active site and to activate the molecule. In contrast, generation of t-PA activity does not depend on its conversion in a two-chain form, but on its interaction with a specific substrate, in particular fibrin. Once bound to this substrate, both the single-chain form and the two-chain form of t-PA are active.

The actual activity of the PAs is regulated not only by their concentration, but also by their interaction with PA inhibitors (PAIs),[7-9] cellular receptors[10,11] and, as indicated above, matrix proteins. Table 1 summarizes some properties of the proteins involved in plasminogen activation.

Table 1. Properties of protein involved in plasminogen activation.

Protein	MW (kD)	Number of amino acids	Plasma concentration (mg/l)	Produced by
Plasminogen	92	791[a]	200	hepatocytes
t-PA	68	530[a]	0.005	endothelium, mesothelium
scu-PA	54	411	0.008[b]	monocytes/mø, renal tubuli, activated endothelial cells
α2-Antiplasmin	70	452	70	hepatocytes
PAI-1	50	379	0.05	smooth muscle cells, activated endothelium, ?liver
PAI-2	46/56[c]	393	< 0.005	macrophages, placenta, mesothelial cells
u-PA Receptor	55	313/282-284[d]	-	present on monocytes-mø, endothelial cells and other cells

[a] The numbering of amino acid residues is usually based on a total of 790 residues for plasminogen (Lys-plasminogen) and of 527 residues for t-PA.
[b] Sum of scu-PA and u-PA: inhibitor complexes.
[c] 46 kD: intracellular non-glycosylated form, 60 kD excreted glycosyated form.
[d] 313: postulated transmembrane form; 282-284: PI-linked form, which has been processed on its COOH-terminal side.
Abbreviations: mø: macrophages; MW: molecular weight.

The three proteases of the fibrinolytic system are counterregulated by potent inhibitors, which are members of the serine protease inhibitor (serpin) superfamily. Plasmin is instantaneously inhibited by α2-antiplasmin,[12] but this reaction is attenuated when plasmin is bound to fibrin. The predominant regulators of t-PA and u-PA are PAI-1 and PAI-2.[8] PAI-1 is a 50 kD glycoprotein present in blood platelets and synthesized by endothelial cells, smooth muscle cells and many other cell types in culture.[9] PAI activity in human plasma is normally exclusively PAI-1. PAI-2 is produced by monocytes/macrophages and can be found as a glycosylated secreted molecule and as an intracellular molecule.[13]

Local direction of fibrinolytic activity can occur by cellular receptors. High affinity binding sites for plasminogen,[14,15] t-PA[16] and u-PA[11] are found on various types of cells including endothelial cells. A specific u-PA receptor has been identified and cloned, which binds both single-chain u-PA and two-chain u-PA via their growth factor domains.[17] The u-PA receptor is heavily glycosylated and proteolytically converted at its carboxy-terminal end; the receptor with the new carboxy-terminus is anchored in the membrane by a phosphatidyl group.[18] The nature of the t-PA receptor and the plasmin(ogen) receptor(s) on endothelial cells is less clear. The lipoprotein Lp(a), which has strong structural homology with a large part of the plasminogen molecule, can compete for plasminogen binding to endothelial cells.[15,19] In addition, clearance receptors exist on liver hepatocytes and liver endothelial cells, which can clear t-PA[20] and plasmin-α2-antiplasmin and PA:PAI-1 complexes[21,22] from the circulation.

REGULATION OF t-PA PRODUCTION: PREVENTION OF INTRAVASCULAR FIBRIN DEPOSITION

The fibrinolytic activity in blood is largely determined by the concentration of t-PA, which is synthesized in the endothelium.[23] The concentration of t-PA in the circulation can change rapidly. This is due to the short half life time of t-PA in the circulation, which is 5 to 10 minutes in man, and to the ability of endothelial cells to release rapidly a relatively large amount of t-PA. Clearance of t-PA occurs in the liver. Consequently, changes in the liver blood flow affect t-PA clearance and the plasma t-PA concentration. The acute release of t-PA from a storage pool in the vessel wall can be induced by vasoactive substances, such as bradykinin, platelet activating factor and thrombin.[24] This mechanism makes it possible to enhance the t-PA concentration exclusively at those places where fibrin generation occurs. Hence, it contributes to the local protection against an emerging thrombus. If a generalized stimulation of the endothelium occurs, for example by catecholamines, the acute release mechanism causes a rapid temporary increase in the blood t-PA concentration. The release of t-PA by endothelial cells also depends on the t-PA synthesis rate in the cells. The t-PA synthesis rate is different in various types of blood vessels, e.g. veins produce more t-PA than arteries. Furthermore, t-PA synthesis can be enhanced pharmacologically or physiologically by various mediators.

During recent years insight has been gained regarding the regulation of the synthesis of t-PA by using endothelial cells in vitro. Activation of protein kinase C has been implicated in the regulation of t-PA transcription and synthesis in human endothelial cells. The stimulation of t-PA production by histamine and thrombin is caused by this process. The induction of t-PA by protein kinase C activation is potentiated by a simultaneous increase of the cellular cAMP concentration.[25,26] It has been suggested that the proto-oncogenes c-fos and c-jun, which can form a heterodimer called AP1, are involved in the regulation of the t-PA gene in endothelial cells by interacting with one or more AP1-binding site(s) of the t-PA promoter.[26] In favour of this suggestion is an experiment of nature, in which a single mutation in a AP-1 binding site of the t-PA promoter (-TGACATCA-) alter this PMA-responsive element in a cAMP-responsive element (-TGACGTCA-).[27] In contrast to human cells, rat endothelial cells enhance their t-PA synthesis markedly upon stimulation of the cAMP generation alone, whereas rat t-PA synthesis does not respond to the sole addition of PMA.[28]

Other - pharmacologically interesting - agents that can enhance t-PA synthesis are retinoids[29,30] and benzodiazepines.[31] Retinoic acid and vitamin A have been demonstrated also to enhance t-PA synthesis in the rat, an elevation which is still present after a feeding period of six weeks.[32] As these components have little effect on the production of PAI-1, vitamin A derivatives are interesting candidates for pharmacological enhancement of t-PA synthesis.

INFLAMMATORY ACTIVATION OF THE ENDOTHELIUM: EFFECT ON PAI-1 AND u-PA SYNTHESIS

In vivo, changes of the plasma levels of t-PA and its main inhibitor PAI-1 often occur in the same direction. Nevertheless, the synthesis of PAI-1 is independently regulated from that of t-PA, albeit that certain mediators, such as thrombin, can induce both t-PA and PAI-1 synthesis. In a number of diseases predominantly PAI-1 is elevated in the blood (sepsis, maturity onset diabetes, postoperative thrombosis) and/or in tissues (arteriosclerosis, sepsis). Several factors involved in inflammatory and vascular diseases, such as the cytokines tumor necrosis factor-α (TNFα) and interleukin-1 (IL-1), endotoxin (LPS), transforming growth factor-β (TGF-β), oxidized lipoproteins and thrombin, can stimulate PAI-1 production by endothelial cells in vitro. Also in vivo, administration of TNFα, IL-1, LPS or thrombin causes an increase in circulating PAI-1. After infusion of LPS in animals, PAI-1 mRNA increased in vascularized tissues and PAI-1 mRNA was elevated in the endothelium of various organs.[33,34] Administration of LPS or TNFα to patients or healthy volunteers caused after about 2 hours a large increase in circulating PAI-1, which was preceded by a rapid and sustained increase in circulating t-PA.[35-37] The mechanism underlying the stimulation of t-PA synthesis in vivo by LPS or TNFα is still unresolved, and may be the indirect result of the LPS- or TNFα-infusion by the generation of another mediator. The large increase in PAI-1 induced two hours after TNFα- or LPS-administration far exceeds the production of t-PA.[35-37] This may result - after an initial raise in fibrinolytic activity - in a prolonged attenuation of the fibrinolysis process. It is generally believed that induction of PAI-1 by inflammatory mediators may contribute to the thrombotic complications in endotoxinemia and sepsis.

However, the effect of inflammatory mediators TNFα, IL-1 and LPS on plasminogen activation is probably more complex. Simultaneous with the increase in PAI-1, these inflammatory mediators induce the synthesis of u-PA in human endothelial cells.[38] Induction of u-PA by TNFα is associated by an increased degradation of matrix proteins.[39] The enhanced secretion of u-PA occurs entirely towards the basolateral side of the cell, whereas the secretion of t-PA and PAI-1 proceed equally to the luminal and basolateral sides of the cell.[38] The polar secretion of u-PA underlines the suggestion that u-PA may be involved in local processes causing the remodelling of the basal matrix of the cell. Therefore the increase of PAI-1 induced by inflammatory mediators may represent, in addition to a role in the modulation of fibrinolysis, a protective mechanism of the cell against uncontrolled u-PA action.

THE ROLE OF THE u-PA RECEPTOR

The site, where local u-PA activity occurs, is probably directed by the u-PA receptor, a GPI-anchored glycoprotein of about 45 kD.[11,18] Human endothelial cells in vitro contain about 40,000 u-PA receptors per cell. Upon secretion, single-chain u-PA binds via its growth factor domain to the receptor, and can subsequently be converted to two-chain u-PA, by which it becomes proteolytically active. As the endothelial cell contains also plasmin(ogen) receptors, an interplay between receptor-bound u-PA and receptor-bound plasmin is likely to happen. The then generated plasmin can degrade a number of matrix proteins. On the other hand, a direct plasmin-independent action of u-PA on matrix proteins may also occur, as has been reported by Quigley et al.[40] for u-PA-dependent degradation of avian fibronectin. In this respect it is of interest to note that in various cell types the u-PA receptor has been localized in the focal attachment sites, which host integrin-matrix interactions, and in cell-cell contact areas.[41] The time that is allowed for u-PA to act as a protease it probably rather short. Receptor-bound two-chain u-PA is also subjected to inhibition by PAI-1. The then formed u-PA:PAI-1 complex is internalized,[42] in contrast to receptor-bound free u-PA. This occurs probably after interaction with another receptor.[43] After internalization, the u-PA:PAI-1 complex is dissociated from the receptor and degraded in the lysosomes, whereas the u-PA receptor returns to the plasma membrane.

The number of u-PA receptors is enhanced by activation of protein kinase C and elevation of the cellular cAMP concentration,[4,44] as well as by several angiogenic growth factors, including b-FGF and VEGF (ref. 45; van Hinsbergh, unpublished data). Although b-FGF enhances u-PA production in bovine endothelial cells, it is unable to induce u-PA production in various types of human endothelial cells. Therefore, the simultaneous exposure of human endothelial cells to TNFα, which induces u-PA synthesis, and to b-FGF and VEGF, which enhance the expression of u-PA receptors, are needed for a maximal increase in local u-PA activity on human endothelial cells.

PAs IN LOCAL PROTEOLYSIS: PUTATIVE ROLE IN ANGIOGENESIS

Migrating and invading cells, such as monocytes and tumor cells, express u-PA activity bound to u-PA receptors on their cellular protrusions and on focal attachment sites, which suggest that PA activity is involved in cellular migration and invasion. Such a mechanism is also likely to be involved in endothelial cell migration and in the formation of new blood vessels (angiogenesis). Proteolysis of the basement membrane of endothelial cells is a prerequisite for these processes. A direct correlation between the expression of PA activity and the migration and formation of capillary sprouts by bovine microvascular endothelial cells was demonstrated in vitro by Pepper and Montesano.[46-48] The outgrowth of tubular structures was increased by b-FGF which increases in bovine endothelial cells u-PA activity, and counteracted by TGF-β, which enhances predominantly PAI-1 and inhibits PA activity. VEGF can also stimulate bovine adrenal microvascular endothelial cells to form tubular structures, and acts cooperatively with b-FGF in this induction.[49] Preliminary studies in our laboratory (Koolwijk et al., in preparation) indicate that no or a limited number of capillary sprouts grow from a monolayer of human endothelial cells into a fibrin matrix after exposure to b-FGF and VEGF. However, many tubular structures are obtained after the simultaneous exposure of the cell monolayers to TNFα, b-FGF and VEGF. Whether this marked proliferation of tubular structures is due to an enhanced local u-PA activity by the induction of u-PA (by TNFα) and u-PA receptors (by b-FGF and VEGF) remains to be elucidated. Alternative or additional mechanisms may be involved. However, in favour of the hypothesis that u-PA is involved in or associated with angiogenesis are in vivo observations, which demonstrate the expression of u-PA during neovascularization of ovarian follicles, the corpus luteum and the maternal decidua.[50]

PAs IN MATRIX REMODELLING: COOPERATION WITH MATRIX METALLOPROTEINASES

The regulation of proteolytic activation and activity is probably more complex than depicted above. It is not known whether interaction of u-PA with its receptor may affect cell metabolism. This may occur by a direct signal transduction into the cell or by an indirect signal generated by a proteolytic event in the focal attachment site, which is connected via a complex of proteins with the actin cytoskeleton of the cell. Furthermore, the local proteolytic activity is not limited to plasminogen activation. Endothelial cells in vitro can produce a number of matrix metalloproteinases (MMPs)[51,52] as summarized in Table 2. They are all secreted in a zymogen form and depend on zinc and calcium ions for

Table 2. Expression of matrix metalloproteinases in human endothelial cells in vitro.

Name	Size (kDa)	Other names (and EC No.)	Substrate specificity	Expression	
				basal	inducible[a]
Collagenases					
MMP-1	55	Interstitial collagenase Type I collagenase Fibroblast collagenase (EC 3.4.24.7)	collagen type I, II, III collagen type VII, X gelatins (?)	yes[b]	yes
Gelatinases					
MMP-2	72	Gelatinase-A 72-kDa gelatinase 72-kDa type IV collagenase (EC 3.4.24.24)	collagen type IV, V, VII collagen type X, XI gelatins, fibronectin elastin	yes	no
MMP-9	92	Gelatinase-B 92-kDa gelatinase 92-kDa type IV collagenase (EC 2.4.24.35)	collagen type IV, V, VII collagen type X, XI gelatins fibronectin	no	yes
Stromelysins					
MMP-3	57	Stromelysin-1 Transin Proteoglycanase (EC 3.4.24.17)	collagen type I, II, III, collagen type IV, IX, X, laminin, fibronectin gelatins, proteoglycans	yes[c]	yes
MMP-7	28	Matrilysin PUMP-1 (EC 3.4.24.23)	gelatins, fibronectin proteoglycans, elastin	no	no

(a) Induction was achieved by PMA or TNFα or a combination of these mediators.
(b) Not detectable in microvascular endothelial cells.
(c) Not detectable at mRNA level. Low protein levels detectable.

activity. Activation of these proteins can proceed by the action of proteases, of which plasmin and stromelysin (MMP-3) have received particular attention.[52] This results in the loss of a pro-peptide and the reduction of the molecular weight by about 8 to 10 kD. The MMPs are inhibited by tissue inhibitors of metalloproteinases, the so-called TIMPs.[53,54] Two mammalian TIMPs have been characterized: TIMP-1 and TIMP-2, the mRNAs of which are both found in human endothelial cells in vitro. Hence, MMP activities are regulated by (a) the activation of the zymogen, (b) the presence and interaction with TIMPs and (c) by the presence of binding and competing substrates. Data in our laboratory[55] have indicated that the synthesis of several MMPs, including stromelysin, is enhanced or induced by the inflammatory mediator TNFα, the same mediator that induce the synthesis of u-PA. This sets the scene for a coordinate expression of matrix remodelling proteases in endothelial cells in inflammation.

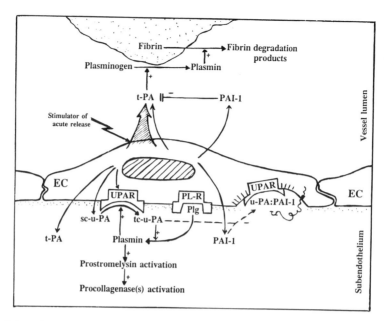

Figure 3. Schematic representation of postulated aspects of the involvement of endothelial cell plasminogen activators in fibrinolysis and local proteolysis. Abbreviations: PA: plasminogen activator; t-PA: tissue-type PA; u-PA urokinase-type PA; sc-u-PA: single-chain u-PA; tc-u-PA: two-chain u-PA; Plg: plasminogen; UPAR: u-PA receptor; PL-R: plasminogen receptor; PAI-1: PA inhibitor-1; EC: endothelial cell. +: stimulation; -: inhibition.

SUMMARY

The endothelial cell uses PAs for several functions (Figure 3). The endothelium is able to respond to emerging fibrin deposits in the blood stream by regulating the production of t-PA, the main fibrinolysis regulator in blood. In addition it can fine-tune fibrinolytic activity by the simultaneous production of PAI-1, so that re-bleeding of a wound is prevented. Furthermore, the endothelium uses PAs, in particular u-PA, for proteolytically changing its interaction with its underlying matrix and for remodelling of its basement membrane, processes which are necessary for cell migration and angiogenesis. This process is limited in space and time by interaction of u-PA with its specific receptor and by the presence of PAI-1. As the expression of u-PA and u-PA receptor are under the control of the leukocyte-derived cytokines TNFα and IL-1 and by angiogenic growth factors, respectively, leukocytes may play an important role in the control of angiogenesis. In addition the induction and activation of stromelysin and other matrix metalloproteinases by the monokine TNFα and the u-PA plasmin system further contribute to the complex regulation of the local proteolytic events associated with endothelial matrix remodelling.

REFERENCES

1. H.F. Dvorak, J.A. Nagy, B. Berse, L.F. Brown, K.-T. Yeo, T.-K. Yeo, A.M. Dvorak, L. Van de Water, T.M. Sioussat, and D.R. Senger, Vascular Permeability factor, fibrin, and the pathogenesis of tumor stroma formation, *Ann. N. Y. Acad. Sci.* 667:101 (1992).
2. H.C. Kwaan, Tissue fibrinolytic activity studied by a histochemical method, *Fed. Proc.* 25:52 (1966).

3. V.W.M. Van Hinsbergh, and P. Koolwijk, Production of Plasminogen Activators and Matrix Metalloproteinases by Endothelial Cells: Their Role in Fibrinolysis and Local Proteolysis, *in*: "Angiogenesis in Health and Disease," M.E. Maragoudakis, P. Gullino, and P.I. Lelkes, eds., NATO ASI Series A Volume 227, Plenum Press, New York (1992).

4. V.W.M. Van Hinsbergh, Impact of endothelial activation on fibrinolysis and local proteolysis in tissue repair, *Ann. N. Y. Acad. Sci.* 667:151 (1992).

5. F. Bachmann, Fibrinolysis, *in*: "Thrombosis and Haemostasis 1987," M. Verstraete, J. Vermylen, R. Lijnen, and J. Arnout, eds., Leuven University Press, Leuven (1987).

6. P. Wallén, Structure and Function of Tissue Plasminogen Activator and Urokinase, *in*: "Fundamental and Clinical Fibrinolysis," P.J. Castellino, P.J. Gaffney, M.M. Samama, and A. Takada, eds., Elsevier, Amsterdam (1987).

7. E.D. Sprengers, and C. Kluft, Plasminogen activator inhibitors, *Blood* 69:381 (1987).

8. E.K.O. Kruithof, Plasminogen activator inhibitor type 1: biochemical, biological and clinical aspects, *Fibrinolysis* 2, Suppl.2:59 (1988).

9. D.J. Loskutoff, Regulation of PAI-1 gene expression, *Fibrinolysis* 5:197 (1991).

10. L.A. Miles, and E.F. Plow, Plasminogen receptors: ubiquitous sites for cellular regulation of fibrinolysis, *Fibrinolysis* 2:61 (1988).

11. E.S. Barnathan, Characterization and regulation of the urokinase receptor on human endothelial cells, *Fibrinolysis* 6, Suppl.1:1 (1992).

12. W.E. Holmes, L. Nelles, H.R. Lijnen, and D. Collen, Primary structure of human α_2-antiplasmin, a serine protease inhibitor (Serpin), *J. Biol. Chem.* 262:1659 (1987).

13. A. Wohlwend, D. Belin, and J.-D. Vassalli, Plasminogen activator-specific inhibitors produced by human monocytes/macrophages, *J. Exp. Med.* 165:320 (1987).

14. L.A. Miles, E.G. Levin, J. Plescia, D. Collen, and E.F. Plow, Plasminogen receptors, urokinase receptors, and their modulation on human endothelial cells, *Blood* 72:628 (1988).

15. R.L. Nachman, Thrombosis and atherogenesis: molecular connections. *Blood* 79:1897 (1992).

16. K.A. Hajjar, The endothelial cell tissue plasminogen activator receptor. Specific interaction with plasminogen, *J. Biol. Chem.* 266:21962 (1991).

17. A.L. Roldan, M.V. Cubellis, M.T. Masucci, N. Behrendt, L.R. Lund, K. Danø, E. Appella, and F. Blasi, Cloning and expression of the receptor for human urokinase plasminogen activator, a central molecule in cell surface, plasmin dependent proteolysis. *EMBO J.* 9:467 (1990).

18. M. Plough, N. Behrendt, D. Løber, and K. Danø, Protein structure and membrane anchorage of the cellular receptor for urokinase-type plasminogen activator, *Semin. Thrombos. Hemostas.* 17:183 (1992).

19. L.A. Miles, G.M. Fless, E.G. Levin, A.M. Scanu, and E.F. Plow, A potential basis for the thrombotic risks associated with lipoprotein(a), *Nature* 339:301 (1989).

20. J. Kuiper, M. Otter, D.C. Rijken, and T.J.C. Van Berkel, Characterization of the interaction in vivo of tissue-type plasminogen activator with liver cells, *J. Biol. Chem.* 263:18220 (1988).

21. K. Orth, E.L. Madison, M.-J. Gething, J.F. Sambrook, and J. Herz, Complexes of tissue-type plasminogen activator and its serpin inhibitor plasminogen-activator inhibitor type 1 are internalized by means of the low density lipoprotein receptor-related protein/α_2-macroglobulin receptor, *Proc. Natl. Acad. Sci. U.S.A.* 89:7422 (1992).

22. G. Bu, S. Williams, D.K. Strickland, and A.L. Schwartz, Low density lipoprotein receptor-related protein/α_2-macroglobulin receptor is an hepatic receptor for tissue-type plasminogen activator, *Proc. Natl. Acad. Sci. U.S.A.* 89:7427 (1992).

23. T.-C. Wun, and A. Capuano, Spontaneous fibrinolysis in whole human plasma. Identification of tissue activator-related protein as the major plasminogen activator causing spontaneous activity in vitro, *J. Biol. Chem.* 260:5061 (1985).

24. J.J. Emeis, Regulation of the acute release of tissue-type plasminogen activator from the endothelium by coagulation activation products, *Ann. N. Y. Acad. Sci.* 667:249 (1992).

25. E.G. Levin, K.R. Marotti, and L. Santell, Protein kinase C and the stimulation of tissue plasminogen activator release from human endothelial cells. Dependence on the elevation of messenger RNA, *J. Biol. Chem.* 264:16030 (1989).

26. T. Kooistra, P.J. Bosma, K. Toet, L.H. Cohen, M. Griffioen, E. Van den Berg, L. Le Clercq, and V.W.M. Van Hinsbergh, Role of protein kinase C and cyclic adenosine monophosphate in the regulation of tissue-type plasminogen activator, plasminogen activator inhibitor-1, and platelet-derived growth factor mRNA levels in human endothelial cells. Possible involvement of proto-oncogenes c-jun and c-fos, *Arterioscler. Thrombos.* 11:1042 (1991).

27. P. Feng, M. Ohlsson, and T. Ny, The structure of the TATA-less rat tissue-type plasminogen activator gene. Species-specific sequence divergences in the promoter predict differences in regulation of gene expression, *J. Biol. Chem.* 265:2022 (1990).

28. J.J. Emeis, and T. Kooistra, Animal models and experimental procedures to study the synthesis and acute release of tissue-type plasminogen activator, *Fibrinolysis* 7, Suppl.1:31 (1993).

29. T. Kooistra, J.P. Opdenberg, K. Toet, H.F.J. Hendriks, R.M. Van den Hoogen, and J.J. Emeis, Stimulation of tissue-type plasminogen activators synthesis by retinoids in cultured human endothelial cells and rat tissues in vivo, *Thromb. Haemostas.* 65:565 (1991).

30. E.A. Thompson, L. Nelles, and D. Collen, Effect of retinoic acid on the synthesis of tissue-type plasminogen activator and plasminogen activator inhibitor-1 in human endothelial cells, *Eur. J. Biochem.* 201:627 (1991).

31. T. Kooistra, K. Toet, C. Kluft, P.F. Von Voigtlander, M.D. Ennis, J.W. Aiken, J.A. Boadt, and L.A. Erickson, Triazolobenzodiazepines: a new class of stimulators of tissue-type plasminogen activator synthesis in human endothelial cells, *Biochem. Pharmacol.* 45: in press (1993).

32. A.M. Van Bennekum, J.J. Emeis, T. Kooistra, and H.F.J. Hendriks, Modulation of tissue-type plasminogen activator by retinoids in rat plasma and tissues, *Am. J. Physiol.* 264:R931 (1993).

33. P.H.A. Quax, C.R. Van den Hoogen, J.H. Verheijen, T. Padró, R. Zeheb, T.D. Gelehrter, T.J.C. Van Berkel, J. Kuiper, and J.J. Emeis, Endotoxin induction of plasminogen activator in plasminogen activator inhibitor type 1 mRNA in rat tissues in vivo, *J. Biol. Chem.* 265:15560 (1990).

34. M. Keeton, Y. Eguchi, M. Swadey, C. Ahn, and D. Loskutoff, Cellular localization of type 1 plasminogen activator inhibitor messenger RNA and protein in murine renal tissue, *Am. J. Pathol.* 142:59 (1993).

35. A.F. Suffredini, P.C. Harpel, and J.E. Parrillo, Promotion and subsequent inhibition of plasminogen activation after administration of intravenous endotoxin to normal subjects, *N. Engl. J. Med.* 320:1165 (1989)

36. V.W.M. Van Hinsbergh, K.A. Bauer, T. Kooistra, C. Kluft, G. Dooijewaard, M.L. Sherman, and W. Nieuwenhuizen, Progress of fibrinolysis during tumor necrosis factor infusion in humans. Concomitant increase of tissue-type plasminogen activator, plasminogen activator inhibitor type-1, and fibrin(ogen) degradation products, *Blood* 76:2284 (1990).

37. S.J.H. Van Deventer, H.R. Büller, J.W. Ten Cate, L.A. Aarden, E. Hack, and A. Sturk, Experimental endotoxemia in humans: analysis of cytokine release and coagulation, fibrinolytic, and complement pathways, *Blood* 76:2520 (1990).

38. V.W.M. Van Hinsbergh, E.A. Van den Berg, W. Fiers, and G. Dooijewaard, Tumor necrosis factor induces the production urokinase-type plasminogen activator by human endothelial cells, *Blood* 75:1991 (1990).

39. M.J. Niedbala, and M. Stein Picarella, Tumor necrosis factor induction of endothelial cell urokinase-type plasminogen activator mediated proteolysis of extracellular matrix and its antagonism by γ-interferon, *Blood* 79:678 (1992).

40. J.P. Quigley, L.I. Gold, R. Schwimmer, and L.M. Sullivan, Limited cleavage of cellular fibronectin by plasminogen activator purified from transformed cells, *Proc. Natl. Acad. Sci. U.S.A.* 84:2776 (1987).

41. J. Pöllänen, K. Hedman, L.S. Nielsen, K. Danø, and A. Vaheri, Ultrastructural localization of plasma membrane-associated urokinase-type plasminogen activator at focal contacts, *J. Cell Biol.* 106:87 (1988).

42. D. Olson, J. Pöllänen, G. Høyer-Hansen, E. Rønne, K. Sakaguchi, T.-C. Wun, E. Appella, K. Danø, and F. Blasi, Internalization of the urokinase-plasminogen activator inhibitor type-1 complex is mediated by the urokinase receptor, *J. Biol. Chem.* 267:9129 (1992).

43. A. Nykjær, C.M. Petersen, B. Møller, P.H. Jensen, S.K. Moestrup, T.L. Holtet, M. Etzerodt, H.C. Thøgersen, M. Munch, P.A. Andreasen, and J. Gliemann, Purified α_2-macroglobulin receptor/LDL receptor-related protein binds urokinase•plasminogen activator inhibitor type-1 complex. Evidence that the α_2-macroglobulin receptor mediates cellular degradation of urokinase receptor-bound complexes, *J. Biol. Chem.* 267:14543 (1992).

44. D.J. Langer, A. Kuo, K. Kariko, M. Ahuja, B.D. Klugherz, K.M. Ivanics, J.A. Hoxie, W.V. Williams, B.T. Liang, D.B. Cines, and E.S. Barnathan, Regulation of the endothelial cell urokinase-type plasminogen activator receptor. Evidence for cyclic AMP-dependent and protein kinase C-dependent pathways, *Circ. Res.* 72:330 (1993).

45. P. Mignatti, R. Mazzieri, and D.B. Rifkin, Expression of the urokinase receptor in vascular endothelial cells is stimulated by basic fibroblast growth factor, *J. Cell Biol.* 113:1193 (1991).

180

46. M.S. Pepper, J.-D. Vassalli, R. Montesano, and L. Orci, Urokinase-type plasminogen activator is induced in migrating capillary endothelial cells, *J. Cell Biol.* 105:2535 (1987).

47. M.S. Pepper, D. Belin, R. Montesano, L. Orci, and J.-D. Vassalli, Transforming growth factor-beta 1 modulates basic fibroblast growth factor-induced proteolytic and angiogenic properties of endothelial cells in vitro, *J. Cell Biol.* 111:743 (1990).

48. R. Montesano, Regulation of angiogenesis in vitro. *Eur. J. Clin. Invest.* 22:504 (1992).

49. M.S. Pepper, N. Ferrara, L. Orci, and R. Montesano, Potent synergism between vascular endothelial growth factor and basic fibroblast growth factor in the induction of angiogenesis in vitro, *Biochem. Biophys. Res. Commun.* 189:824 (1992).

50. E. Bacharach, A. Itin, and E. Keshet, In vivo patterns of expression of urokinase and its inhibitor PAI-1 suggest a concerted role in regulating physiological angiogenesis, *Proc. Natl. Acad. Sci. U.S.A.* 89:10686 (1992).

51. L.M. Matrisian, The matrix-degrading metalloproteinases, *BioEssays* 14:455 (1992).

52. G. Murphy, S. Atkinson, R. Ward, J. Gavrilovic, and J.J. Reynolds, The role of plasminogen activators in the regulation of connective tissue metalloproteinases, *Ann. N. Y. Acad. Sci.* 667:1 (1992).

53. L.A. Liotta, P.S. Steeg, and W.G. Stetler-Stevenson, Cancer metastasis and angiogenesis: an imbalance of positive and negative regulation, *Cell* 64:327 (1991).

54. A.J.P. Docherty, and G. Murphy, The tissue metalloproteinase family and the inhibitor TIMP: a study using cDNAs and recombinant proteins, *Ann. Rheumatic Diseases* 49:469 (1990).

55. R. Hanemaaijer, P. Koolwijk, L. Le Clercq, W.J.A. De Vree, and V.W.M. Van Hinsbergh, Regulation of matrix-degrading metalloproteinases (MMPs) expression in human vein and microvascular endothelial cells. Effects of TNFα, IL-1 and phorbol ester, *Biochem. J.* submitted for publication.

TUMOR ANTIANGIOGENESIS:WHICH WAY?

Pietro M. Gullino

Department of Biomedical Sciences
University of Torino
Via Santena, 7
10126 Torino, Italy

The presence of neoplastic cells in a normal tissue elicits from the host vascular network a large number of microvessels that sustain tumor growth. Angiogenic capacity is peculiar but not restricted to neoplastic cells;[1,2] it appears in the course of neoplastic transformation before the cell population can form a tumor upon transplantation;[3,4] it suggests an increased risk of neoplastic transformation when acquired by a cell population normally not angiogenic.[5,6] Present knowledge of tumor angiogenesis sustains the belief of many investigators that antiangiogenesis treatment of a tumor bearing host might arrest or effectively impair growth of primary as well as metastatic neoplasia. Fig. 1 depicts one observation that supports this belief. In this model a very rapid growth consistently occurs as soon as the vessels colonize the neoplastic cell population, thus, it seems reasonable to expect growth arrest or impairment if angiogenesis is blocked. What guide does present knowledge of angiogenesis offer in searching for an effective antiangiogenesis treatment? An effort to address this question is made here in an environment most suitable for speculation.

The simplest suggestion derived from the observation reported in fig. 1 is that neoplastic cells produce "angiogenesis factors". Indeed, most neoplastic tissues were found to be angiogenic in several model systems and neoplastic cells released angiogenic molecules in the fluid surrounding them in vivo as well as in vitro. The search for "angiogenesis factors" has produced so far a relatively long list of molecules all with angiogenic capacity but so different for chemical characteristics and biological properties as to justify the assumption that they act like "triggers" of a multifactorial event. If so, two facts have to be kept in mind: (a) under physiological conditions angiogenesis is practically absent in the adult organism (the reconstruction of the uterine mucosa and the corpus luteus formation are in fact wound-repair events and (b) angiogenesis occurs in a number of pathological conditions as different as diabetic retinopathy, psoriasis, rheumatoid arthritis, wound healing, tumor growth, etc. Thus, either angiogenesis is caused by one "factor"

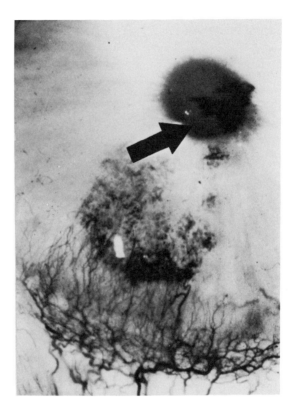

Fig. 1. Melanosarcoma cells were injected in the centre of a rabbit cornea (arrow). As long as the distance between neoplastic cells and limbal vessels at the periphery of the cornea was more than 2 mm, the cornea remained avascular, transparent and without morphologic signs of alterations. As soon as the neoplastic cells migrated at a distance of 2 mm or less from the limbus, a large number of microvessels originated from it within a few days. Angiogenesis was present in all tumors we tested and appeared a few days after neoplastic cells were located at an appropriate distance (< 2 mm) from the microvessels of the limbus. Rapid tumor growth occurred as soon as the cell population was colonized by the newly-formed vessels.

peculiar to each of these conditions or the liberation of a common factor, stored as inactive molecule, occurs in each of these processes. Examples sustaining both possibilities are found in the literature.

The first possibility is illustrated by work of Bouck, Polverini and coworkers.[7,8,9]

BHK21/clone 13 cells are neoplastically transformed and their culture media is angiogenic. When fused with normal human fibroblasts, these cells lose the neoplastic phenotype and their culture media loses angiogenic capacity. Revertants from the hamster-human hybrids resume the neoplastic phenotype and their culture medium becomes angiogenic again. The accepted interpretation is that fusion adds to the BHK cells an active cancer suppressor gene and angiogenesis inhibition is one of the effects of its presence. Indeed, a polypeptide similar to thrombospondin, a protein present in platelets and matrix, was found in conditioned media of the BHK cells that lost the neoplastic phenotype after fusion. Both molecules inhibited neovascularization in vivo and migration of endothelium in vitro. Moreover, when microvascular endothelium in culture started to produce "cords" i.e. to mimic angiogenesis, secretion of thrombospondin was reduced, and increment of cord formation was observed if anti-thrombospondin IgG was added to endothelial cell cultures to limit thrombospondin availability during the cord-forming period.[10] Moreover, thrombospondin is known to limit endothelial cell proliferation.[11] Thus, arrest of tumor angiogenesis could require blocking the effect of a gene product closely associated to the genomic alteration sustaining the neoplastic phenotype.[12,13]

The possibility that one or a few "factors" stored in the adult tissue are actually involved in the angiogenic response is suggested by the work of Vlodavsky and coworkers.[14,15,16] Basic fibroblastic growth factor (bFGF) is an angiogenic polypeptide, practically ubiquitous in adult tissues, normally stored in inactive form bound to the heparans of the stroma and released by heparanases or heparitinases. Indeed, vascular endothelium suffers structural and functional alterations in the absence of bFGF,[17,18] which has been shown to accumulate into the sub-endothelial matrix. Moreover, lymphoma cells as well as platelets, neutrophils and mast cells express heparanase activity suggesting that FGF release can occur in a variety of conditions.[19] The angiogenesis-angiostasis equilibrium could therefore be modified by any event determining the release of angiogenic factors. If so, the angiogenic response should be preceeded by a change in the compositions of the tissue to be colonized by the newly formed vessels.

To test this possibility we induced angiogenesis in the rabbit cornea, normally avascular, utilizing two triggers, prostaglandin E1 (PGE1) and bFGF. Before any vessel was formed we observed in the cornea an increment in copper and sialic acid concentrations. In particular, the ganglioside content doubled while the ratio GM3:GD3 gangliosides was sharply enhanced. Experimental reproduction of local increment in ganglioside concentration showed that doses of angiogenic molecules too small to induce neovascularization became angiogenic when the cornea was enriched with gangliosides GD3 or GM1. On the contrary, doses of bFGF or PGE1 sufficient to induce angiogenesis were prevented to do so by local enrichment of the corneal tissue with GM3 ganglioside.[20,21,22] Since at the doses used the gangliosides were not angiogenic, we concluded that they could modulate angiogenesis induced by PGE1 or bFGF, both molecules normally present in adult tissues where angiogenesis is absent. The modulation depended on total concentration as well as relative proportion among the gangliosides normally present in the rabbit cornea. Thus the equilibrium between angiogenesis and angiostasis could be modified by a change in the microenvironment due to non-angiogenic molecules.

The formation of a vessel requires mobilization and growth of endothelium and the production of a basement membrane. Impairment of any of these events could arrest neovascularization. If the hypothesis is accepted, an obvious approach to block angiogenesis is the utilization of molecules that alter endothelial cell behaviour. To this end several avenues are being followed. On the assumption that fibroblastic growth factors are indispensable for endothelial cell growth, a variety of molecules have been tested to alter FGF bioactivity: Chimeric toxins composed of acidic fibroblastic growth factor fused to mutant forms of pseudomonas exotoxin able to enzymatically inhibit protein synthesis,[23] interleukin 1, α and β,[24] recombinant human platelet factor 4[25] or analogs thereof,[26] gold compounds[27,28] high oxygen tension,[29] transforming growth factor β[30], a 16k-fragment of prolactin,[31] were all described to have an antiangiogenic effect often secondary to an "anti'FGF" activity, mostly due to interference with the receptor. This approach is also comforted by the observation that removal of FGF from the culture medium of human umbilical vein endothelium results in cell death by apoptosis.[18]

Interference with endothelial cell proliferation as an approach to block angiogenesis, was pursued through other routes besides the anti-FGF action. Limitation or block of endothelial cell proliferation were ascribed to products of the lipoxygenase pathway,[32] to D-penicillamine in the presence of copper sulfate,[33] to depletion of polyamines,[34] and, more recently, to a fraction of conditioned media of adherent macrophage cultures.[35]

Experiments analyzing the possible antiangiogenic effect of molecules able to impair endothelial cell motility have also been reported. A polymeric peptide based on the Arg-Gly-Asp (RGD) core-sequence of fibronectin when coinjected with tumor cells or given i.v. on day 1,2,3 of the tumor innoculation, significantly reduced tumor neovascularization. The inhibitory effect was dose-dependent and in vitro Poly - (RGD) prevented haptotactic migration.[36] Protamine, known for its capacity to bind heparin, was able to block migration of capillary endothelium stimulated by mast cells or heparin. Indeed, protamine prevented neovascularization in chronic inflammation of the cornea, in the developing embryo and in tumors.[37]

Following the hypothesis that antiangiogenesis could be achieved by impairing vascular morphogenesis, treatments were devised to alter the extracellular matrix, basement membrane in particular. Promising results were obtained using so called angiostatic steroids, tetrahydrocortisol in particular. Improvement on the steroids ability to alter basement membrane turnover was obtained in the presence of heparin or heparin-like molecules[38,39,40,41] and with combined treatment with a sulfated polysaccharide-pepidoglycan isolated from a culture of an Arthrobacter species.[42] The damage to the basement membrane appears dependent upon disruption of laminin sub-units synthesis by the endothelium.[43] This observation is corroborated by the inhibition of angiogenesis and tumor growth obtained with a synthetic laminin peptide.[44] As expected, molecules that interfere with collagen metabolism like proline derivatives or β-aminopropionitrile, can enhance the antiangiogenic action of steroids.[45,46]

During the antiangiogenic response the role of heparin, heparin-like molecules and polysaccharides of the stroma in general, appears to be very important. Heparin injected i.v. (1000 u/rat) concentrated in endothelium[47] and induced increment of FGF-like molecules in plasma.[48] Heparin is incorporated into the complex fibronectin + collagen in a saturable manner and the combination heparin + fibronectin have strong chemotactic effect on microvascular endothelium.[49] Thus, heparin is involved in all 3 major events related to angiogenesis: proliferation, motility and matrix dependence of endothelium, however, the mechanism of its action is obscure.

Microbial products with angiostatic effects have been described and particular attention has been given to a naturally secreted antibiotic of Aspergillus fumigatus fresenius.[50] Somatostatin analogues have been reported to possess inhibitory effects on angiogenesis in the chick chorioallantoic membrane[51] and the antitumor action of interferons has been in part ascribed to suppression of angiogenesis[52] possibly related to the ability of interferon-y secreted by the tumor to induce expression of histocompatibility complex class II antigens on tumor endothelium. This may be a promising approach for an antibody directed to targeting of tumor microvessels.[53] These observations, however, are still in embryonal stage.

In conclusion, the angiogenesis-angiostasis equilibrium can be considered either a result achieved by "factors" produced ad hoc during a variety of biological processes or the consequence of a balance among molecules normally present in the microenvironment but shifting in concentration according to different conditions of a tissue. In the first case induction of angiostasis requires knowledge of these "factors" and capacity of producing them and their counterparts in pharmacologically useful preparations. In the second case one should identify the pertinent molecules of the microenvironment able to influence the equilibrium. Central position is occupied by the "responsiveness" of the microvascular endothelium to these molecules and the tools available to assess it. Since formation of a new vessel is a morphogenetic event, in vitro evaluation of endothelial cell growth, motility and matrix penetration as well as "cord" formation are very useful parameters but not identifiable with the in vivo angiogenesis or angiostasis. Angiogenesis in a solid tumor occurs in an organoid structure with irregular and erratic blood supply sustained by a relatively small fraction of cardiac output. This has to be kept in mind when molecules with short half-life are utilized in the attempt to modify the tumor microenvironment. Finally, angiostasis is experimentally identified as absence of angiogenesis when expected to occur under angiogenic stimulation. The angiogenesis trigger as well as the "locus" of action are obviously important components of the experimental system particularly because induction of angiogenesis, or lack of it, is different in the chicken chorioallantoic membrane, in the rabbit cornea or in the subcutaneous sponge. In particular, if tumor growth is impaired by a drug and a cytotoxic effect of this drug is not directly demonstrable, it is wise not to conclude that the growth limitation is "probably due to an antiangiogenic effect" of this drug.

REFERENCES

1. P.M. Gullino, Angiogenesis factors, in: "Handbook of Experimental Pharmacology," R. Baserga, Springer Verlag, New York, 57:427 (1981).
2. J. Folkman, Tumor angiogenesis, Adv. Cancer Res. 19:331(1974) and 32:1975(1985).
3. M. Ziche and P.M. Gullino, Angiogenesis and prediction of sarcoma formation, Cancer Res. 41:5060(1981).
4. M. Ziche and P.M. Gullino, Angiogenesis and neoplastic progression in vitro, J. Nat. Cancer Inst. 69:483(1982).
5. S.S. Brem, P.M. Gullino, and D. Medina, Angiogenesis: a marker for neoplastic transformation of mammary papillary hyperplasia, Science 195:880(1977).
6. S.S. Brem, Jensen H, and P.M. Gullino, Angiogenesis as marker of preneoplastic lesions of the human breast, Cancer 41:239(1978).
7. F.Rastinejad, P.J. Polverini, and N.P. Bouck, Regulation of the activity of a new inhibitor of angiogenesis by a cancer suppressor gene, Cell 56:345(1989).
8. D.J. Good, P.J. Polverini, F. Rastinejad, M.M. Le-Beau, R.S Lemons, W.A. Frazier, and N.P. Bouck, A tumor suppressor-dependent inhibitor of angiogenesis is immunologically and functionally indistinguishable from a fragment of thrombospondin, Proc. Nat. Acad. Sci. USA 87:6624(1990).
9. N.P. Bouck, Tumor angiogenesis: The role of oncogenes and tumor suppressors genes, Cancer Cells 2:179(1990) Cold Spring Harbor Laboratory Press ISSN.
10. M.L. Iruela-Arispa, P. Bornstein, and H. Sage, Thrombospondin exerts an antiangiogenic effect on cord formation by endothelial cells in vitro, Proc. Nat. Acad. Sci. USA 88:5026(1991).
11. P. Bagavandoss and J.W. Wilks, Specific inhibition of endothelial cells proliferation by thrombospondin, Biochem. Biophys. Res. Commun. 170:867(1990).
12. T. Ichikawa, Y. Ichikawa, and J. Isaacs, Genetic factors and

suppression of metastatic ability of prostatic cancer, Cancer Res. 51:3788(1991).

13. T.Ichikawa, Y. Ichikawa, J. Dong, A.L. Hawkins, C.A. Griffin, W.B. Isaacs, M. Oshimura, J.C. Barret, and J.T.Isaacs, Localization of metastasis suppressor gene(s) for prostatic cancer to the short arm of human chromosome 11, Cancer Res. 52:3486(1992).

14. I. Vlodavsky,J. Folkman, R. Sullivan, R. Fridman, R. Ishai-Michaeli, J. Sasse J, and M. Klagsbrun, Endothelial cell-derived basic fibroblastic growth factor:Synthesis and deposition into subendothelial extracellular matrix, Proc. Nat. Acad. Sci. USA 84:2292(1987).

15. I. Vlodavsky, R. Fridman, R. Sullivan, J. Sasse, and M. Klagsbrun, Aortic endothelial cells synthesize basic fibroblastic growth factor which remains cell associated and platelet-derived growth factor-like protein which is secreted, J. Cell. Physiol. 131:402(1987).

16. I. Vlodavsky, Z.Fues, R. Ishai-Michaeli, P. Bashkin, E.Levi, G. Korner, R. Bar-Shavit, and M. Klagsbrun, Extracellular matrix-resident basic fibroblastic growth factor: Implication for the control of angiogenesis, J. Cell. Biochem. 45:167(1991).

17. I. Vlodavsky, L.K. Johnson, G. Greenburg, and D. Gospodarowicz, Vascular endothelial cells maintained in the absence of fibro-blastic growth factor undergo structural alterations that are incompatible with their in vivo differentiated properties, J. Cell. Biol. 83:468(1979).

18. S. Araki, Y. Shimada, K. Kaji, and H. Hayashi, Apoptosis of vascular endothelial cells by fibroblast growth factor deprivation, Biochem. Biophys. Res. Comm. 168:1194(1990).

19. R. Ishai-Michaeli, A. Eldor, and I. Vlodavsky I, Heparanase activity expressed by platelets, neutrophils and lymphoma cells releases active fibroblastic growth factor from extracellular matrix, Cell Regulation 1:833(1990).

20. M. Ziche, G. Alessandri, and P.M. Gullino, Gangliosides promote the angiogenic response, Lab. Invest. 61:629(1989).

21. G. Alessandri, G. De Cristan, M. Ziche, A.P.M. Cappa, and P.M. Gullino, Growth and motility of microvascular endothelium are modulated by the relative concentration of gangliosides in the medium, J. Cell. Physiol. 151:23(1992).

22. M.Ziche, L. Morbidelli, G. Alessandri, and P.M. Gullino, Angiogenesis can be stimulated or repressed in vivo by a change in the GM3:GD3 ganglioside ratio, Lab. Invest. 67:711(1992).

23. J.R. Merwin, M.J. Lynch, J.A. Madri, I. Pastan, and C.B. Siegall, Acidic fibroblastic growth factor - Pseudomonas exotoxin chimeric protein elicits antiangiogenic effects on endothelial cells, Cancer Res. 52:4995(1992).

24. K. Norioka, M. Hara, A. Kitani, T. Hirose, W. Hirose, M. Harigai, K. Suzuki, M. Kawakam, H. Tabata, M. Kawagoe, and H. Nakamura, Inhibitory effect of human recombinant interleukin 1α and β on growth of human vascular endothelial cells, Biochem. Biophys. Res. Comm. 145:969(1987).

25. T.E. Maione, G.S. Gray, S. Petro, A.J. Hunt, A.L. Donner, S.I. Bauer,

H.F. Carson, and R.J. Sharpe, Inhibition of angiogenesis by recombinant human platelet factor 4 and related peptides, Science 247(1990).

26. T.E. Maione, G.S. Gray, A.J. Hunt, and R.J. Sharpe, Inhibition of tumor growth in mice by an analogue of platelet factor 4 that lacks affinity for heparin and retains potent angiostatic activity, Cancer Res. 51:2077(1991).

27. T. Matsubara, and M. Ziff, Inhibition of human endothelial cell proliferation by gold compounds, J. Clin. Invest. 79:1440(1987).

28. A.E. Koch, M. Cho, J. Burrows, S.J. Leibovich, and P.J. Polverini, Inhibition of production of macrophage-derived angiogenic activity by the anti-rheumatic agents gold-sodium thiomalate and auranofin, Biochem. Biophys. Res. Comm. 154:205(1988).

29. M.M. Grant, H.C. Koo, W. Rosenfeld, Oxygen affects human endothelial cell proliferation by inactivation of fibroblast growth factors, Am. J. Physiol. 263:L370(1987).

30. G. Muller, J. Beherens, U. Nussbaumer, P. Bohlen, and W. Bircmeier, Inhibitory action of transforming growth factor β on endo-thelial cells, Proc. Nat. Acad. Sci. USA 84:5600(1987).

31. N. Ferrara, L. Clapp, and R. Weiner, The 16K fragment of prolactin specifically inhibits basal or fibroblastic growth factor stimulated growth of capillary endothelial cells, Endocrinology 129:986(1991).

32. B.N. Yamaja-Setty, R.L. Dubowy, and M.J. Stuart, Endothelial cell proliferation may be mediated via the production of endogenous lipo-oxygenase metabolites, Biochem. Biophys. Res. Comm. 144:345(1987).

33. T. Matsubara, R. Saura, K. Hirohata, and M. Ziff, Inhibition of human endothelial cell proliferation in vitro and neovascularization in vivo by D-penicillamine, J. Clin. Invest. 83:158(1989).

34. M. Takigawa, M. Enomoto, Y. Nishida, H.O. Pan, A. Kinoshita, and F. Suzuki, Tumor angiogenesis and polyamines: α difluoromethyl-ornithine, an irreversible inhibitor of ornithine dicarboxylase, inhibits B16 melanoma-induced angiogenesis in ovo and the proliferation of vascular endothelial cells in vitro, Cancer Res. 50:4131(1990).

35. G.E. Besner, and M. Klagsbrun, Macrophages secrete a heparin-binding inhibitor of endothelial cell growth, Microvascular Res. 42:187(1991).

36. I. Saiki, J. Murata, T. Makabe, N. Nishi, S. Tokura, and I. Azuma, Inhibition of tumor angiogenesis by a synthetic cell-adhesive polypeptide containing the Arg-Gly-Asp (RGD) sequence of fibro-nectin poly (RGD) Jpn. J. Cancer Res. 81:668(1990).

37. S. Taylor, and J. Folkman, Protamine is an inhibitor of angiogenesis, Nature 297:307(1982).

38. J. Folkman, R. Langer, Linhardt, C. Haudenschild, and S. Taylor, Angiogenesis inhibition and tumor regression caused by heparin or a heparin fragment in the presence of cortisone, Science 221:719(1983).

39. R. Crum, S. Szabo, and J. Folkman, A new class of steroids inhibits angiogenesis in the presence of heparin or a heparin fragment, Science 230:1375(1985).

40. J. Folkman J, and D.E. Ingberg, Angiostatic steroids, Amer. Surg. 206:374(1987).
41. D.E. Ingberg, and J. Folkman, Inhibition of angiogenesis through modulation of collagen metabolism, Lab. Invest. 59:44(1988).
42. N.G. Tanaka, N. Sakamoto, K. Inoue, H. Korenaga, S. Kadoya, H. Ogawa, and Y. Osada, Antitumor effect of an antiangiogenic poly-saccharide from an arthrobacter species with and without a steroid, Cancer Res. 49:6727(1989).
43. Y. Tokida, Aratani, A. Morita, and Y. Kitagawa, Production of two variant laminin forms by endothelial cells and shift in their relative levels by angiostatic steroids, J. Biol. Chem. 265:18123(1990).
44. N. Sakamoto, M. Iwahana, N.G. Tanaka, and Y. Osada, Inhibition of angiogenesis and tumor growth by a synthetic laminin peptide CDPGYIGSR-NH$_2$, Cancer Res. 51:903(1991).
45. M.E. Maragoudakis, M. Sarmonika, and M. Panoutsacopoulou, Inhibition of basement membrane biosynthesis prevents angiogenesis, J. Pharmacol. Exp. Therapeutics 244:729(1988).
46. W.M. Leuko, L. Liotta, M.S. Wicha, Vonderhaar, and W.R. Kidwell, Sensitivity of N-Nitrosomethyl urea-induced rat mammary tumors to cis-hydroxypraline, an inhibitor of collagen production, Cancer Res. 41:2855(1981).
47. L.M. Hiebert, and L.B. Jaques, Heparin concentration in endothelium, Thrombosis Res 8:195(1976).
48. P.A. D'Amore, Heparin-endothelial cell interaction, Haemostasis 20 suppl. 1:159(1990).
49. S. Ungari, K.S. Raju, G. Alessandri, and P.M. Gullino, Cooperation between fibronectin and heparin in the mobilization of capillary endothelium, Invasion and metastasis 5:193(1985).
50. D. Ingberg, T. Fujita, S. Kishimoto, K. Sudo, T.Kanamaru, H. Brem, and J. Folkman, Synthetic analogues of fumagillin that inhibit angiogenesis and suppress tumor growth, Nature 348:555(1990).
51. E.A. Woltering, R. Barrie, T.M. O'Dorisio, D. Arce, T. Ure, A. Cramer, D. Holmes, J. Robertson, and J. Fassler, Somatostatin analogues inhibit angiogenesis in the chick chorioallantoic membrane, J. Surg. Res. 50:245(1991).
52. Y.A. Sidky, and E.C. Borden, Inhibition of angiogenesis by interferons: effects on tumor and lymphocyte-induced vascular responses, Cancer Res. 47:5155(1987).
53. F.J. Burrows, Y. Watanabe, and P.E. Thorpe, A murine model for antibody-directed targeting of vascular endothelial cells in solid tumors, Cancer Res. 52:5954(1992).

THERAPEUTIC ANGIOGENESIS IN SURGERY AND ONCOLOGY

Michael Höckel, Karlheinz Schlenger, Renate Frischmann-Berger, Sabine Berger and Peter Vaupel[1]

Department of Obstetrics and Gynecology and [1]Institute of Physiology and Pathophysiology
University of Mainz Medical Center
6500 Mainz, Germany

INTRODUCTION

The aim of this presentation is to demonstrate the pathological importance of microenvironmental tissue hypoxia and to elucidate a general treatment concept for this situation which we have termed *therapeutic angiogenesis*[1]. Hypoxia not only represents an insufficient oxygen supply for the cells of a given tissue area but is also regarded as an indicator for their metabolic deprivation and the concomitant accumulation of waste products. Therapeutic angiogenesis applied either with clinically established methods or using novel ways, which are the objectives of laboratory research and clinical trials at present, or in so far hypothetical forms, should lead to an expansion of the functional microvascular space resulting in an increased nutritive blood flow. Thus microregional oxygen availability should be elevated and directly counteract local tissue hypoxia. The problems of nutritional deprivation and waste product accumulation are also treated by therapeutic angiogenesis.

MICROENVIRONMENTAL TISSUE HYPOXIA

For an analytical approach, a tissue compartment is further subdivided into micro-regions which are composed of several cells having direct contact with each other and which are surrounded by a common microenvironment (Fig.1). The microenvironment is determined by the extracellular matrix - which may be in a dynamic state - and a complex flow equilibrium of chemical and physical factors including metabolic substrates, biologic response modifiers, interstitial hydrostatic pressure, pH, pO_2.

The microenvironmental pO_2 is the result of O_2 availability and cellular O_2 consumption. In the case of homeostasis, the microenvironmental pO_2 represents normoxia irrespective of its measured value. Normoxic pO_2 values can be low in certain tissues (e.g. parous uterine cervix, $pO_2 < 5$ mm Hg) and much higher in other tissues (e.g. subcutaneous fat, $pO_2 \sim 60$ mm Hg).

The occurrence of microenvironmental hypoxia despite normal systemic oxygenation parameters is caused by (i) vascular pathology, (ii) any form of tissue damage including surgical wounding, (iii) solid tumor growth.

Angiogenesis: Molecular Biology, Clinical Aspects
Edited by M.E. Maragoudakis *et al*, Plenum Press, New York 1994

The consequences of hypoxia for *normal tissue* are compiled in Fig. 2. Whereas the switch to an anaerobic metabolic state and the induction of angiogenesis are compensatory mechanisms probably without adverse effects for the tissue function, the further cellular events in the case of severe and prolonged hypoxia finally result in tissue damage. In wounded normal tissue, zones of hypoxia always exist[2-4]. A variety of conditions may exaggerate wound hypoxia (which will be discussed later in this article) leading to impaired and prolonged wound healing and tissue regeneration. It has been clearly shown that hypoxia increases the susceptibility for local infections which is also important for the surgical wound[5,6].

In a *solid tumor,* normoxia and hypoxia cannot be defined on the basis of tissue homeostasis. Nevertheless, a great number of investigations have confirmed correlations between microregional pO_2 gradients and proliferation status and viability of tumor cells

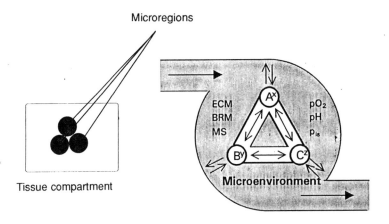

Figure 1. Schematic representation of the cellular microenvironment. Cells in different functional states (A^x, B^y, C^z) interact directly with each other and with a common microenvironment. ECM indicates extracellular matrix; BRM, biologic response modifiers; MS, metabolic substrates; P_{is}, interstitial pressure.

within solid malignancies. Further evidence of hypoxia-induced effects on tumor cells has been gathered from in vitro experiments at reduced oxygen tensions and with clamped rodent tumors. Although the growth rate of a solid tumor is reduced under hypoxic conditions, the overall effect from a clinical standpoint is a dramatic increase of its aggressiveness (Fig.2). Most important is the elevated metastatic potential and increased resistance towards conventional radiotherapy which may render a neoplastic disease incurable[7-19]. In a prospective clinical trial of locally advanced cancer of the uterine cervix we have recently found tumor oxygenation to be the most powerful predictor of treatment outcome and survival in this tumor entity which is treated with standard radiation[20,21].

Because of these considerations we have theoretical reservations against any adjuvant angiostatic treatment in primary solid tumors which are prone to increase "tumor hypoxia".

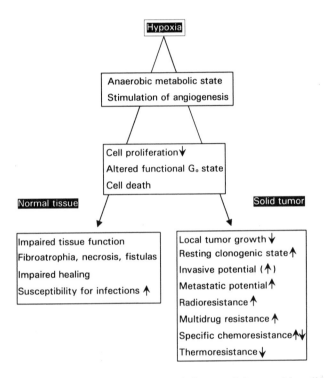

Figure 2. Effects of microenvironmental hypoxia in normal tissue and in solid tumors.

MODES OF THERAPEUTIC ANGIOGENESIS

The goal of therapeutic angiogenesis is the controlled induction or stimulation of new blood vessel formation in order to reduce hypoxia-related unfavorable tissue effects or to enhance tissue repair in normal tissues, and to sensitize a (residual) tumor for salvage radiotherapy or chemotherapy. Therapeutic angiogenesis can be achieved by surgical methods, i.e. transposition or transfer of autologous tissues with uncompromized vasculature and high angiogenic potential in close proximity to the site of the desired neovascularization (e.g. omentum majus flaps, musculo/fascio/cutaneous flaps, detubularized seromuscular bowel flaps).

Experimental and clinical experience demonstrate the initiation of a strong neovascularization reaction providing microvascular connections between the transferred tissue and the recipient bed at day 2 to 3. At days 7 to 10 these vascular connections usually suffice for complete perfusion of the flap and ensure its survival after ligation of the nourishing pedicle. In the case of a hypoxic bed, new blood vessels originating from the flap will continue to grow into the recipient region until a "normal" vascularization is ultimately obtained[22-27].

Therapeutic angiogenesis by use of autologous tissue transfer is most effective if the flap has an axial macrovascular pattern, a healthy microvascularization and releases factors capable of stimulating angiogenesis. These prerequisites are best fulfilled with omentum majus flaps or muscle flaps with one or two dominant pedicles such as the rectus abdominis muscle flap[28-32].

These classical surgical ways of therapeutic angiogenesis might be supplemented in near future by two further strategies: (i) pharmaceutical methods, i.e. the local application of angiogenesis factors, (ii) cell biological methods, i.e. the implantation of autologous capillary endothelial cells cultured ex vivo. Although successful in situ tissue formation by use of cultured autologous cells has been achieved with keratinocytes[33] and urothelial cells[34], the feasibility of therapeutic angiogenesis with endothelial cell implants is speculative at present and needs experimental confirmation. However, considerable experimental results as well as some preliminary clinical data exist to support the usefulness of angiogenic factors for therapeutic angiogenesis.

THERAPEUTIC ANGIOGENESIS BY USE OF ANGIOGENESIS FACTORS

A great number of substances have been identified as showing positive results in one or several angiogenesis assay systems. These substances differ widely in origin, molecular weight and chemical composition as well as biological activity. Most of the factors are at present only partly characterized in these respects. The majority of the angiogenesis factors of higher molecular weight (for review see[35]) belong to the three substance classes which are potentially involved in the regulation of the process of angiogenesis in vivo, i.e. (i) growth factors/cytokines, (ii) cell surface and extracellular matrix components, and (iii) proteases. However, a definitive physiological role in angiogenesis control has not been established for any of these substances. The function of the large group of low molecular weight angiogenesis factors (for review see[36]) is mostly uncertain, some may act as chemotactic signals or enzymatic cofactors. Another principle to classify angiogenesis factors is their direct (receptor-mediated) action on endothelial cells. Many potent angiogenesis-inducing substances appear to act indirectly by modifying the extracellular matrix or by affecting other cells. Independently from their potential physiological role and their action in the assay systems, angiogenesis factors suitable for therapeutic angiogenesis should fulfill the following criteria:

(i) induction or stimulation of controlled neovascularization in mature vascularized tissues,
(ii) defined tissue reactions associated with angiogenesis, neglectable local and systemic side effects,
(iii) effective doses in the nano/picomolar range and dose-effect relationship,
(iv) chemically defined substances, easy to handle,
(v) large scale availability.

Table 1: Characterization of cytokines potentially applicable for therapeutic angiogenesis. (From: Höckel et al.[1] with permission).

Cytokine	Chemistry	Availability for use in humans	Angiogenic dose (assay)	Tissues tested	Associated tissue reactions
Acidic Fibroblast Growth Factor (aFGF)	Polypeptide 16kd	yes	1 μg (chick chorioallantoic membrane)	Subcutis Peritoneum Peripheral nerve	Granulation tissue Nerve regeneration
Basic Fibroblast Growth Factor (bFGF)	Polypeptide 18kd	yes	10 - 100 ng (rabbit cornea)	Dermis Gastric/duodenal submucosa Brain Bone	Inflammation (?) Granulation tissue
Epidermal Growth Factor (EGF)	Polypeptide 6kd	yes	10 ng - 1 μg (chick chorioallantoic membrane)	Dermis	Epidermal proliferation Granulation tissue
Transforming Growth Factor Alpha (TGFα)	Polypeptide 5,5kd	yes	0,3 - 1 μg (hamster cheek pouch)	Dermis	Epidermal proliferation
Transforming Growth Factor Beta (TGFß$_1$)	Homodimeric Polypeptide 25kd	yes	50 ng (rabbit cornea)	Subcutis	Inflammation Granulation tissue Epidermal metaplasia
Tumor Necrosis Factor (TNFα)	Polypeptide 17kd	yes	5 ng - 5 μg (rat cornea)	Dermis Uterus	Inflammation Granulation tissue Fibrosis Necrosis
Vascular Endothelial Growth Factor (VEGF) Vascular Permeability Factor (VPF)	Glycosylated dimeric polypeptide 46 - 48kd	Potential	1 μg (rabbit cornea) 20 - 200 ng (rat cornea)	Bone	Inflammation (?)
Platelet-derived Endothelial Cell Growth Factor (PD-ECGF)	Polypeptide 45kd	Potential	50 ng (chick chorioallantoic membrane)	Subcutis	Not known
Angiogenin	Polypeptide 14kd	Potential	50 - 500 ng (rabbit cornea)	Cartilage	Not known
Angiotropin	Polyribonuc-1leopolypep-tide (4,5)$_n$kd	No	0,5 ng (rabbit cornea)	Dermis	Inflammation Granulation tissue Epidermal proliferation Hair growth

From our present knowledge, angiogenic growth factors/cytokines available by recombinant biotechnology are most appropriate with regard to these prerequisites. A selection of polypeptide angiogenesis factors are listed and characterized in table 1.

Systemic side effects should not be of concern in the case of local application of angiogenic cytokines. However, local overstimulation of angiogenesis might lead to hemorrhagic necrosis. Neovascularization stimulated in the vasa vasorum of the vessel wall may facilitate thrombus formation. Angiogenic factors which induce inflammatory angiogenesis like angiotropin, TNF-alpha or TGF-beta may also cause local tissue injury through the action of proteolytic enzymes and/or peroxides. Unwanted functional effects could occur as a consequence of local edema formation. Local pain may possibly arise. Pharmacologically stimulated angiogenesis is likely to enhance tumor growth if the angiogenic substances are administered at the tumor bed without simultaneous antitumor treatment. Topical application of TGF-beta to partial thickness wounds in pigs caused abnormal epithelial differentiation and decreased epithelial volume concomitant with increased angiogenesis and dermal regeneration[37]. As with all recombinant biomolecules, angiogenic cytokines may induce antibody formation in individuals.

Most important for the clinical concept of therapeutic angiogenesis with cytokines is the development of *therapeutic systems* for their defined local delivery. Experimental or clinical data concerning local delivery of polypeptide growth factors/cytokines are scarce. Up to now cream formulations, hydrogels, multilamellar liposomes and biodegradable sponges made from natural and synthetic polymers have been empirically used[38-44]. A promising approach to the controlled delivery of angiogenic factors could be realized with microencapsulation. As matrix materials a variety of biocompatible and biodegradable polymers can be used, either synthetic (e.g. polylactic-co-glycolic acid) or natural ones (e.g. albumin, collagen[45-48]). Microparticles releasing their drug at a controlled rate would allow different angiogenic factors to be combined giving optimal local concentrations. The microspheres could be administered in several different ways such as powder, suspension or in a tissue adhesive allowing maximum flexibility in their application under surgical conditions.

THERAPEUTIC ANGIOGENESIS IN SURGERY

Therapeutic angiogenesis is applied in the surgical field as *treatment* and as *adjuvant* for a variety of indications, of which the treatment of *chronically ischemic tissues, chronically infected tissues* and *non-healing wounds* are probably the most important ones. There is a broad overlap between these three entities because tissue hypoxia leads to susceptibility to infections and healing is impaired by both hypoxia and infection. If occlusion of the small blood vessels and reduced angiogenic potential as a consequence of late radiation damage, diabetes mellitus, connective tissue diseases, burns, smoking are the underlying reasons for tissue necrosis, chronic infection and non-healing, surgical therapeutic angiogenesis offers the best chance for successful treatment. By use of flaps from the omentum majus, detubularized seromuscular bowel or adjuvant muscles such as the gracilis or rectus abdominis muscle, radiation-induced soft tissue necrosis, osteonecrosis and fistulas can be often repaired[49].

However, even in the absence of vascular occlusive disease chronically infected tissues, such as resistent osteomyelitis, perineal sinuses following proctocolectomy, pressure sores, ulcers due to venous insufficiency, infected vascular allografts, should be treated with surgical therapeutic angiogenesis delivered with various flap techniques since microenvironmental hypoxia is the common problem in these situations[50,51].

Although we have successfully treated a radiation-induced skin ulcer with topical TNF-alpha, it is not clear at present if angiogenic hyperstimulation by use of angiogenesis factors can overcome the impaired angiogenic potential in cases of microvascular disease. Optimistic results are reported from the first clinical studies employing angiogenic factors topically on non-healing skin wounds[52,53]. The most clinical experience has been gathered with the so-called "Platelet-Derived Wound Healing Formula (PDWHF)", a blend of platelet-derived growth factors containing platelet factor 4, TGF-beta, PDGF, platelet-derived angiogenesis factor and platelet-derived epidermal growth factor. This growth factor cocktail is an autologous product from the patient's own blood. Its impressive effect in the treatment of chronic non-healing cutaneous wounds has been demonstrated by Knighton et al.[54]. It must be stressed that the clinical benefit of PDWHF may be in part due to the stimulation of other aspects of wound healing/tissue regeneration besides angiogenesis which may represent either direct effects on other target cells (fibroblasts, epithelial cells, mononuclear cells) or indirect effects induced by the angiogenesis reactions.

Healing of experimental gastric ulcers in the rat is promoted by intragastric administration of EGF at least partly through its angiogenic action[55]. Recently, Folkman et al. showed that an acid-stable form of basic FGF administered orally to rats with duodenal ulcer promoted a ninefold increase of angiogenesis in the ulcer bed and accelerated ulcer healing more potently than cimetidine[56]. Clinical studies using angiogenesis factors in gastric ulcer patients have been initiated[57].

Many more clinical situations characterized by local tissue ischemia might in principle be treated by surgical and pharmaceutical therapeutic angiogenesis. Examples are aseptic bone necrosis and heart, brain, or limb ischemia due to (partial) occlusion of smaller arteries[22-27]. By the creation of anastomoses between an extracardiac artery and the coronary circulation the prerequisites for myocardial revascularization with angiogenesis factors in ischemia of the heart have been recently achieved in a dog model[58].

Whereas the concept of therapeutic angiogenesis by use of surgical methods is at present more adequate for the treatment of sequelae of local tissue hypoxia, pharmaceutical therapeutic angiogenesis shows promise for the prevention of hypoxia-related problems in certain surgical situations, i.e.

- tissue transfer and grafting,
- suboptimal healing conditions,
- surgical high risk wounds,
- repair with alloplastic materials.

Moreover, angiogenesis factors applied in adjuvant fashion may accelerate the repair process in wounds with long healing time.

All *tissue transfers* using either free (vessel anastomosed) or pedicled flaps are necessarily associated with a reduction of the functional vasculature which brings about anatomical restrictions and the constant threat of flap necrosis[59]. In animal experiments, Höckel and Burke[60] could prevent skin flap necrosis by treating the donor and receptor sites with angiotropin, an angiogenic substance isolated from porcine monocytes. The concept of local blood vessel activation and hypervascularization of donor and/or receptor sites before transplantation surgery might substitute for delayed flap procedures and may extend the vascular territories[61,62].

Since the revascularization process is critical for the survival and subsequent function of free autologous *tissue grafts* (e.g. skin, nipple, muscle, bone, nerve transplants), these surgical procedures may also benefit from therapeutic angiogenesis. Eppley et al.[63]

showed in a rabbit study that free bone grafting to an irradiated bed was significantly more successful if the recipient sites were pretreated with basic FGF prior to placement of the graft. The fate of free autologous nerve grafts is determined by the time course and extent of revascularization[64]. In a rat experiment, transplanted superior cervical ganglia survived better when pretreated with basic FGF and nerve growth factor. The density of capillaries in these grafts was significantly higher compared to the controls[65]. Rat peripheral nerve regeneration could be greatly ameliorated when acidic FGF was added to a collagen-filled nerve guide by an obvious increase in the vascular supply[66]. Basic FGF given intraventricularly promoted cerebral angiogenesis in a rat model of mild chronic forebrain ischemia[67].

Wound healing and/or tissue regeneration can be *suboptimal* due to a variety of etiologic factors such as diabetes mellitus, connective tissue diseases, chronic venous insufficiency, cachexia, paraneoplastic syndrome, jaundice, alcoholism, uremia, smoking, previous radiation therapy, cytotoxic therapy, steroids, contamination and infection, allografts, burns. Several studies, using either chamber wound healing models or animals treated with the cytostatic agent doxorubicine to delay healing of incisions, have shown that impaired wound healing is reversed to near normal levels with angiogenic cytokines. In such studies, the effective topical preparations have included TGF-beta, TNF-alpha, or combinations of TGF-beta with EGF and PDGF[68-70]. EGF can restore healing in a wound chamber model in the presence of methyl prednisolone[71].

It has been shown that the quality of wound healing (i.e. tensile strength resistance) is associated with early vascularization[72]. Angiogenesis factors may therefore be applied as novel adjuvant means in *high risk wounds* such as digestive tract and urinary tract anastomoses or in uterotomies of cesarean sections to avoid complications of wound breaking which might be life-threatening. In a rat uterotomy model we found a significant early increase in uterine bursting pressure compared to the controls when TNF-alpha was incorporated into the wound before closure[73].

Angiogenic cytokines have been successfully used in vascular prosthesis in order to speed up their endothelial lining[74,75]. Angiogenesis factors might support the *repair with alloplastic materials* by improving the tissue fixation of a variety of artificial supporting or functional items like slings, patches, joint prothesis, heart valves. Moreover, the risk of infection might be reduced. However, this has not yet been proven experimentally.

Hard tissue (i.e. bone cartilage, tendon) *healing* and regeneration as well as *secondary healing* of large soft tissue defects produces considerable morbidity due to the long immobilization and hospitalization. In addition, treatment costs are high in these cases. The results of recent animal studies showed that healing in general could be accelerated by use of exogenous growth and/or angiogenic factors[37,76,77]. Angiogenesis is tightly linked to cartilage mineralization and bone formation[78]. Indeed, bone repair could be significantly increased by means of an omental angiogenic lipid fraction in the rat[29]. An 80% overall increase in bone density was obtained compared to the controls in this model and this increase was associated with a twofold rise in regional blood perfusion. Thus therapeutic angiogenesis might also become an interesting perspective for bone healing/regeneration in traumatology and orthopedics. Osseous defects, high-risk fractures, bone grafting, and arthrodeses may benefit from local implantation of angiogenesis factors. Certain ocular tissues heal slowly since they have a poor blood supply.

Angiogenic cytokines seem to be promising for the treatment of ocular wounds in supporting ocular surgery[79]. The effect of EGF on deep corneal wound healing has been proven in animal studies[80] and is currently undergoing investigation in clinical trials as well.

EGF has also been successfully applied to partial thickness skin defects and second degree burns in animal and clinical studies[42,81]. Epidermal regeneration could be enhanced by this angiogenic recombinant human growth factor. The biological mechanisms responsible for the results have not yet been established. Although a direct mitogenic effect of EGF on keratinocytes and fibroblasts is suggested, stimulation of angiogenesis could also have contributed to the accelerated epidermal regeneration as more granulation tissue subjacent to the epidermis has been observed as well[82]. A dose-dependent enhancement of the dermis regeneration has been achieved with angiotropin in artificial skin matrices implanted into rabbit skin defects[60].

THERAPEUTIC ANGIOGENESIS IN ONCOLOGY

We have outlined in the introductory sections that a great amount of experimental and clinical evidence supports the view that "tumor hypoxia" may reduce the chance of a solid neoplasm to be cured. Although therapeutic angiogenesis alone accelerates local tumor growth, its integration in multimodality cancer treatment concepts appears to be a worthwhile consideration.

Theoretically, therapeutic angiogenesis might be applied to sensitize low pO_2 tumors for chemotherapy and radiotherapy. Indeed, two multimodality treatment approaches in gynecologic oncology involve the promotion of neovascularization which may contribute to their clinical success (i) surgical debulking followed by chemotherapy for stage III ovarian carcinoma and (ii) the CORT procedure for pelvic sidewall recurrences of gynecologic malignancies.

The intraabdominal tumor spread in *FIGO stage III ovarian cancer* renders complete surgical resection impossible. The large tumor masses usually do not show long-term responses if treated by chemotherapy alone. However, the combination of *surgical debulking* - optimally to a microscopic residual tumor - and subsequent platinum based chemotherapy reaches long term complete remission and 5-year survival rates of approximately 20 %[83].

The proposed chemosensitizing effect of surgical debulking is multifactorial including the reduction of the total number of viable tumor cells and the intratumoral interstitial pressure. The proportion of the remaining cells within the proliferative stage is increased[84,85].

The surgical trauma produced by the debulking process stimulates the angiogenesis in the tumor bed and the tumor stroma which may lead to better oxygenation of the remaining tumor cells[86].

The *CORT procedure* (Combined Operative and Radiotherapeutic Treatment) has been developed to treat recurrent gynecologic malignancies infiltrating the pelvic wall[87,88]. Patients with pelvic wall recurrences have an extremely poor prognosis: with conventional treatment, median survival is less than 10 months.

The surgical part of CORT consists of (i) staging laparotomy/ lymphadenectomy, (ii) maximum tumor resection at the pelvic wall and exenteration of infiltrated central pelvic organs, (iii) implantation of guide tubes on the residual tumor/tumor bed at the pelvic wall, (iv) pelvic wall plasty with muscle, musculocutaneous and omentum flaps, (v) operative reconstruction of bowel, bladder and perineo-vulvo-vaginal functions. Radiation is performed as postoperative interstitial high dose rate brachytherapy through the implanted tubes. Patients without prior pelvic irradiation receive in addition preoperative whole pelvis teletherapy.

Therapeutic angiogenesis is accomplished in the tumor bed at the pelvic wall containing residual tumor cells through the process of subtotal tumor resection and the overlay of the tissue flaps transferred from a primarily non-irradiated region. Postoperative MRI investigations have shown an up to fourfold increase in local blood perfusion at the infiltrated pelvic wall as a consequence of the pelvic wall plasty. Increased oxygen availability is thought to sensitize the residual tumor cells to the tube-guided radiation. In addition, late tissue reactions of the highly irradiated pelvic wall which are at least in part due to microvascular rarefaction should be reduced through the angiogenic potential of the far less irradiated pelvic wall plasty.

CORT has been evaluated in a prospective trial at the University of Mainz. Within a 4 year period 32 patients have been treated. Twenty-five patients have been irradiated as part of the previous therapy with a median total midpelvic dose of 65 Gy (range 40-100 Gy). In 23 patients, local tumor control has been achieved. After a median follow-up period of 22 months (range 6-46 months) Kaplan-Meier life table analysis revealed a 3 year survival probability of 60 % (recurrence-free 55 %). Therapeutic angiogenesis may have contributed to the attainment of these encouraging results.

SUMMARY AND CONCLUSIONS

Microenvironmental hypoxia in normal tissue may result in the loss of its functional and structural integrity ultimately leading to fibroatrophia and necrosis. Prolonged hypoxia in wounds impairs healing and increases susceptibility to infections. "Hypoxia" in solid tumors can exaggerate their aggressiveness and resistance towards nonsurgical treatment modalities such as radiation and chemotherapy.

We have proposed the term therapeutic angiogenesis to elucidate a general treatment concept which should reduce or prevent unfavorable sequelae caused by local tissue hypoxia. Therapeutic angiogenesis can be achieved by surgical methods, i.e. the transfer (or transposition) of autologous tissues with uncompromised vasculature and high angiogenic potential, such as omentum majus flaps and muscle flaps. Recent advances in the understanding of the biological process of neovascularization and the availability of biosubstances inducing or stimulating angiogenesis may add a new clinical tool: therapeutic angiogenesis by pharmaceutical and cell biological methods.

The efficacy of surgical therapeutic angiogenesis in the treatment of chronically ischemic tissues, chronically infected tissues and non-healing wounds is clinically established. A great deal of evidence from animal experiments and preliminary clinical trials show that pharmaceutical therapeutic angiogenesis might be effective in certain surgical situations, such as tissue transfer and grafting, suboptimal healing conditions, high-risk wounds, repair with alloplastic materials, to prevent hypoxia-related problems. Moreover, angiogenesis factors applied in an adjuvant fashion may accelerate the repair process in wounds with long healing time.

However, further pharmacological characterization of the angiogenic biomolecules and the development of specific therapeutic systems for the programmed local delivery are mandatory before pharmaceutical therapeutic angiogenesis may get entry into the clinical practice.

Although therapeutic angiogenesis alone accelerates local tumor growth, its integration in multimodality cancer treatment concepts with respect to sensitizing "hypoxic" tumors for radiotherapy and chemotherapy appears to be clinically successful. This is demonstrated in gynecologic oncology with surgical debulking followed by chemotherapy for stage III ovarian carcinoma and in the CORT procedure for isolated pelvic side wall recurrences.

ACKNOWLEDGEMENTS

The work on therapeutic angiogenesis and tumor hypoxnia is supported by grants from

LTS Lohmann and from Deutsche Krebshilfe (M40/91 Val). The authors thank Dr. Debra Bickes-Kelleher for her valuable suggestions in preparing the manuscript.

REFERENCES

1. Höckel, M., Schlenger, K., Doctrow, S., Kissel, T., and Vaupel, P. Therapeutic angiogenesis. *Arch. Surg., 128:* 423-429, 1993.

2. Jonsson, K., Jensen, J.A., Goodson, W.H., and Hunt, T.K. Wound healing in subcutaneous tissue of surgical patients in relation to oxygen availability. *Surg. Forum, 37:* 86-88, 1986.

3. Hunt, T.K., and Pai, M.P. The effect of varying ambient oxygen tensions on wound metabolism and collagen synthesis. *Surg. Gynecol. Obstet., 135:* 561-567, 1972.

4. Pai, M.P., and Hunt, T.K. Effect of varying oxygen tensions on healing of open wounds. *Surg. Gynecol. Obstet., 135:* 756-758, 1972.

5. Jonsson, K., Hunt, T.K., and Mathes, S.J. Effect of environmental oxygen on bacterial induced tissue necrosis in flaps . *Surg. Forum, 35:* 589-591, 1984.

6. Knighton, D.R., Halliday, B., and Hunt, T.K. Oxygen as an antibiotic: The effect of inspired oxygen on infection. *Arch. Surg., 119:* 199-204, 1984.

7. Brown, J.M. Tumor hypoxia, drug resistance, and metastases. *J. Natl. Cancer Inst., 82:* 338-339, 1990.

8. Coleman, C.N. Hypoxia in tumors: a paradigm for the approach to biochemical and physiologic heterogeneity. *J. Natl. Cancer Inst., 80:* 310-317, 1988.

9. Moulder, J.E., and Rockwell, S. Hypoxic fractions of solid tumors: experimental techniques, methods of analysis, and a survey of existing data. *Int. J. Radiat. Oncol. Biol. Phys., 10:* 695-712, 1984.

10. Powers, W.E., and Tolmach, L.J. A multicomponent X-ray survival curve for mouse lymphosarcoma cells irradiated in vivo. *Nature, 197:* 710-711, 1963.

11. Rice, G.C., Hoy, C., and Schimke, R.T. Transient hypoxia enhances the frequency of dihydrofolate reductase gene amplification in chinese hamster ovary cells. *Proc. Natl. Acad. Sci. USA, 83:* 5978-5982, 1986.

12. Teicher, B.A., Holden, S.A., Al-Achi, A., and Herman, T.S. Classification of antineoplastic treatments by their differential toxicity toward putative oxygenated and hypoxic tumor subpopulations in vivo in the FSaIIC murine fibrosarcoma. *Cancer Res., 50:* 3339-3344, 1990.

13. Gerweck, L.E., Nygaard, T.G., and Burlett, M. Response of cells to hyperthermia under acute and chronic hypoxic conditions. *Cancer Res., 39:* 966-972, 1979.

14. Overgaard, J. Effect of hyperthermia on the hypoxic fraction in an experimental mammary carcinoma in vivo. *Br. J. Radiol., 54:* 245-249, 1981.

15. Chapman, J.D. The detection and measurement of hypoxic cells in solid tumors. *Cancer, 54:* 2441-2449, 1984.

16. Gray, L.H., Conger, A.D., Ebert, M., Hornsey, S., and Scott, O.C.A. The concentration of oxygen dissolved in tissues at the time of irradiation as a factor in radiotherapy. *Br. J. Radiol., 26:* 638-648, 1953.

17. Drescher, E.E., and Gray, L.H. Influence of oxygen tension on X-ray induced damage in Ehrlich ascites tumor cells irradiated in vitro and in vivo. *Radiol. Res., 11:* 115-146, 1959.

18. Durand, R.E. Keynote address: The influence of microenvironmental factors on the activity of radiation and drugs. *Int. J. Rad. Onc. Biol. Phys., 20:* 253-258, 1991.

19. Young, S.D., Marshall, R.S., and Hill, R.P. Hypoxia induces DNA overreplication and enhances metastatic potential of murine tumor cells. *Proc. Natl. Acad. Sci., 85:* 9533-9537, 1988.

20. Höckel, M., Schlenger, K., Knoop, C., and Vaupel, P. Oxygenation of carcinomas of the uterine cervix: Evaluation by computerized O₂ tension measurements. *Cancer Res., 51:* 6098-6102, 1991.

21. Höckel, M., Knoop, C., Schlenger, K., Vorndran, B., Bauβmann, E., Mitze, M., Knapstein, P.G., and Vaupel, P. Intratumoral pO₂ predicts survival in advanced cancer of the uterine cervix. *Radiotherapy and Oncology, 26:* 45-50, 1993.

22. Hoshino, S., Hamada, O., Iwaya, F., Takahira, H., and Honda, K. Omental transplantation for chronic occlusive arterial diseases. *Int. Surg., 64:* 21-29, 1979.

23. Goldsmith, H.S. Salvage of end stage ischemic extremities by intact omentum. *Surgery, 88:* 732-736, 1980.

24. Hoshino, S., Nakayama, K., Igari, T., and Honda, K. Long-term results of omental transplantation for chronic occlusive arterial diseases. *Int. Surg., 68:* 47-50, 1983.

25. In: "The Greater Omentum," Liebermann-Meffert, D.; White, H.,ed., Springer, Berlin (1983).

26. Maurya, S.D., Singhal, S., Gupta, H.C., Elhence, I.P., and Sharma, B.D. Pedicled omental grafts in the revascularization of ischemic lower limbs in Buerger's disease. *Int. Surg., 70:* 253-255, 1985.

27. Pevec, W.C., Hendricks, D., Rosenthal, M.S., Shestak, K.C., Steed, D.L., and Webster, M.W. Revascularization of an ischemic limb by use of a muscle pedicle flap: A rabbit model. *J. Vasc. Surg., 13:* 385-390, 1991.

28. Mathes, N. Classification of the vascular anatomy of muscles: Experimental and clinical correlation. *Plast. Reconstr. Surg., 67:* 177-187, 1981.

29. Nottebeart, M., Lane, J.M., Juhn, A., Burstein, A., Schneider, R., Klein, C., Sinn, R.S., Dowling, C., Cornell, C., and Catsimpoolas, N. Omental angiogenic lipid fraction and bone repair. An experimental study in the rat. *J. Orthop. Res., 7:* 157-169, 1989.

30. Anthony, J.P., Mathes, S.J., and Alpert, B.S. The muscle flap in the treatment of chronic lower extremity osteomyelitis: Results in patients over 5 years after treatment. *Plast. Reconstr. Surg., 88:* 311 1991.

31. Jones, N.F., Eadie, P., Johnson, P.C., and Mears, D.C. Treatment of chronic infected hip arthroplasty wounds by radical debridement and obliteration with pedicled and free muscle flaps. *Plast. Reconstr. Surg., 88:* 95 1991.

32. Phillips, G.D., and Knighton, D.R. Angiogenic activity in damaged skeletal muscle (43025). *P.S.E.B.M., 193:* 197-202, 1990.

33. Green, H., Kehinde, O., and Thomas, J. Growth of cultured human epidermal cells into multiple epithelia suitable for grafting. *Proc. Natl. Acad. Sci. USA, 76:* 5665-5668, 1979.

34. Romagnoli, G., De Luca, M., Faranda, F., Bandelloni, R., Franzi, A.T., Cataliotti, F., and Cancedda, R. Treatment of posterior hypospadias by the autologous graft of cultured urethral epithelium. *N. Engl. J. Med., 323:* 527-530, 1990.

35. Eisenstein, R. Angiogenesis in arteries: Review. *Pharmac. Ther. , 49:* 1-19, 1991.

36. Odedra, R., and Weiss, J.B. Low molecular weight angiogenesis factors. *Pharmac. Ther., 49:* 111-124, 1991.

37. Lynch, S.E., Colvin, R.B., and Antoniades, H.N. Growth factors in wound healing. Single and synergistic effects on partial thickness porcine wounds. *J. Clin. Invest., 84:* 640-646, 1989.

38. Laato, M., Niinikoski, J., Lebel, L., and Gerdin, B. Stimulation of wound healing by epidermal growth factor (EGF): A dose dependent effect. *Ann. Surg., 203:* 379-381, 1986.

39. Buckley, A., Davidson, J.M., Kamerath, C.D., and Woodward, S.C. Epidermal growth factor increases granulation tissue formation dose dependently. *J. Surg. Res., 43:* 322-328, 1987.

40. Broadley, K.N., Aquino, A.M., and Hicks, B. Growth factors -FGF and TGF-beta- accelerate the rate of wound repair in normal and in diabetic rats. *Int. J. Tiss. Reac. , 10:* 345-353, 1988.

41. Brown, G.L., Curtsinger, L.J., White, M., Mitchell, R.D., Pietsch, J., Nordquist, R., von Fraunhofer, A., and Schultz, G.S. Acceleration of tensile strength of incisions treated with EGF and TGF-beta. *Ann. Surg., 208:* 788-794, 1988.

42. Brown, G.L., Nanney, L.B., Griffen, J., Cramer, A.B., Yancey, J.M., Curtsinger L.J., Holtzin, L., Schultz, G.S., Jurkiewicz, M.J., and Lynch, J.B. Enhancement of wound healing by topical treatment with epidermal growth factor. *N. Engl. J. Med., 321:* 76-79, 1989.

43. Langer, R., and Moses, M. Biocompatible controlled release polymers for delivery of polypeptides and growth factors. *J. Cell. Biochem., 45:* 340-345, 1991.

44. Chu, G.H., Ogawa, Y., and McPherson, J.M. Collagen wound healing matrices and process for their production. *Collagen Corp.* 990; WO 90/00060.

45. Kopecek, J., and Ulbrich, K. Biodegradation of biomedical polymers. *Prog. Polym. Sci., 9:* 1-58, 1983.

46. Höckel, M., Ott, S., Siemann, U., and Kissel, T. Prevention of peritoneal adhesions in the rat with sustained intraperitoneal dexamethasone delivered by a novel therapeutic system. *Ann. Chirurg. Gynaecol., 76:* 306-313, 1987.

47. In: "Novel drug delivery," Prescott, L.F.; Nimmo, W.S.,ed., Wiley and Sons, Chichester (UK) (1989).

48. Kissel, T., Brich, Z., Bantle, S., Lancranjan, I., Nimmerfall, F., and Vit, P. Parenteral depot systems on the basis of biodegradable polyesters. *J. Contr. Rel.,* in press.

49. Edington, H.D., Sugarbaker, P.H., and McDonald, H.D. Management of the surgically traumatized, irradiated, and infected pelvis. *Surgery, 103:* 690-697, 1987.

50. Mathes, S. J., Feng, L.J., and Hunt, T. Coverage of the infected wound. *Ann. Surg., 198:* 420-426, 1983.

51. Eshima, I., Mathes, S.J., and Paty, P. Comparison of the intracellular bacterial killing activity of leukocytes in musculocutaneous and random-pattern flaps. *Plast. Reconstr. Surg., 86:* 541-547, 1990.

52. Knighton, D.R., Ciresi, K.F., Fiegel, B.S., Austin, L.L., and Butler, E.L. Classification and treatment of chronic nonhealing wounds. *Ann. Surg., 204:* 322-330, 1986.

53. Burgos, H., Herd, A., and Bennett, J.P. Placental angiogenic and growth factors in the treatment of chronic varicose ulcers: preliminary communication. *J. Royal Soc. Med., 82:* 598-599, 1989.

54. Knighton, D.R., Ciresi, K.F., Fiegel, V.D., Schumerth, S., Butler, E., and Cerra, F. Stimulation of repair in chronic, nonhealing, cutaneous ulcers using platelet derived wound healing formula. *Surg., Gynecol. and Obstet. 170:* 56-60, 1990.

55. Hase, S., Nakazawa, S., Tsukamoto, Y., and Segawa, K. Effects of prednisolone and human epidermal growth factor on angiogenesis in granulation tissue of gastric ulcer induced by acetic acid. *Digestion, 42:* 135-142, 1989.

56. Folkman, J., Szabo, S., Stovroff, M., McNeil, P., Li, W., and Shing, Y. Duodenal ulcer. Discovery of a new mechanism and development of angiogenic therapy that accelerates healing. *Ann. Surg., 214:* 414-427, 1991.

57. tenDijke, P., and Iwata, K.K. Growth factors for wound healing. *Bio/Technology, 7:* 793-798, 1989.

58. Unger, E.F., Sheffield, C.D., and Epstein, S.E. Creation of anastomoses between an extracardiac artery and the coronary circulation. Proof that myocardial angiogenesis occurs and can provide nutritional blood flow to the myocardium. *Circulation, 82:* 1449-1466, 1990.

59. Myers, B. Understanding flap necrosis. *Plast. Reconstr. Surg., 77:* 813-814, 1986.

60. Höckel, M., and Burke, J.F. Angiotropin treatment prevents flap necrosis and enhances dermal regeneration in rabbits. *Arch. Surg., 124:* 693-698, 1989.

61. McGregor, I.A., and Morgan, G. Axial and random pattern flaps. *Br. J. Plast. Surg., 26:* 202-213, 1973.

62. in: "The Arterial Anatomy of Skin Flaps," Cormack, G.C.; Lamberty, B.G.H.,ed., Churchill Livingstone, New York (1986).

63. Eppley, B.L., Connolly, D.T., Winkelmann, T., Sadove, A.M., Heuvelman, D., and Feder, J. Free bone graft reconstruction of irradiated facial tissue: Experimental effects of basic fibroblast growth factor stimulation. *Plast. Reconstr. Surg., 88:* 1 1991.

64. Penkert, G., Bini, W., and Samii, M. Revascularization of nerve grafts: An experimental study. *J. Reconstr. Microsurg., 4:* 319-325, 1988.

65. Ohta, H., Ishiyama, J., Saito, H., and Nishiyama, N. Effects of pretreatment with basic fibroblast growth factor, epidermal growth factor and nerve growth factor on neuron survival and neovascularization of superior cervical ganglion transplanted into the third ventricle in rats. *Japan. J. Pharmacol., 55:* 255-262, 1991.

66. Cordeiro, P.G., Seckel, B.R., Lipton, S.A., D'Amore, P.A., Wagner, J., and Madison, R. Acidic fibroblast growth factor enhances peripheral nerve regeneration in vivo. *Plast. Reconstr. Surg., 83:* 1013-1019, 1989.

67. Lyons, M.K., Anderson, R.E., and Meyer, F.B. Basic fibroblast growth factor promotes in vivo cerebral angiogenesis in chronic forebrain ischemia. *Brain Res., 558:* 315-320, 1991.

68. Lawrence, T.W., Sporn, M.B., Gorschboth, C., Norton, J.A., and Grotendorst, G.R. The reversal of an adriamycin induced healing impairment with chemoattractants and growth factors. *Ann. Surg., 203:* 142-147, 1986.

69. Mooney, D.P., Gamelli, R.L., and O'Reilly, M. Improved wound healing through the local delivery of tumor necrosis factor. *Surg. Forum, 39:* 77-79, 1988.

70. Curtsinger, L.J., Pietsch, J.D., Brown, G.L., von Fraunhofer, A., Ackerman, D., Polk, H.C. J., and Schultz, G.S. Reversal of adriamycin-impaired wound healing by transforming growth factor-beta. *Surg. Gynecol. Obstet., 168:* 517-522, 1989.

71. Latoo, M., Jyrki, H., Veli, M.K., Niinikkoski, J., and Gerdin, B. Epidermal growth factor (EGF) prevents methylprednisolone induced inhibition of wound healing. *J. Surg. Res., 47:* 354-359, 1989.

72. Eliseenko, V.I., Skobelkin, O.K., Chegin, V.M., and Degtyarev, M.K. Microcirculation and angiogenesis during wound healing by first and second intention. *Bull. Exp. Biol. Med., 105:* 289-292, 1988.

73. Schlenger, K., Höckel, M., Schwab, R., and Frischmann-Berger, R. How to improve the uterotomy healing. I Effects of fibrin and tumor necrosis factor alpha in the rat uterotomy model. *J. Surg. Res., in press*

74. Greisler, H.P., Klosak, J.J., Dennis, J.W., Karesh, S.M., Ellinger, J., and Kim, D.U. Biomaterial pretreatment with ECGF to augment endothelial cell proliferation. *J. Vasc. Surg., 5:* 393-402, 1987.

75. Clowes, A.W., and Kohler, T. Graft endothelialization: The role of angiogenic mechanisms. *J. Vasc. Surg., 13:* 734-736, 1991.

76. Buntrock, P., Jentzsch, K.D., and Heder, G. Stimulation of wound healing, using brain extract with fibroblast growth factor (FGF) activity. *Exptl. Path., 21:* 46-53, 1982.

77. Buckley, A., Davidson, J.M., Kamerath, C.D., Wolt, T.B., and Woodward, S.C. Sustained release of epidermal growth factor accelerates wound repair. *Proc. Natl. Acad. Sci. USA, 82:* 7340-7344, 1985.

78. Winet, H., Bao, J.Y., and Moffat, R. A control model for tibial cortex neovascularization in the bone chamber. *J. Bone Mineral Res., 5:* 19-30, 1990.

79. Gills, J.P., and McIntyre, L.G. Growth factors and their promising future. *J. Amer. Optom. Assoc., 60:* 442-445, 1989.

80. Woost, P.G., Brightwell, J., Eiferman, A., and Schultz, G. Effect of growth factors with dexamethasone on healing of rabbit corneal stromal incisions. *Exptl. Eye Res., 40:* 47-60, 1985.

81. Schultz, G.S., White, M., Mitchell, R., Brown, G., Lynch, J., Twardzik, D.R., and Todaro, G.J. Epithelial wound healing enhanced by transforming growth factor alpha and vaccinia growth factor. *Science, 235:* 350-352, 1987.

82. Nanney, L.B. Epidermal and dermal effects of epidermal growth factor during wound repair. *J. Invest. Dermatol.* , *94:* 624-629, 1990.

83. Hoskins, W., and Rubin, S. Surgery in the treatment of patients with advanced ovarian cancer. *Semin. Oncol., 18:* 213-221, 1991.

84. Simpson-Herren, L., Sanford, A.H., and Holmquist, J.P. Effects of surgery on the cell kinetics of residual tumor. *Cancer Treat. Rep., 60:* 1749-1760, 1976.

85. Gunduz, N., Fisher, B., and Saffer, E.A. Effect of surgical removal on the growth and kinetics of residual tumor. *Cancer Res., 39:* 3861-3865, 1979.

86. Wong, R.J., and DeCosse, J.J. Cytoreductive surgery. *Gynecol. and Obstet., 170:* 279-281, 1990.

87. Höckel, M., Knapstein, P.G., and Kutzner, J. A novel combined operative and radiotherapeutic treatment approach for recurrent gynecologic malignant lesions infiltrating the pelvic wall. *Surg., Gynecol. and Obstet., 173:* 297-302, 1991.

88. Höckel, M., and Knapstein, P.G. The combined operative and radiotherapeutic treatment (CORT) of recurrent tumors infiltrating the pelvic wall: First experience with 18 patients. *Gynecol. Oncol., 46:* 20-28, 1992.

ENDOTHELIAL CELL STIMULATING ANGIOGENESIS FACTOR IN RELATION

TO DISEASE PROCESSES

Jacqueline B. Weiss

Wolfson Angiogenesis Unit
University of Manchester
Rheumatic Diseases Centre
Hope Hospital
Manchester M6 8HD
England

INTRODUCTION

In this chapter I will describe work which has been carried out with many collaborators some of which is published and some of which is not yet published. I have made reference to these co-workers at the appropriate parts of the chapter.

Endothelial cell stimulating angiogenesis factor (ESAF) has been described elsewhere in this book by McLaughlin and Weiss and its ability to activate the prometalloproteinase system has been described (Weiss JB and McLaughlin, 1993).

ESAF is mitogenic for microvessel cells only, that is for endothelial cells and pericytes and has no effect on aortic cells or on smooth muscle cells or on fibroblasts. In fact, it is arguable that the latter two cell types maybe a source of this factor and we have certainly observed ESAF production in early passage cultures of skin fibroblasts and significant increase in ESAF production also occurs in stimulated muscles.

The action of ESAF on microvessel endothelial cells always produces a bell-shaped curve suggesting that there is an autocrine control in the endothelial system, preventing its over-stimulation by ESAF (Odedra and Weiss 1991). This would suggest that the stimulation of the microvasculature by the small molecule may be strictly regulated and unlike the effect of some other growth factors such as basic fibroblast growth factor, uncontrolled growth of the microvasculature could not occur. It is worth noting in Figure 1 that ESAF derived from either serum or synovial fluids, at the same unit activity, displays the same bell-shaped curve.

ASSAYING FOR ESAF ACTIVITY

The ability of ESAF to activate the three major prometalloproteinases; progelatinase

A and prostromelysin and to de-inhibit the TIMP enzyme complex may explain some aspect of its angiogenic function. On a more mundane level this property is very useful in the measurement of ESAF in biological fluids. In fact, we use as a standard assay the activation of procollagenase. Such an assay has the advantage that it is very sensitive and indeed very reproducible: a unit of ESAF represents a percentage activation of available procollagenase.

Early work published in 1983 (Brown et al), using the chick chorioallantoic membrane as a test system, showed that synovial fluid from knees of patients with both osteoarthritis and ankylosing spondylitis had far greater ESAF activity than did that from patients with rheumatoid arthritis and this led us to wonder whether ESAF was not involved in situations in which inflammatory cells were also present. In other words, we considered that natural angiogenesis which would occur during growth, particularly bone, might constitute a different system from that occurring in inflammatory conditions such as in rheumatoid arthritis and probably wound healing.

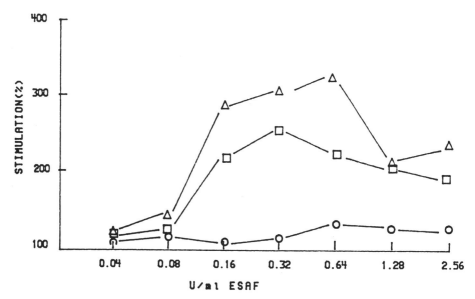

Figure 1. Mitogenic effect of ESAF from serum ⊡ and synovial fluid △ of an osteoarthritic patient, on bovine adrenal microvessel endothelial cells grown on a collagen substratum. There was no effect when cells were grown on plastic ‑ ○ ‑

Since ESAF is a small factor we had to be sure that when we were assaying for it in serum and tissues we were actually assaying for the correct low molecular mass factor. This proved difficult at first but we devised a method (Fig. 2) in which we were able to extract ESAF from tissues or serum in a standardised manner. ESAF is always bound to carrier protein(s) in the body and so it is necessary to homogenise the tissue with a chaotrophic agent such as magnesium chloride to dissociate it from carrier proteins. We use 2M magnesium chloride for this purpose. Although ethanol will also dissociate ESAF from its carrier, we find the magnesium chloride method more satisfactory and more reproducible. The extraction method, as can be seen from the figure, is tedious and requires the use of individual filters for each assay, which have a cut-off of 3000 Daltons and we use (Microsep concentrators; cut off 3000Da. Flowgen, Kent, U.K) for this purpose. After removing high molecular mass material, we can desalt the low molecular mass ESAF fraction on a reverse phase column and elute the majority of salt with water washes. Subsequently, we can elute ESAF at an apropriate dilution using a methanol water gradient (Fig.2).

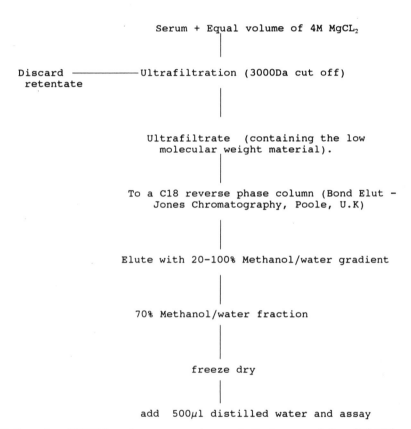

Figure 2. Extraction of ESAF from tissues or serum in a standardized manner; elution of ESAF through the appropriate dilution using a methanol water gradient.

Strangely enough, although ESAF is carrier bound in all adult tissues and serum, this is not so in foetal calf serum where we can dialyse most of it out directly. This is an important point because "free" (in two senses of the word) ESAF is available to all workers in the tissue culture field who are using foetal calf serum.

Originally, we attempted to purify ESAF from tissue culture medium of transformed malignant cells. However, this was not satisfactory due to the large number of other small factors produced in these circumstances from which it was difficult, indeed impossible, to separate ESAF. We therefore used the pineal gland as a relatively clean source of ESAF since the pineal gland does not contain the products of inflammatory cells and contains remarkably high levels of ESAF (Taylor et al 1988). The reason for this relatively high concentration level of ESAF in the pineal gland is not clear but the gland does have a very rich vascular bed (Hodde 1979) and undergoes some degree of remodelling (Tapp 1979, Blask 1984). The high vascularity of the pineal is probably not in itself the explanation for the high levels of ESAF but it is noteworthy that the pineal contains substances which appear to be highly resistant to tumourogenesis and there are reports of inhibition of carcinosarcomas in rats. Serotonin is a product of the pineal gland which has inhibitory effects on solid tumours but not on ascites tumours and this does suggest that it impairs vascularisation. Increased concentrations of ESAF therefore may be present in an attempt to overcome the vascular inhibitory activities that reside normally in the pineal gland (Odedra and Weiss 1991).

TISSUE ESAF LEVELS

A rich source of ESAF is the foetal growth plate and we find this to be very significant since this presently avascular tissue is soon to become transformed into highly vascular bone. It is perhaps worth mentioning here that although we can show the relative amounts of ESAF in growth plate (and pineal glands and serum from patients where neovascularisation is known to be occurring), the actual amounts are very small indeed. It is necessary to stresss this point, because we get a considerable amount of "stick" for not having obtained the structure of ESAF. However, it is not only a very small molecule but it also appears to be produced on demand. The amount circulating in normal serum is negligible. It's potency is such that it would be extremely dangerous to have too much of it. The foetal growth plate has the highest level of all the foetal tissues tested and this is an order of magnitude greater than in other tissue (Table 1). Rather interestingly foetal bone ESAF levels rise much earlier than do those in the diaphysis and brain while the concentration of ESAF in foetal growth plate rises much later (Fig. 3) than all other tissues. Growth plate chondrocytes in megaculture produce very low levels of ESAF. However when a substrate, such as calcium β glycerophosphate, is added to the medium which enables them to calcify the matrix which they have synthesised, levels of ESAF rise to very significant levels (Brown et al 1987). This perhaps explains why levels of ESAF are high in synovial fluids and serum from patients with osteoarthritis and probably reflect the re-opening of the growth plate which will occur in the formation of osteophytes (Brown and Weiss, 1988). An interesting observation has been that in ankylosing spondylitis, where endochondral ossification is occurring, very high levels of ESAF can be measured in the serum. In two studies of patients with this condition we have shown that serum levels of ESAF relate, although not quite attaining significance, to the degree of sacro-iliitis and also that there is a relationship between the enthesitis index and levels of ESAF in the serum. Bearing in mind that both the sacro-iliitis grade and the enthesitis index are somewhat subjective then it is perhaps not surprising that although the levels of ESAF relate, they do not necessarily become truly significant (Jones P, Makki R and Weiss JB in press, Taylor G, Taylor CM and Weiss JB in press).

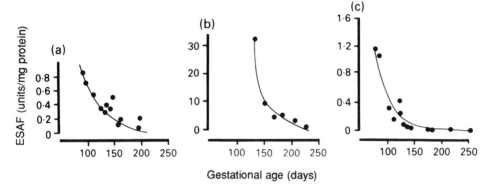

Figure 3. Concentration of ESAF on developing bovine tissues in relation to gestational age of the foetus a)brain b)growthn plate c)diaphysis (from Taylor CM, Mclaughlin B, Weiss JB and Maroudas N. 1992, J. Reprod. and Fertil. 94: 445-9 with permission of Portland Press).

Table 1. Relative concentrations of ESAF in adult and foetal tissues and in malignant brain tumors.

Tissue	Units of ESAF
Brain	0.4
Vitreous	0.3
Retina	0.5
Kidney	0.2
Adrenal	0.2
Liver	0.2
Malignant Brain Tumour	3.9
Pineal Gland	5.8
Foetal Growth Plate (140 days)	30.0

In previous work we have shown a relationship between concentrations of ESAF in brain tumour and extent of malignancy (Taylor et al 1991). It is important to stress however that the levels of ESAF in these tumours although high were less than that of the pineal gland and the growth plate (Table 1) suggesting that ESAF may not have a primary role in tumour angiogenesis.

In an early study on diabetic patients we were unable to find a significant difference between the serum levels of ESAF in normal control subjects or in patients with diabetes. When we divided the patients with diabetes into those with and without retinopathy, we found a significant difference in the concentrations of ESAF between those patients with retinopathy and those without (Taylor and Weiss 1989). We have just recently embarked on a new study of ESAF levels in diabetic patients in which we looked for a relationship between age and ESAF and fasting blood glucose levels and ESAF. There was none. In this study new patients were allocated randomly to either diet, diet and tablets or insulin. Those on diet alone had lower levels of ESAF. This may reflect the fact that patients who were initially entered onto a diet regimen and who were not responding satisfactorily were changed to a more therapeutic regime. We have also looked at ten patients with no complication of diabetes other than proliferative retinopathy and found that their serum concentrations of ESAF were very high. The concentration of ESAF dropped after laser treatment and returned to normal after three months if the treatment had proved successful (Makki, Ologuko, Dornan, Boulton, Weiss in preparation).

In another study on patients with sickle cell anaemias, carried out in collaboration with G.R. Sergeant a the Medical Research Council laboratory in Jamaica, we found in contrast that, although patients with no haemoglobinopathy or with the sickle cell trait had normal serum concentrations of ESAF, patients with either sickle cell disease or SC disease had very high levels irrespective of whether or not they had retinopathy. This suggests that angiogenic processes in these patients may have been occurring elsewhere in the body perhaps in joint spaces. In previous work we reported that the concentration of ESAF in serum of patients with tibial fractures was higher than the normal values for age and sex matched controls and that the values appeared to decrease with time after the fracture. Concentrations of ESAF were relatively higher in patients who had their tibial fractures held with external fixation rather than in patients with tibial fractures held with intra-medullary nails (Wallace et al 1991). We have continued this research and have studied two groups of four sheep with externally fixed tibial osteotomy, where in one group the soft tissues were carefully protected with minimal extra-periosteal distubance prior to osteotomy. In the other group osteotomy was followed by circumferential stripping and excision of the periosteum for 20mm proximally and distally after which a 40mm sleeve of silicone was placed over both fragments to prevent revascularisation of the underlying cortex and the surrounding soft tissue. Serum samples were taken pre-oseotomy and at intervals up to 42 days post operatively. The concentration of ESAF rose measurably in the well vascularised group within the first hour and increased steadily for six days whereafter it fell sharply, and rose again and then declined steadily towards 42 days when it returned to normal. The devascularised group on the other hand, did not show any significant increase until six days when an increased concentration was observed, not quite as high as in the vascularised group, but nevertheless very significant and this fell to normal within 42 days (Fig. 4). We suppose that the first peak of activity reflects healing from the periosteum and perhaps that the second peak of ESAF concentration could have arisen from the endosteal blood supply (Wallace, Makki and Weiss unpublished observations).

In another recent, and also as yet unpublished, study we have looked at the effect of chronic electrical stimulation of rat muscles on their production of ESAF (Brown, Hudlicka, Makki, Weiss in preparation). We compared ESAF values in these muscles with values of

Table 2.   Total ESAF in conditioned media from cultured psoriatic skin
            fibroblasts

		Mann Whitney
Passage 2	Normal versus involved	0.001
8	Normal versus involved	0.001
9-11	Normal versus involved	0.007

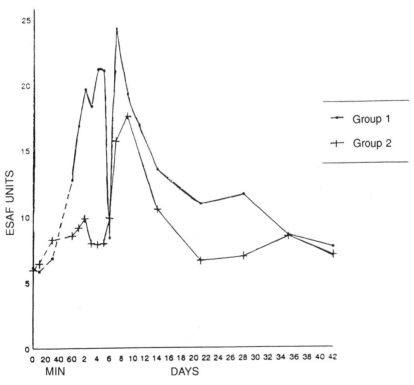

Figure 4. Serum levels of ESAF in sheep after tibial osteotomy. Group 1 normal healing. Group 2 healing
following stripping of periosteum (see text).

muscles treated with the vasodilator prazosin. We compared the capillary formation and ESAF concentration levels in both treatments. We found a very significant association between the capillary density and ESAF concentrations in the electrically stimulated muscles but there was no correlation whatsoever between capillary density and ESAF concentrations in the prazosin treated animals. In fact in these animals ESAF levels did not rise above normal although capillary density increased (Hudlicka, Brown, Makki and Weiss, in preparation). This is a further indication of our proposition that distinct methods of achieving neovascularisation are controlled by different cytokine groups.

TISSUE CULTURE EXPERIMENTS

In a study of conditioned media from skin fibroblasts obtained from patients with psoriasis, we cultured cells taken from involved and uninvolved sites and showed significant increases in ESAF concentrations compared to media from sex, age matched and site matched controls. However, these differences were only seen in media from cells in early and late passages. After passage 3 and until passage 8 no differences were observed (Table 2) (McLaughin, Weiss, Rowland-Payne, Staughton and Zemelman unpublished observtions). This may be of relevance to the general comparison of tissue culture studies between groups, where passage number of cells used in experiments is frquently not included in results. It is worth mentioning that many studies on endothelial cell motility and endothelial cell proliferation are carried out at different passage numbers not necessarily at the primary or first passage but they are all expected to relate to each other and this may not be the case. Similarly, age of individuals (human or animal) from which cells are taken may be of vital importance and may explain some of the differences reported in the literature between animal and human endothelial cells in culture. Sadly, much research on angiogenesis utilises aortic cells which are unlikely to be programmed for normal angiogenesis. It is worth pointing out again that ESAF is only active against microvascular cells and not against aortic cells or human umbilical vein endothelial cells.

DISCUSSION

From the experiments described above, it is clear that ESAF has a real involvement in diseases in which uncomplicated angiogenesis may be occurring but it is not involved, or may not be involved to any great extent, in diseases where inflammatory cells are probably the conveyors of the angiogenic cytokines. Thus even though we can show higher levels of ESAF in brain tumours compared to normal brain, high values are lower than what we would normally expect in a normal pineal gland and very considerably lower than those found in the bovine foetal growth plate. It must be assumed therefore that ESAF does not play a major part in tumour angiogenesis but does play a major part in the angiogenesis of growth and as we have indicated in muscle activity, fracture healing and in the proliferative retinopathy of diabetes. It would seem that there are several types of angiogenesis relating to difference diseases processes or in some instances to normal processes such as growth, and that these different types will be controlled by different cytokines or by different cocktails of cytokines.

In some instances such as tumour angiogenesis, the cytokine cocktail may be so intoxicating that it may not be possible to pinpoint accurately the actual angiogenic components. By contrast in the angiogenesis of new bone formation the pattern of events is laid out much more clearly and the actors involved are more defined. Therefore, we do not claim a vital role for ESAF in all the known angiogenic processes but we do think that in "normal" angiogenesis such as occurs during growth it is probably of major importance. The physical amounts of ESAF which are produced in any of these processes are immensely small. They can be measured easily because of the ability of ESAF to activate proenzymes and this gives rise to the false impression that we are dealing with large amounts of the compound. Sadly, this is not the case. It is very unlikely that there is a real source of ESAF in any tissue in the body and it is almost certainly made on demand. Even in the pineal gland only trace amounts of ESAF in physical quantities are found.

However, we have been able to establish a physiological connection between angiogenesis and ESAF production which we believe warrants its place high amongst the plethora of "angiogenic factors".

REFERENCES

Blash, D.E., 1984, The pineal gland: an oncostatic gland? in: "The Pineal Gland," R.J. Reiter ed., Raven Press, New York.

Brown, R.A., Tomlinson, I.W., Hill, C.R., and Weiss, J.B., 1983, Relationship of angiogenesis factor in synovial fluid to various joint diseases, *Ann. Rheum. Dis.* 42:301.

Brown, R.A., Taylor, C., McLaughlin, B., McFarland, C.D., Weiss, J.B., and Ali, S.Y., 1987, Epiphyseal growth plate cartilage and chondrocytes in mineralising culture produce a low molecular mass angiogenic procollagenase activator, *Bone Miner.* 3(2):143.

Brown, R.A., and Weiss, J.B., 1988, Neovascularisation and its role in the osteoarthritic process, *Ann. Rheum. Dis.* 47(11):881

Hodde, K.C., 1979, The vascularisastion of the rat pineal organ, in: "Progress in Brain Research" Vol. 52: "The Pineal Gland of Vertebrates Including Man," J.A. Kapper, and P. Pevet, eds., Elsevier, North Holland, Amsterdam.

Odedra, R., and Weiss J.B., 1991, Low molecular weight angiogenesis factors, *Pharmac. Ther.* 49:111

Tapp, E., 1979, The histology and pathology of the pineal gland, in: "Progress in Brain Research" Vol. 52: "The Pineal Gland of Vertebrates Including Man," J.A. Kapper, and P. Pevet, eds., Elsevier, North Holland, Amsterdam.

Taylor, C.M., McLaughlin, B., Weiss, J.B., and Smith, I., 1988, Bovine and human pineal glands contain substantial quantities of endothelial cell stimulating angiogenesis factor, *J. Neural. Transm.* 71:79

Taylor, C.M., and Weiss, J.B., 1989, Raised endothelial cell stimulating angiogenesis factor in diabetic retinopathy, *Lancet* 2(8675):1329

Taylor, C.M., Weiss, J.B., and Lye, R.H., 1991, Raised levels of latent collagenase activating angiogenesis factor (ESAF) are present in actively growing human intracranial tumours, *Br. J. Cancer* 64:164

Wallace, A.L., McLaughlin, B., Weiss, J.B., and Hughes, S.P.F., 1991, Increased endothelial cell stimulating angiogenesis factor in patients with tibial fractures, *Injury* 22(4):375

Weiss, J.B., and McLaughlin B., 1993, Activation of gelatinase A and re-activation of the gelatinase A inhibitor complex by endothelial cell stimulating angiogenesis factor (ESAF), *J. Physiol.* 467:49P

HYALURONAN AND ANGIOGENESIS: MOLECULAR MECHANISMS AND CLINICAL APPLICATIONS

[1]S Kumar, [1]J Ponting, [2]P Rooney, [3]P Kumar, [1]D Pye and [1]M Wang

[1]Christie Hospital NHS Trust
Manchester, M20 9BX, UK
[2]University of Manchester, UK
[3]Manchester Metropolitan University, UK

INTRODUCTION

The extracellular matrix (ECM) is found in contact with most cells of a multicellular organism and can be thought of as an extension of the cell surface. It provides mechanical support for cells, may act as a reservoir for growth factors, is the environment through which nutrients are passed and, by virtue of the fact that many of its constituents interact with each

Table 1 A selective list of putative functions of HA

Stromal component, lubricant
Organogenesis, cell adhesion, migration, proliferation
Angiogenesis
Wound healing
Immune response
Tumour development,desmoplasia
Activation of early response genes

other and with the cell surface via specific receptors, it can affect and participate in functions of the cell including adhesion, migration, proliferation, differentiation and gene regulation (Hay, 1991; Rooney and Kumar 1993). A major and important component of the ECM is hyaluronic acid (HA, also known as hyaluronan and hyaluronate). HA plays many roles (Table 1); in particular we have shown that in angiogenesis it has a dual function i.e. dependent on its molecular mass it can promote or inhibit angiogenesis.

(1) STRUCTURE, DISTRIBUTION, SYNTHESIS AND DEGRADATION OF HA

HA was first isolated from vitreous humor in 1934 by Meyer and Palmer. It is present to some degree in all connective tissues and is one of the very few molecules found in both eukaryotes and prokaryotes. HA is a glycosaminoglycan (GAG) and chemically, is built up from a repeating disaccharide unit comprising D-glucuronic acid and N-acetyl D-glucosamine (Fig 1). In its native state it has a molecular mass of up to 1×10^7 Da. The richest source of HA is skin and it is found in significant amounts in body fluids like lymph, serum and synovial fluid (Laurent and Fraser, 1992).

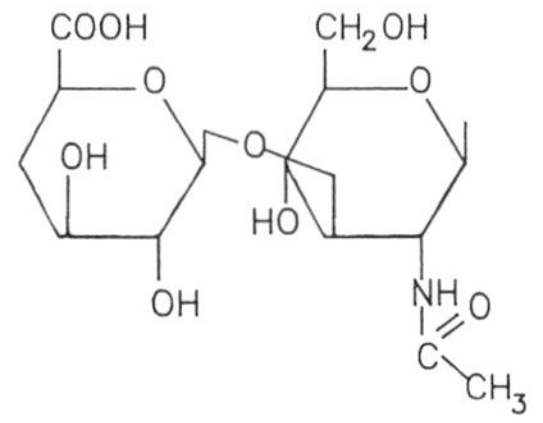

Fig 1 Chemical structure of disaccharide unit of HA.

HA is synthesized by the enzyme, HA synthase, on the inner surface of the plasma membrane and is extruded across the membrane into the ECM. The regulation of HA synthesis is under the influence of a variety of factors. There appears to be a feedback mechanism whereby the nascent HA chains suppress their own elongation (Prehm 1990). Conversely, removal of HA from either B6 or oligodendroglioma cells stimulated the production of HA in skin fibroblast cultures (Larnier *et al*, 1989). Increased synthesis is also activated by cell transformation with Rous Sarcoma Virus. Several growth factors are known to stimulate HA synthesis, transforming growth factor-β (TGF-β) having attracted the most attention (Green and Morales, 1991). TGF-α, basic fibroblast growth factor, epidermal growth factor, platelet derived growth factor and cartilage growth factor also influence HA synthesis (Lembach, 1976; Hamerman *et al*, 1986; Heldin *et al*, 1986). Some cytokines are also stimulatory e.g. interleukin1 (Yaron *et al*, 1987). HA synthesis is influenced by sex hormones and specific HA stimulatory factors have been identified from breast tumour derived fibroblasts (Schor *et al* 1989) and sera of patients with mammary carcinomas (Decker *et al* 1989). In co-culture with tumour cells, fibroblasts have been noted to produce increased amounts of HA. In several physiological and pathological conditions HA synthesis increases.

The half life of HA varies from tissue to tissue, 90 minutes in the aqueous humor, 12 hours in the skin, 45 days in the brain and 70 days in the vitreous humor (Laurent and Fraser, 1992). While it is difficult to estimate normal turnover, in plasma it ranges from 0.3 - 1.0 µg/min/kg body weight i.e. approximately 25% of plasma HA is removed every minute (Fraser and Laurent, 1989). Rodén and co-workers (1989) found that when [^{3}H]-labelled HA was injected into rats, ^3H_2O could be detected in the plasma within 10-20 minutes. This rapid degradation is mediated by hyaluronidases and other enzymes.

(2)   HA AND ANGIOGENESIS

HA is found in all tissues in which neovascularisation is occurring and, as mentioned above, depending on its size it can either promote or inhibit angiogenesis *in vivo* and *in vitro*.

(A)   HA and *in vivo* angiogenesis

(i)   Chicken chorioallantoic membrane assay. For this assay chicken chorioallantoic membranes (CAM) on day 10 are treated with oligosaccharides of HA (o-HA) or native macromolecular HA (n-HA). Four days after the application of HA, CAMs are fixed and examined for the evidence of neovascularisation. o-HA induced a marked angiogenic response, as indicated by the presence of whorls of blood vessels surrounding the site of application. A striking feature of the o-HA treated CAMs at the electron microscope level was the deposition of a large amount of collagen fibrils (Rooney *et al*, 1993). In contrast neither n-HA nor sham treated CAMs showed any angiogenic response or deposition of collagen fibrils. The role of collagens in angiogenesis is unclear. Ingber and Folkman (1988) have shown that metabolic reduction of collagen synthesis inhibits microvessel formation on the CAM, suggesting that collagen might be a necessary substratum for the migration of endothelial cells. Furthermore, Maragoudakis *et al* (1991) reported that the addition of a specific inhibitor of a basement membrane collagen biosynthesis (GPA 1734) prevented o-HA induced angiogenesis on CAMs. Type VIII collagen has also been reported to play a role in angiogenesis (Iruela-Arispe and Sage, 1991 see also section Bii).

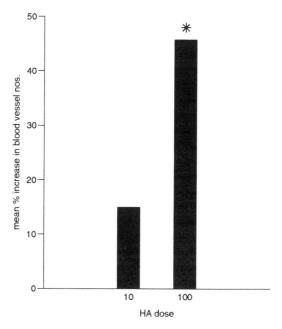

Fig 2 Subcutaneous implants of methylcellulose discs containing HA, induced marked vascularisation.

(ii)   HA and Skin Angiogenesis. Methylcellulose discs containing 0, 10 or 100 µg of o-HA were surgically implanted subcutaneously in both sides of the backs of rabbits. 5 days later skin biopsies were fixed in formalin and processed for routine histology. Implantation of HA resulted in a dose dependent increase in the number of blood vessels (Fig 2).

In a related investigation, daily topical application of o-HA for 5 days on the back of rats induced a significant increase in the number of blood vessels in 6 of 11 rats compared with control sites which were treated with buffer alone (Fig 3). This action of o-HA may prove useful in retarding blood vessel paucity and degeneration observed during the ageing process and following radiotherapy. Since vascularity of skin is the single most important determinant in the loss of hair, application of o-HA may be useful in the restoration of hair growth.

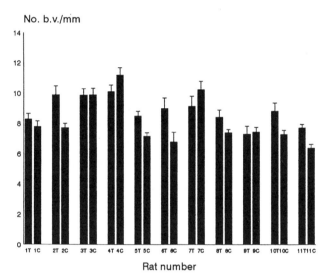

Fig 3   Topical application of HA on rat skin for 5 days induced statistically significant increase (p ≤ 0.05) in the number of blood vessels in rats (nos. 2, 5, 6, 8, 10, 11; Sattar *et al* manuscript submitted)

In a collaborative study with Dr T Fan's group in Cambridge, the effects of o-HA on blood flow in subcutaneously implanted sponges were examined. Daily injections of o-HA into the sponges in rats for 5 days significantly promoted neovascularisation which was confirmed by an increased clearance of locally injected ^{133}Xe and concomitant increase in protein, and DNA synthesis, and more importantly in blood vessel numbers (Smither *et al*, manuscript in preparation).

(iii)   HA and Angiogenesis in myocardial infarcts. The possible relevance of HA in myocardial infarcts was investigated using two different animal experimental models. Professor W Schaper has pioneered a method to induce progressive occlusion of a coronary artery in pig heart, which results in angiogenesis. In general, infarcted and adjacent normal-looking heart tissue (5/6) induced angiogenesis in the CAM assay, whereas normal non-infarcted heart tissue contained little or no angiogenic activity. More relevant to the present review is the fact that low molecular mass HA was found in the angiogenic but not in normal non-angiogenic heart tissue. More recently, in collaboration with Dr D Smith we have shown that the intracardiac infusion of o-HA resulted in an increase in the actual number of blood vessels per unit area in treated compared with control heart tissue (our unpublished data).

(iv) Topical application of HA in healing and non-healing wounds. A striking change in the ECM of normally healing wounds is the replacement of non-sulphated GAG, HA, with sulphated GAGs. In non-healing venous leg ulcers this transition does not occur (Patrick, personal communication). In two patients with venous leg ulcers and in a pig with full thickness wounds, topical application of o-HA failed to induce any apparent improvement in healing. However, this failure may be due to limitations imposed by dose and mode of application. It is possible that aqueous HA solution as was used in these cases may not be an ideal method of application of o-HA.

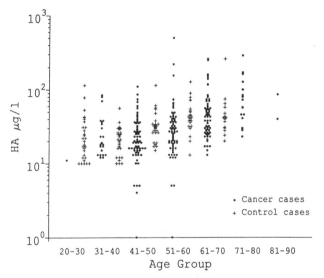

Fig 4   The serum HA levels in 238 women with breast carcinomas, when compared with 120 normal donors, were of no diagnostic or prognostic value (for details see Ponting *et al*, 1992).

(v) HA and Tumour Angiogenesis. The role of HA in tumour development may be inferred, if not directly demonstrated, from its known effects on cell behaviour and roles in other physiological events. Increased levels of HA have been found in human and animal tumours. Metastatic variants of a melanoma cell line (Turley and Tretiak, 1985) produce more HA than do their less aggressive counterparts and highly metastatic variants of mouse mammary carcinoma cells have been shown to possess a large pericellular HA coat compared with lesser non-metastatic ones (Kimata *et al*, 1987). It should be noted that despite the increase in production, the carcinoma cells still produce less HA than do normal fibrobla

We have examined the occurrence of HA in the sera of patients with mammary carcinoma and renal tumours using a specific radiometric assay. The serum HA levels of 238 women with mammary carcinoma showed no increase compared with 120 control, sex and age matched, sera (Fig 4). A number of prognostic factors were evaluated including stage of disease, lymph node involvement, tumour size, histology and presence of oestrogen and progesterone receptors in the tumours. No correlation was found with serum HA concentration and thus serum HA level is of no prognostic significance in breast cancer (Ponting et al, 1992). In another investigation a radiometric assay was used to measure HA levels in 57 children with renal tumours (55 Wilms' tumours and 2 bone-metastasing renal tumours of childhood-BMRTC) and 20 normal siblings (Kumar *et al*, 1989; Fig 5). The HA levels in the sera of normal children were barely detectable and had a molecular mass of 1-5 x 10^5 Da. In both Wilms' and BMRTC patients, very high levels of HA were found in pre-operative serum samples; these fell dramatically following surgical excisions of the tumours. A novel finding was the presence of low molecular mass HA (similar to angiogenic o-HA) in the sera only of BMRTC patients. Therefore the molecular mass of HA rather than its serum concentrations alone may be a hallmark of certain tynes of malignancy, especially bone metastasis (Kumar et al, 1989).

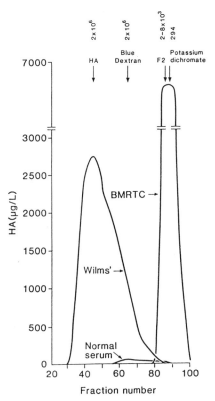

Fig 5 While sera from children with both Wilms' tumours and bone metastasizing renal tumour of childhood - BMRTC, contained very high levels of HA, the presence of low molecular mass HA was the hallmark of BMRTC (for details see Kumar *et al*, 1989).

Using specific staining methods, we have studied the distribution of HA and one of its binding proteins, hyaluronectin (HN) in breast carcinomas (Ponting et al, 1993). The two components of the ECM were apparently co-localised in all tissues. The levels of both were observed to be increased in neoplastic compared to normal tissues. In particular the staining was strongest immediately adjacent to the invading part of the tumours (Fig 6).

(B)   HA and *in vitro* Angiogenesis

(i)   Endothelial cell proliferation and migration. Since in the above *in vivo* test systems o-HA induced angiogenesis, its effects were also examined using tissue cultured endothelial cells. These findings have been published and can be summarised as follows (West and Kumar, 1989; Sattar *et al*, 1992; Rooney *et al*, 1993) For both large and microvessel derived endothelial cells o-HA had no effect on cell morphology, but DNA synthesis and cell proliferation was induced by o-HA (maximal effect 0.1 - 1.0 µg/ml). Macromolecular HA caused inhibition of DNA synthesis and cell proliferation (~ 100 µg/ml) in bovine aortic endothelial cells (BAEC). Effects are specific, i.e. not seen with fibroblasts or smooth muscle cells by us, although other workers have shown stimulation of fibroblasts by high molecular mass HA (Knudson, 1989). The stimulatory effect was not due to contaminants of HA e.g.

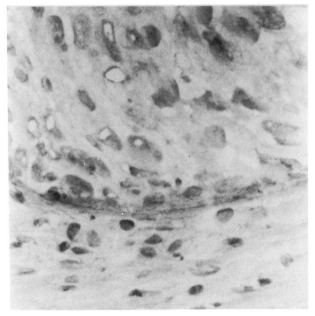

Fig 6   In breast cancer tissue the strongest staining for HA was seen immediately adjacent to the invading front of the tumour (for details see Ponting *et al*, 1993).

chondroitin sulphate. Fluorescein isothiocyanate conjugated HA was specifically taken up by endothelial cells. [³H]-labelled HA bound to HA receptors on endothelial cells - it was calculated that there were approximately 2000 receptors sites per cell (Kd 10⁻¹⁰M). The HA receptor is thought to be part of the CD44 family (Underhill, 1992). In cell migration experiments o-HA selectively stimulated BAECs migration through millipore filters in a dose-dependent manner yet had little or no effect on fibroblasts or smooth muscle cells. It is important to highlight the fact that o-HA failed to stimulate proliferation and migration of brain tumour derived endothelial cells. This is yet another example of endothelial cell heterogeneity.

(ii)   Collagen biosynthesis.   As mentioned earlier the application of o-HA induced deposition of collagen fibrils in the *in vivo* CAM assay. We wondered whether endothelial cells *in vitro* would also show enhanced synthesis of collagen. The treatment of sub-confluent cultures of BAEC with o-HA (1µg/ml) increased the uptake of [³H] proline by approximately 60%. SDS-polyacrylamide gel electrophoresis of treated cultures demonstrated the enhanced synthesis of type I and type VIII collagens. The production of the type VIII collagen was further confirmed by Western blotting and immunocytochemistry using specific antibodies kindly provided by Drs E H Sage and R Kittleberger (Rooney *et al*, 1993; Fig 7).

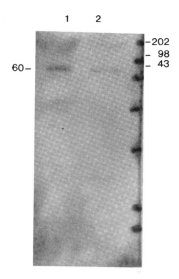

Fig 7   Treatment of aortic endothelial cells with o-HA induced the enhanced synthesis of type I and type VIII collagens. This figure is a Western blot showing the presence of type VIII collagen (for details see Rooney *et al*, 1993

(3)   HA AND GENE REGULATION

In a collaborative unpublished study with Drs R Deed and J Norton, we have studied the nature of signalling pathways involved in the stimulation of BAEC by HA. As a first step we have used the BAEC to investigate the effects of HA on the induction of immediate early response genes (ERG): Krox 20, Krox 24, c-fos, c-jun and jun-B. The treatment with o-HA resulted in a rapid, transient induction in expression of all the ERGs, which is consistent with a mitogenic signalling cascade typical of that elicited by ligand-receptor interaction in other cell types. These early signalling pathways alone were not sufficient to evoke a proliferative response in BAEC, an effect only brought about by continuous exposure to o-HA. Treatment of BAEC with non-angiogenic/anti-angiogenic high molecular mass n-HA did not induce expression of ERGs. Moreover, n-HA was able to inhibit o-HA induced ERG expression in a dose-dependent manner. The clear implication of these findings is that anti-angiogenic and angiogenic signals elicited by n-HA and o-HA respectively are integrated at a very early stage in cellular signalling pathways either at or immediately downstream, of the HA receptor. It is known that ERG upregulation results in an increased transcription of ERG dependent genes, for example of collagenase and stromelysin (Fig 8; Curran and Franza, 1988). How the

HA binds to HA receptor on cell

↓

Increased transcription of ERGs
jun B, c-jun, c-fos, Krox-20, Krox-24

↓

Increased transcription of ERG-dependent genes
(e.g. collagenase, stromelysin)

↓

Cell growth

Fig 8   Flow chart to show how binding of HA to cell receptor can lead to cell growth.

induction of collagenase and increased synthesis of collagen occurs in the same cells remains to be resolved. There is probably a spatial and temporal compartmentalisation of these events. Whether protein kinase C dependent or independent pathways are involved in the induction of o-HA induced angiogenesis is not known.

(4)   PRACTICAL/THERAPEUTIC USES OF HA

HA has been the subject of numerous patents (Balazs and Denlinger, 1989). The following is a short account of clinical situations where HA has been of practical value. HA has rheological properties which are not matched by any other natural or synthetic polymers.

(A)   HA and Disease States

The normal adult human plasma contains ~ 50µg HA/litre. Highly increased levels of HA have been found in the sera of other body fluids of patients with inflammatory rheumatic diseases, liver cirrhosis, Crohn's and coeliac disease, mesothelioma, and renal tumours. In these disorders HA levels correlated with the extent of disease (for review see Engstrom-Laurent, 1989). Patients with septicaemia and shock with fatal outcome had higher serum levels than patients who survived (Berg *et al*, 1988). The practical value of estimation of HA levels in the sera of individuals who are at an increased risk of developing mesothelioma has been demonstrated with some success (Pluygers *et al*, 1991).

(B)   Viscosurgery and HA

During various surgical procedures, an elastoviscous coloured solution or wafer of high molecular mass HA is used as a space marker to separate tissue and create the space for surgical manipulations and for the insertion of implants (Lieseg-ang, 1990; Assia *et al*, 1992; Smith and Burt, 1992). Apart from acting as tissue protector or shock absorber and lubricant, when left at the site of surgery it can reduce adhesion. Viscosurgical implants can also minimise post-operative bleeding and exhudation and therefore promote healing. HA has been used as an implant and surgical tool for the reattachment of detached retina. Perhaps the widest use of HA is in the removal of cataractous lens. More recently, both in animals and humans, HA was reported to improve the healing process and reduce the formation of fibrous tissue on application to perforated tympanic membranes (Rivas-Lacarte *et al*, 1992 and Spandow and Hellstrom, 1992).

(C)   HA and artificial tears

HA is the only ingredient in some artificial tear drops used by patients who cannot produce sufficient lubricant naturally. Using quantitative scintography, the effectiveness of hydroprophyldmethylcellulase, polyvinyl alcohol and HA were compared. The half life of HA was nearly 6-fold greater than the other two solution ($P \leq 0.01$). This was attributable to high viscosities and low shear rates of HA (Balazs and Baird, 1984; Snibson *et al*, 1992).

(D)   HA and Cosmetics

One of the oldest uses of HA is in medical and personal care products such as hand creams and body lotions (Baird, 1985). Its possible ability to improve skin vascularity and ameliorate radiation damage to vasculature are other examples of the potential uses of HA which are at an early exploratory stages.

(E)   HA and joint disorders

Bertolami et al (1993) assessed the efficacy of high molecular mass HA as a treatment for certain intracapsular temporomandibular disorders in 121 patients. Their conclusions were that twice as many patients treated with HA showed improvement compared to those given placebo. Furthermore, only 3% of patients who were treated with HA relapsed in contrast with 31% given placebo.

(5)   CONCLUSIONS

As a major component of the extracellular matrix, HA is known to influence a variety of physiological events. One of the events central to tumour growth is angiogenesis and HA has been demonstrated *in vivo* to influence this process in a size and concentration dependent manner. Its increased deposition in breast carcinoma compared to normal tissue has been demonstrated, although its concentration in sera showed that it has no diagnostic or prognostic value. *In vitro* addition of O-HA has been found to promote vascular endothelial cell migration and proliferation. This is likely to be mediated via cell surface HA receptors. Indeed a receptor has been identified in vascular endothelial cells and shown to mediate the uptake of HA. A significant finding is the stimulation of early response genes in endothelial cells by O-HA but not native HA, indicating a role for angiogenic fragments of HA early in the stimulation of endothelial and tumour cell metastasis. One such change may be the observed increase in deposition of collagen type VIII by O-HA-treated endothelial cells.

Existing clinical uses of HA exploiting its physical properties include viscosurgery and lubrication. An increased knowledge of the mechanisms by which HA affects cell behaviour will aid our understanding of the roles played in a variety of physiological events and may lead to an ability to target HA metabolism in order to influence cell behaviour.

REFERENCES

Assia EA, Apple DJ, Lim ES, Morgan RC and Tsai JC. Removal of viscoelastic materials after experimental cataract surgery *in vitro*. J Cataract Refract Surg 18, 3-6 (1992).

Balazs EA and Denlinger JL. Clinical uses of hyaluronan. Ciba Symposium, 143, Wiley, Chichester, 265-280 (1989).

Band P. Effective use of hyaluronic acid. Drugs and Cosmetics, 53, 54-57 (1985).

Berg S, Brodin B, Hesselvik F, Laurent TC and Maller R. Elevated levels of plasma hyaluronan in septicaemia. Scand J Clin Lab Invest 48, 727-732 (1988).

Bertolami CN, Gay T, Clark GT, Rendell J, Shetty V, Liu C and Swann DA. Use of sodium hyaluronate in treating temporomandibular joint disorders: a randomised, double blind, placebo-controlled clinical trial. J Oral-Maxillofac Surg 51, 232-242 (1993).

Curran T and Franza BR. Fos and Jan: the AP-1 connection. Cell 55, 1009-1011.

Decker M, Chiu ES, Dollbaum C, Moiin A, Hall J, Spenlove R, Longaker MT and Stern R. Hyaluronic acid-stimulating activity in sera from the bovine fetus and from breast cancer patients. Cancer Res 49, 3499-3505 (1989).

Engstrom-Laurent A. Changes in hyaluronan concentration in tissues and body fluids in disease states. Ciba Symposium 143, Wiley, Chichester pp 233-247 (1989).

Fraser JRE and Laurent TC. Turnover and metabolism of hyaluronan. Ciba Symposium No. 143, Wiley, Chichester pp41-59.(1989).

Iruela-Arispe ML and Sage EH. Expression of type VIII collagen during morphogenesis of the chicken and mouse heart. Dev Biol 144, 107-118 (1991).

Ingber DE and Folkman J. Inhibition of angiogenesis through modulation of collagen metabolism. Lab Invest 59, 44-51 (1988).

Hamerman D, Sasse J, Klagsbrun M. A cartilage-derived growth factor enhances hyaluronate synthesis and diminishes sulphated glycosaminoglycan synthesis in chondrocytes. J Cell Physil 127, 317-322 (1986).

Hay ED. The cell biology of extracellular matrix. Plenum Press, NY (1991).

Heldin P, Laurent TC, Heldin CH. Effect of growth factors on hyaluronan synthesis in cultured human fibroblasts. Biochem J 258, 919-922 (1989).

Kimata K, Yoneda M and Morita H. Interaction of extracellular matrix macromolecules between host and metastatic tumour cells. Gan To Kagaku Ryoho 14, 2025-32 (1987).

Kumar S, West DC, Ponting JM and Gattamaneni HR. Sera of children with renal tumours contain low-molecular-mass hyaluronic acid. Int J Cancer <u>44</u>, 445-448 (1989).

Larnier, C., Kerneur, C., Robert, L., Moczar, M. (1989). Effect of testicular hyaluronidase on hyaluronate synthesis by human skin fibroblasts in culture. Biochim. Biophys. Acta, 1014: 145-152.

Laurent TC, Hellstrom S and Fellenius E. Hyaluronan improves the healing of experimental tympanic membrane perforations. A comparison of preparations with different rheologic properties. Arch Otolaryngol Head Neck Surg <u>114</u>, 1435-1441 (1988).

Laurent TC and Fraser JRE. Hyaluronan. FASEB J <u>6</u>, 2397-2404 (1992).

Lembach KJ. Enhanced synthesis and extracellular accumulation of hyaluronic acid during stimulation of quiescent human fibroblasts by mouse epidermal growth factor. J Cell Physiol <u>89</u>, 277-288 (1976).

Liesegang TJ. Viscoelastic substances in ophthalmology. Surv Ophthalmology <u>34</u>, 268-293 (1990).

Maragoudakis ME, Missirlis E, Karakiulakis G, Bastiki M and Isopanoglou N. Basement membrane biosynthesis as a target for developing inhibitors of angiogenesis with anti-tumour activity. Int J Radiat Biol <u>60</u>. 54-59 (1991).

Morale TI. Transforming growth factor-beta 1 stimulates synthesis of proteoglycan aggregates in calf articular cartilage organ cultures. Arch Biochem Biophys <u>286</u>, 99-106 (1991).

Pluygers E, Baldewyns P, Minnte P, Beauduin M, Gourdin P and Robinet P. Biomarker assessments in asbestos-exposed workers as indicators for selective prevention of mesothelioma or broncogenic carcinoma: ratio rank and practical implementations.Eur J Cancer Prev <u>1</u>, 57-68 (1991).

Ponting J, Howell A, Pye D and Kumar S. Prognostic relevance of serum hyaluronan levels in patients with breast cancer. Int J Cancer <u>52</u>, 873-876 (1992).

Ponting J, Kumar S and Pye D. Co-localisation of hyaluronan and hyaluronectin in normal and tumour breast tissues. Int J Onclogy <u>2</u>, 889-893 (1993).

Prehm P. Release of hyaluronate from eukaryotic cells. Biochem J <u>267</u>, 185-189 (1990).

Rivas-Lecarte MP, Casnasin T and Alonso A. Effects of sodium hyaluronate on tympanic membrane perforations. J Int Med Res <u>20</u>, 353-359 (1992).

Rodén L, Campbell P, Fraser JR, Laurent TC, Pertoft H and Thompson JN. Enzymic pathways of hyaluronan catabolism. Ciba Symposium 143, Wiley, Chicester pp60-86,(1989).

Rooney P and Kumar S. Inverse relationship between hyaluronan and collagens in development and angiogenesis. Differentiation <u>54</u>, 1-9 (1993)

Rooney P, Wang M, Kumar P and Kumar S. Angiogenic oligosaccharides of hyaluronan enhance the production of collagens by endothelial cells. J Cell Sci <u>105</u>, 213-218 (1993).

Sattar A, Kumar S, West D. Does hyaluronan have a role in endothelial cell proliferation of the synovium? Sem Arth Rheum <u>22</u>, 37-43 (1992).

Schor SL, Schor AM, Grey AM, Chen J, Rushton G, Grant ME and Ellis I. Mechanism of action of the migration stimulating factor produced by fetal and cancer patient fibroblasts: effect on hyaluronic acid synthesis. In Vitro Cell Dev Biol <u>25</u>, 737-746 (1989).

Smith KD and Burt WL. Fluorescent viscoelastic enhancement. J Cataract-Refract-Surg <u>18</u>, 572-576 (1992).

Snibson GR, Greaves JL, Soper ND, Tiffany JM, Wilson CG and Bron AJ. Ocular surface residence of artificial tear solutions. Cornea <u>11</u>, 288-293 (1992).

Spandow O and Hellstrom S. Healing of tympanic membrane perforation - a complex process influenced by variety of factors. Acta Otolaryngeal-Suppl-Stockh <u>492</u>, 90-93 (1992).

Turley EA and Tretiak M. Glycosaminoglycan production by murine melanoma variants *in vivo* and *in vitro*. Cancer Res <u>45</u>, 5098-5105 (1985).

Underhill CB. CD44: The hyaluronan receptor. J Cell Sci <u>104</u>, 293-298 (1992).

West DC and Kumar S. The effect of hyaluronate and its oligosaccharides on endothelial cell proliferation and monolayer integrity. Exp Cell Res <u>183</u>, 179-196 (1989).

West DC and Kumar S. Hyaluronan and angiogenesis. Ciba Symposium <u>143</u>,Wiley,Chichester pp187-207 (1989).

Yaron I, Meyer FA, Dayer JM, Yaron M. Human recombinant interleukin-1 beta stimulates glycosaminoglycan production in human synovial fibroblast cultures. Arthritis Rheum <u>30</u>, 424-430 (1987).

CHRONIC CELLULAR INJURY AND HUMAN PROLIFERATIVE DISORDERS

Harry N. Antoniades

The Center for Blood Research and
the Department of Cancer Biology and Nutrition
Harvard School of Public Health
Boston, MA 02115

INTRODUCTION

Carcinogenesis and normal tissue regeneration share some common characteristics. In both states there is an active cellular proliferation and induction of the expression of potent mitogenic growth factors and receptors that promote the proliferative activity. Normal tissue regeneration is initiated by acute injury. Carcinogenesis is linked to prolonged chronic cellular injury by a variety of toxic agents including carcinogen. The major difference between the two processes is that in normal tissue repair, both cellular proliferation and growth factor expression are suppressed upon the healing of the wound. In contrast, in cancer and in some nonmalignant proliferative disorders cellular proliferation remains uncontrolled, and the expression of the mitogenic growth factors and receptors remains unsuppressed. It seems that in these disorders, the pathologic lesion is not the induction of growth factor and cytokine expression. Induction of these factors in response to acute or chronic injury appears to be a physiologic process aiming at the repair of the damage caused by the injury. What is abnormal in these pathologic states is a malfunction of the reversal mechanisms that normally control the suppression of the genes encoding for these growth factors and cytokines.

I have proposed the hypothesis that prolonged chronic cellular injury is the common initial event in the development of several human proliferative disorders and neoplasias[1]. This supposition is based on parallel *in vivo* studies aiming at the understanding of the molecular basis of normal tissue repair and of pathologic proliferative disorders. As described below, these investigations support a unifying molecular basis for the development of several proliferative disorders and neoplasias.

ACUTE INJURY INDUCES THE LOCALIZED EXPRESSION OF MITOGENIC GROWTH FACTORS AND RECEPTORS

In vivo studies in swine have shown that acute cutaneous injury induces the expression of selective mitogenic growth factors and receptors. This expression was localized in the skin epithelial and connective tissue cells of the injured tissue[2,3]. For example, injury induced the co-expression of the c-*sis*/ PDGF-B mitogen and its receptor β (PDGF-R β) in the skin epithelial cells at the site of injury[2]. This finding was unexpected because normally epithelial cells do not express PDGF and PDGF receptors. This co-expression was seen only *in vivo* in the malignant epithelial cells of primary human tumors, including lung, gastric, and prostatic carcinomas.

Acute injury was also shown to induce the localized expression of the c-*sis/* PDGF-B mitogen in the connective tissue fibroblasts of the injured tissue[2]. Again, this expression was unexpected since normally fibroblasts express only PDGF receptors but not c-*sis*/PDGF B mitogen. Expression of the PDGF mitogen was seen in human malignant fibroblasts derived from fibrosarcomas, and in malignant astrocytes, and meninges of primary human astrocytomas and meningiomas (reviewed in Ref. 1).

As shown in Figure 1, a similar expression after injury was seen with other mitogenic growth factors including transforming growth factor alpha (TGF-α) and its epidermal growth factor receptor (EGF-R), and acidic and basic fibroblast growth factors (aFGF, bFGF)[3]. Other reports also provided evidence for the expression of growth factors during wound healing[4-10].

The intensity of the expression of the mitogenic growth factors and receptors induced by injury paralleled the proliferative state of the injured tissue[2,3]. The expression was highest at the stages of active cellular proliferation, it declined progressively during the healing phase, and it was totally suppressed upon the healing and re-epithelialization of the injured tissue which in our model occurs at about 5 days after injury[11]. This controlled reversible expression of mitogenic growth factors and their receptors in response to acute injury suggests that, at the cellular level, molecular mechanisms can be activated that contribute to physiologic processes of normal tissue repair. These mechanisms appear to control not only the induction of gene expression in response to cellular injury but also suppression upon healing. As described below, in contrast to these controlled physiologic processes induced by acute injury, prolonged chronic injury may lead to inappropriate, unsuppressible cellular gene expression that contributes to a pathologic state.

We have investigated recently the *in vivo* response of p53 to injury. The wild-type p53 protein is a tumor suppressor gene and this function is mediated by its ability to arrest cell cycle progression and proliferation (for recent reviews, see Refs. 12-14). The time-course of the expression of the p53 mRNA and its protein product in the injured tissue was found to be opposite to that seen with the expression of the mitogenic growth factors. p53 protein expression was suppressed at the stages of active cellular proliferation and it re-emerged at the stages of healing. In contrast, the growth factor expression was highest at the stages of active cellular proliferation and it was suppressed upon healing. This profile of p53 expression is consistent with its role as an anti-proliferative agent. Its suppression at the stages of active cellular proliferation eliminated a potent anti-proliferative factor. Re-emergence of its expression at the stages of healing may contribute to the down regulation of the proliferative process and possibly contribute to the suppression of the mitogenic genes. Wild-type p53 nuclear protein is a transcriptional transactivator and transcriptional suppressor of genes that are involved in the control of cellular growth, including suppression of c-*fos* and c-*jun* (for reviews see Refs. 12-14). Suppression by p53 may involve transcriptional transactivation of other genes which in turn induce transcriptional suppression. The suppressing activity of wild-type p53 protein is removed in malignancies through mutation-inactivation of the p53 protein[12-14].

In summary, normal tissue regeneration in response to acute injury includes the selective localized expression of mitogenic factors that promote cell proliferation, and, the concomitant suppression of p53 which inhibits cell proliferation. At the healing stage there is a decline in the expression of the mitogenic growth factors, and total suppression upon healing. In contrast, at the healing stage there is a strong expression of p53 which serves to downregulate cellular proliferation.

EXPRESSION OF GROWTH FACTORS IN PROLIFERATIVE DISORDERS: SIMILARITIES WITH THE EXPRESSION INDUCED BY ACUTE INJURY

As mentioned above, I have promoted the hypothesis that chronic injury can contribute to the development of proliferative disorders and neoplasias by inducing the

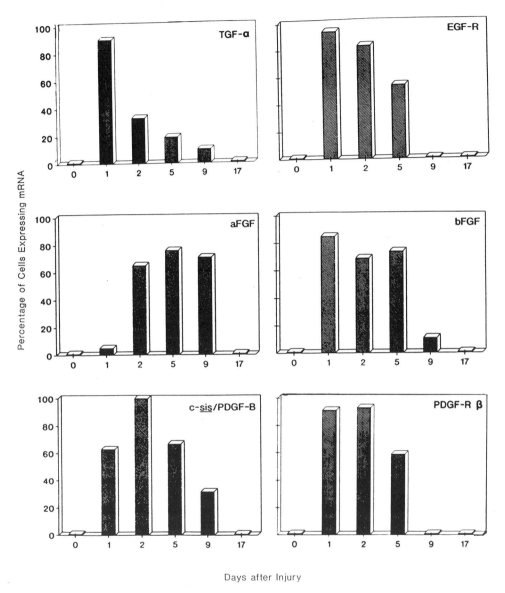

Figure 1. Localized expression of TGF-α, EGF-R, aFGF, bFGF, c-*sis*/PDGF-B, and PDGF-R β in skin epithelial cells before and after acute cutaneous injury. Notice the lack of expression in the epithelial cells of control, uninjured tissue. From Antoniades *et al.*[3].

expression of selective cellular proto-oncogenes and genes encoding for mitogenic and angiogenic growth factors and their receptors and by disrupting the mechanisms that normally control suppression of these genes. This hypothesis is supported by the *in vivo* studies described above, that link acute injury to expression of mitogenic and angiogenic growth factors. This hypothesis provides a common molecular basis for the understanding of the initial event leading to pathologic states. Furthermore, it provides a basis for the understanding of epidemiologic findings linking malignant transformation and nonmalignant proliferative disorders to diverse environmental factors. For example, epidemiologic studies have linked cigarette smoking to lung cancer; inhalation of coal dust, asbestos, beryllium and a variety of organic dusts to lung disorders; dietary factors to gastric cancer; aflatoxins and alcohol to liver cirrhosis and carcinoma; solar radiation to skin cancer. In other cases, chronic injury may result from immune processes causing inflammation and the localized release of toxic agents by inflammatory cells.

"Inappropriate", or "aberrant" expression of mitogenic growth factors has been reported in several pathologic cases. *In vivo* studies have shown the strong expression of c-*sis*/PDGF-B mRNA and its protein product in the lung alveolar epithelial cells and macrophages of human patients with idiopathic pulmonary fibrosis[15]. This disease is characterized by inflammation and fibrosis. The fibrotic process includes excessive fibroblast proliferation and collagen deposition within the alveolar structures leading to progressive respiratory failure. There was no significant expression of PDGF-B expression in the lung epithelial cells of non-IPF subjects. IPF is linked to chronic injury of unknown etiology. The strong c-*sis*/PDGF-B expression seen in the lung epithelial cells of these patients is similar to that seen in the skin epithelial cells following acute injury. However, the expression seen after acute injury is suppressed upon healing, but the expression in IPF remains unsuppressed.

Studies in primary human lung[16], gastric[17], and prostate carcinomas[18] have demonstrated the co-expression of c-*sis*/PDGF-B and its receptor, the PDGF-R β mRNAs and respective protein products in the malignant epithelial cells of the tumors of these patients. There was no significant expression of c-*sis*/PDGF-B and its receptor in nonmalignant lung, gastric, or prostate epithelial cells. The expression in the malignant epithelial cells remained unsuppressed, while a similar co-expression seen in the skin epithelial cells after acute injury was suppressed upon healing. Primary lung carcinoma is linked to cigarette smoking, gastric carcinoma to dietary factors, and prostatic carcinoma to environmental factors.

"Inappropriate" expression of c-*sis*/PDGF-B has been reported in malignant cell lines derived from human fibrosarcomas, and in malignant astrocytes and meninges of primary human astrocytomas and meningiomas (reviewed in Ref. 1). This expression was not present in their nonmalignant cellular counterparts. There is no known direct connection between environmental factors and primary human astrocytomas and meningiomas. However, it has been reported that some tumors may arise in patients who incur scars from head injuries. Cranial irradiation and exposure to some chemicals has been linked to an increased incidence of both astrocytomas and meningiomas. The predominance of meningiomas in women and the effects of menses and pregnancy on the development of meningiomas have implicated hormonal influences in their etiology (reviewed in Ref. 19).

Increased levels of expression of growth factors and cytokines has been reported in human atheromatous plaques compared to uninvolved vessels (reviewed in Refs. 20-21). Chronic vascular injury caused by a variety of factors, including hyperlipidemia, hypertension, and exogenous toxins, is considered to play an important role in the development of this disorder.

EXPRESSION OF GROWTH FACTORS IN PROLIFERATIVE DISORDERS IS A PHYSIOLOGIC RESPONSE TO INJURY. THE PATHOLOGIC LESION IS THE BREAKDOWN OF THE SUPPRESSING MECHANISMS.

The term "inappropriate", or "aberrant" expression has been used to describe growth factor and cytokine expression seen in malignant cells that was not present in their normal counterparts. This "inappropriate" expression was assumed to represent a pathologic lesion, contributing to the uncontrolled growth of the malignant cells. However, as described above, investigations in normal tissue regeneration have shown a similar expression by the regenerating cells following injury. This expression induced by injury, serves to promote the repair of the injured tissue and it was suppressed upon healing. In pathologic cases linked to prolonged chronic injury, the expression of the mitogenic factory remains unsuppressed. Thus, gene induction in response to acute or chronic injury is a physiologic process aiming at the repair of the damage caused by the injury. What is abnormal in pathologic cases is a malfunction of the reversal mechanisms that suppress gene induction. It seems that at some point, a cell abused by prolonged chronic injury has lost the ability to suppress the expression of the mitogenic genes induced by the injury. This abused cell will enter into uncontrolled proliferation, promoted by the expression of mitogenic factors. To borrow the expression of Dvorak[22], this abused cell is a cell "that does not heal".

ACKNOWLEDGMENTS

This research was supported by the National Institutes of Health Grant CA30101, and by the Institute of Molecular Biology. I thank Amal Ghaly for preparation of the manuscript.

REFERENCES

1. H.N. Antoniades, Linking cellular injury to gene expression and human proliferative disorders: examples with the PDGF genes, *Mol. Carcinog.* 6:175 (1992).

2. H.N. Antoniades, T. Galanopoulos, J. Neville-Golden, C.P. Kiritsy, and S.E. Lynch, Injury induces *in vivo* expression of PDGF and PDGF-receptor mRNAs in skin epithelial cells and PDGF mRNA in connective tissue fibroblasts, *Proc. Natl. Acad. Sci. USA* . 88:565 (1991).

3. H.N. Antoniades, T. Galanopoulos, J. Neville-Golden, C.P. Kiritsy, and S.E. Lynch, Expression of growth factor and receptor mRNAs in skin epithelial cells following acute cutaneous injury, *Am. J. Pathol.* 142:1099 (1993).

4. H.-A. Hansson, E. Jennische, and A. Skottner, Regenerating endothelial cell express insulin-like growth factor I immunoreactivity after arterial injury, *Cell Tissue Res.* 250:499 (1987).

5. G.R. Grotendorst, C.A. Grotendorst, and T. Gilman, Production of growth factors (PDGF an TGF-beta) at the site of tissue repair, *Prog. Clin. Biol. Res.* 266:47 (1988).

6. G.R. Grotendorst, Y. Soma, K. Takehara, and M. Charette, EGF and TGF-alpha are potent chemoattractants for endothelial cells and EGF-like peptides are present at sites of tissue regeneration. *J. Cell Physiol.* 139:617 (1989).

7. H. Steefnos, C. Lossing, and H.A. Hansson, Immunohistochemical demonstration of endogenous growth factors in wound healing, *Wounds* 2:218 (1990).

8. C.J.M. Kane, P.C. Hanawait, A.M. Knapp, and J.N. Mansbridge, Transforming growth factor-β1 localization normal and psoriatic epidermal keratinocytes *in situ, J. Cell Physiol.* 144:144 (1990).

9. D.J. Whitby, and M.W. Ferguson, Immunohistochemical localization of growth factors in fetal wound healing, *Dev. Biol.* 147:207 (1991).

10. S. Werner, K.G. Peters, M.T. Longaker, F. Fuller-Pace, M.J. Bander, and L.T. Williams, Large induction of keratinocyte growth factor expression in the dermis during wound healing, *Proc. Natl. Acad. Sci. USA.* 89:6896, 1992.

11. S.E. Lynch, R.B. Colvin, and H.N. Antoniades, Growth factors in wound healing: single and synergistic effects on partial thickness porcine skin wounds, *J. Clin. Invest.* 84:640 (1989).

12. B. Vogelstein, and K.W. Kinzler, p53 function and dysfunction, *Cell* 70:523 (1992).

13. S.J. Ullrich, C.W. Anderson, W.E. Mercer, and E. Appella, The p53 tumor suppressor protein, a modulator of cell proliferation, *J. Biol. Chem.* 267:15259 (1992).

14. G.P. Zambetti, and A.J. Levine, A comparison of the biological activities of wild-type and mutant p53, *FASEB J.* 7:855 (1993).

15. H.N. Antoniades, M.A. Bravo, R.E. Avila, T. Galanopoulos, J. Neville-Golden, M. Maxwell, and M. Selman, Platelet-derived growth factor in idiopathic pulmonary fibrosis, *J. Clin. Invest.* 86:1055 (1990).

16. H.N. Antoniades, T. Galanopoulos, J. Neville-Golden, and C.J. O'Hara, Malignant epithelial cells in primary human lung carcinomas co-express *in vivo* platelet-derived growth factor (PDGF) and PDGF receptor mRNAs and their protein products, *Proc. Natl. Acad. Sci. USA.* 89:3942 (1992).

17. C.C. Chung, and H.N. Antoniades, Expression of c-sis/platelet-derived growth factor B, insulin-like growth factor I, and transforming growth factor α messenger RNAs and their respective receptor messenger RNAs in primary gastric carcinomas: *In vivo* studies with *in situ* hybridization and immunocytochemistry, *Cancer Res.* 52:3453 (1992).

18. M. Xiao, T. Galanopoulos, J. Neville-Golden, J.P. Richie, and H.N. Antoniades, Human prostate adenocarcinomas express *in vivo* mRNA and protein products for platelet-derived growth factor B and its receptor, *Int. J. Oncol.* in press.

19. P. McL Black, Brain tumors, *N. Eng. J. Med.* 324:1471 (1991).

20. S.K. Clinton, and P. Libby, Cytokines and growth factors in atherogenesis, *Arch. Pathol. Lab. Med.* 116:1292 (1992).

21. R. Ross, The pathogenesis of atherosclerosis: a perspective for the 1990s, Nature 362:801 (1993).

22. H.F. Dvorak, Tumors: wounds that do not heal, *N. Eng. J. Med.* 315:1650 (1986).

CELLULAR AND MOLECULAR MECHANISMS OF ANGIOGENESIS IN THE BRAIN

A.M.Tsanaclis, MD, Ph.D.

Laboratory of Experimental Neuropathology. University of
São Paulo School of Medicine. Av. Dr. Arnaldo 455
01246-000 São Paulo, Brasil

The formation of new capillaries occurs in a variety of normal and pathological conditions, which include embryo and organ development, wound healing, atherosclerosis and tumor growth. In embryogenesis and wound healing this process is well regulated and ceases when tissue repair or organ development is completed; in contrast, new vessels in the tumor stroma seems to be continuously induced by the growing tumor.

In the normal brain, which is a well vascularized organ, the turnover of the endothelial cells is extremely low - is it estimated that it takes about 23 years for a normal cerebral endothelial cell to enter the cell cycle and achieve mitosis (1). However, during organogenesis the situation is different, and young brain endothelial cells do proliferate in order to provide the forming organ with the necessary bulk of vessels for its adequate function (2). A modification of the normal status survenes when a tumor develops in the brain parenchyma - the capillary endothelial cells proliferate to a variable extent depending on the individual tumor, from slight to severe degrees of endothelial hyperplasia (glomeruloid formations). Important is the fact that in astrocytomas, the most frequent cerebral tumor in adults, endothelial hyperplasia constitutes a hallmark of malignant transformation (3) .

Endothelial cell proliferation is mediated by specific biochemical factors which are mitogenic for these cells. It has been proposed that the known endothelial cell mitogens fall into two classes: molecules related to endothelial cell growth factor, which are eluted from heparin-sepharose with lM NaCl and have an acidic pI, and molecules related to fibroblast growth factor, which bind more strongly to heparin-sepharose and have a basic pI (4,5). One of the best studied is the basic fibroblast growth factor (bFGF), a heparin-binding peptide with potent angiogenic properties either in vitro or in vivo, in developing or adult endothelial cells. Cells transfected with bFGF, fused to a signal peptide, stimulate neoplastic transformation in vitro and produce vascularized tumors in vivo (6).

Distribution of bFGF in normal tissues was described initially in basement membranes; in arterial endothelium and sub endothelial matrix and in the margins of brain wounds (7,8,9). These observations were the result of immunocytochemical studies using policlonal antisera. Recently, the cloning and production of purified recombinant bFGF have led to the development of high affinity monoclonal antibodies which permitted a more precise mapping of the distribution of bFGF within the tissues. DG2 and DE6 permitted us to localize bFGF to the basement membrane of small cerebral vessels, occasionally to neurons and to the

Angiogenesis: Molecular Biology, Clinical Aspects
Edited by M.E. Maragoudakis *et al*, Plenum Press, New York 1994

microvasculature of human brain tumors (10). bFGF exists in different molecular weight forms and these have separate intracellular localization, either cytoplasmic or nuclear (11).

Many of the biological properties of bFGF are well established by now but the mechanisms underlying the release and the intimate mode of action of bFGF are largely unknown. This is partly due to the fact that a well defined secretory signal peptide sequence associated with the bFGF gene has not been identified to date (12). Yet, the finding of a family of oncogenes that share homology with the bFGF gene associated with a signal peptide sequence suggests that altered forms of bFGF may be involved in the phenotypic switch to a neoplastic state (13,14).

bFGF exists as either a stable, inactive molecule (stored in the basement membranes) or a rapidly degradable, bioactive molecule (active form). The active form acts in the major steps of the angiogenic cascade, that is, endothelial cell mitosis, migration and remodeling of basement membranes. Thus, local stimulus to initiate angiogenesis leads to the production of proteases by the endothelial cell capable of degrade the basement membrane where bFGF is stored, and the peptide is released in the active form. The interaction between the triggering agent and the growth factor is determined by multiple factors - either genetic or epigenetic (7,15).

bFGF appears to be one of the most potent angiogenesis inducers. In vitro, the response of microvascular endothelial cells to bFGF consists of three major steps: increased production of proteolytic enzymes such as plasminogen activator and collagenase (16), an increase in the rate of cell proliferation, and a stimulation of endothelial cell migration along a gradient of angiogenic factor (17). During angiogenesis, a cascade of proteolytic events occurs, which leads to the degradation of the extracellular matrix. The production of proteinases by endothelial cells is believed to be of fundamental importance for the degradation of the perivascular extracelullar matrix and the stroma of the tissue to be vascularized (18). The rate of basement membrane biosynthesis has been used as an index to angiogenesis (19).

bFGF is neurotrophic for cultured neurons and induces plasmin activator production in different cell types including astrocytes (20). Secreted neuronal bFGF might exert a paracrine effect on brain glial and endothelial cells by stimulating their plasmin-generating system. In effect, it has been shown that human fetal neurons synthesize bFGF and neuronal bFGF stimulates plaminogen activator activity when added to cultured endothelial cells (21,22).

In at least some systemic tumors, neovasculariztion is probably mediated, in part, by macrophages. The studies of Polverini and co-workers suggest that the mechanism of tumor associated macrophages-induced neovascularization is via the secretion of a factor or factors that stimulates capillary formation (23). There is no report as to the role of these cells in the angiogenic tumoral process in the brain.

Examination of the brain tissue lying beyond the visible edge of a primary cerebral tumor discloses from essentially normal to grossly infiltrated tissue, in pace with the biological properties of these lesions; other the infiltrative behavior, the small vessels of the neuropile may or not show morphological changes ranging from normal to varying degrees of endothelial hyperplasia. This alteration is entirely independent of neoplastic invasion, and is the result of diffusible bFGF and/or other angiogenic peptides probably synthesized by the neoplastic and/or endothelial cells.

Brain tissue surrounding tumors contains varying numbers of abnormal cells that are thought to have migrated from the tumor. Since blood vessels that grow into a tumor lose their blood-brain barrier and become highly abnormal, it seems likely that blood vessels in surrounding brain might also show some degree of abnormality as a result of tumor cells and their environment.

It is well established now that the growth fraction in the vascular endothelial cells is greater than that of the tumor cells in the proliferating area of human gliomas. However, a frequent observation in this group of tumors is endothelial hyperplasia in the vessels of the neuropile free from tumor cells, in close proximity of the neoplastic mass; this observation is compatible with the hypothesis that a soluble factor diffuses at a distance from the tumor mass to the surrounding tissue, thus stimulating angiogenesis. However, why this phenomenon does not occur in all gliomas, specially in the glioblastoma multiforme variety where the growth fraction is high, is difficult to answer. Even if it is well established that malignancy and angiogenesis are linked, it seems that, at least for the malignant astrocytomas, factors other than angiogenic play an important role in determining malignancy.

In human brain tumors we detected immunocytochemically the presence of bFGF predominantly in the malignant ones. In particular, for astrocytomas, our observations indicate that bFGF can be detected in the malignant variants, suggesting that this peptide may have an important role in the development of a malignant phenotype. In addition our studies point to a link between the cell proliferation index, as evaluated by the Ki-67 monoclonal antibody, and the expression of bFGF - for astrocytomas, bFGF can be detected in the anaplastic forms but not in the benign varieties. It is interesting to note that, in meningiomas the detection of bFGF was restricted to that recurrent or malignant types (10).

bFGF has been implicated in the modulation of tumor invasiveness. To test the invasive potential we implanted NIH-3T3 cells transfected with the bovine brain bFGF gene with (6-1) or without (B-7) an artificial signal peptide sequence in the brains of nude mice and we observed the morphological appearance after application of these cells to a Matrigel. 6-1 cells form extensive branching colonies as early as 24 hours after plating on Matrigel coated dishes; B-7 cells form tight individual colonies. Light and electron microscopy of the experimental tumors showed that intracerebral 6-1 invades largely the neuropile surrounding the main tumoral mass; in contrast, B-7 cell tumors grew within the boundaries of a kind of capsule constituted by an amorphous basement membrane-like substance that separated the tumor from the neuropile. These findings suggest that bFGF has a role in the modulation of tumor invasiveness (24).

The demonstration of bFGF within the tumor microvasculature and surrounding neuropile as well as the modulator effect of the peptide on invasiveness provide a molecular target for the adjuvant therapy by angiosuppression and for anti-invasion. Experimental studies are currently being developed.

Human neurofibrosarcoma implanted in the sub renal capsule of nude mice had retard growth and a slight angiogenic process when the animals were treated with a combination of heparin and hydrocortisone (25) thus corroborating previous observations (26). Heparin alone promotes angiogenesis in vivo and the angiostatic steroids have usually an effect on the induction of new vessels. Potent inhibition of angiogenesis requires the combined effect of both agents - heparin and steroids.

The antiangiogenic effect of low serum copper in experimental animals implanted intracerebrally with VX2 carcinoma or 9L gliosarcoma was well established (27); in this model, invasiveness of the brain parenchyma by the tumor cells was clearly inhibited by low serum copper levels (28).

Suramin is a strongly negatively charged molecule with six sulphonate groups which are believed to bind directly to positively charged bFGF in a manner similar to that of heparin binding (29). In a brain tumor model, suramin administered at specific doses, is able to retard tumor growth by inhibiting the angiogenic process (30). This agent is being under intensive study as to its ultimately antineoplastic properties.

VASCULAR ENDOTHELIAL GROWTH FACTOR (VEGF)   or   VASCULAR PERMEABILTIY FACTOR (VPF)

Another endothelial cell-specific mitogen recently described is the vascular permeability factor (VPF) or vascular endothelial growth factor (VEGF) (31). It was purified from media conditioned by bovine pituitary follicular cells. Levels of VPF mRNA are increased within a few hours of exposing different cell cultures to hypoxia and return to previous levels when normal oxygen supply is resumed; these studies suggest that VPF functions as a hypoxia-inducible angiogenic factor (32).

Hybridiztion of normal brain and glioma tissue in situ with S35-labeled antisense RNA showed that in low grade gliomas VPF is expressed in few tumor cells in contrast with glioblastomas where VPF mRNA is observed in two subsets of cells - the pallisading cells and clusters of cells not associated with necrotic areas. VPF mRNA is not present in the white matter of normal human brain, but is expressed in cells scattered in the cortex; VPF is undetectable in normal brain vasculature (33). VPF is present in the vascular endothelial cells of glioblastomas and experimental tumors but it was not detected in the tumoral cell compartment (34). The important observation that VPF mRNA is present in consistently higher levels in glioblastoma as compared with astrocytoma is of practical interest - the finding of VPF expression in an otherwise phenotipically benign astrocytoma would implicate in the malignant progression of the tumor - early detection of this expression could be of benefit for the patient in terms of planning therapy and establishing prognosis.

Receptors for VPR have been identified and found in astrocytomas and glioblastomas. flt mRNA transcript was found in astrocytomas and glioblastomas by northern analysis; in situ hybridiztion to investigate the cellular localization of flt disclosed expression of flt mRNA in endothelial cells of the vascular component of gliomas. Normal brain endothelial cells do not express VPF receptor mRNA. It is concluded that expression of the VPR receptor is induced in endothelial cells during tumor development.

CONCLUDING REMARKS

Steroids, fumagillin, minociclyne, thiomolybdates, antagonists of bFGF, interferon - alfa, low copper treatment have among others been described as antiangiogenic agents that mey interfere with the receptor, inactivate the gene, prevent the mitogenci effect of bFGF or bind directly to the peptide. Regardless the intrinsic mechanism of action of each, the objective of this search is to establish a reliable and safe antiangiogenic therapy for those unfortunate patients harboring a tumor of the central nervous system. The path from laboratory to the bedside is very long but extremely rewarding when successfully achieved and, if we look to the overall effort currently being done by many researchers all over the world to find an effective antiangiogenic therapy, we can say that the light at the end of the tunnel is extremely pale, but it certainly begins to shine.

BIBLIOGRAPHY

1. Hobson, B. & Denekamp, J. - Endothelial cell proliferation in tumours and normal tissues; continuous labeling studies.Br. J. Cancer 1984; 49: 405.

2. Bar, Th. - The vascular system of the cerebral cortex. Adv. Anat. Embryol. Cell Biol. 1980; 59: 1.

3. Russell, D. & Rubinstein, L.J. - Pathology of Tumours of the Nervous System. Wilkins and Wilkinson, 1989, 5th ed.

4. Schweigerer, L.; Neufeld, G.; Friedman, J.; Abraham, J.A.; Fiddes, J.C.; Gospodarowicz, D. - Capillary endothelial cells express basic fibroblast growth factor, a mitogen that promotes their own growth. Nature 1987; 325: 257.

5. Rifkin,D.B. & Moscatelli, D. - Recent developments in the cell biology of basic fibroblast growth factor. Cell Biol. 1989; 109: 1.

6. Rogeli, S.; Weinberg,R.A.; Fanning,P. & Klagsbrun,M. - Characterization of tumors produced by signal peptide-basic fibroblast growth factor-transformed cells. J. Cell. Biochem. 1989; 39:132.

7. Folkman,J.; Klagsbrun, M.; Sasse,J.; et al. - A heparin-binding angiogenic protein - basic fibroblast growth factor - is stored within the basement membrane. Am.J. Pathol. 1988; 130:393.

8. Finkelstein, S.; Apostolides,P.J.; Cady, C.G., et al. - Increased basic fibroblast growth factor (bFGF) immunoreactivity at the site of focal wounds. Brain Res. 1988; 460:253.

9. Gonzalez, A.M.; Buscaglia, M.; Ong, M. & Baird, A. - Distribution of basic fibroblast growth factor in the 18-day rat fetus: localization in the basement membrane of diverse tissues. J. Cell Biol. 1990; 110: 753.

10. Brem, S.; Tsanaclis, A.M.C.; Gately, S., et al. - Immunolocalization of basic fibroblast growth factor to the microvasculature of human brain tumors. Cancer 1992;70:2673.

11. Bugler, A.; Amalric, F. & Prats, H. - Alternative initiation of translation determines cytoplasmic or nuclear localization of basic fibroblast growth factor. Mol. Cell. Biol. 1991; 11: 573.

12. Baird, A. & Klagsbrun, M. - The fibroblast growth factor family. Cancer Cells 1991; 3:239.

13. Kandel,J.; Bossy-Wetzel,E.; Radvanyi, F.; et al. - Neovascularization is associated with a switch to the export of bFGF in the multistep development of fibrosarcoma. Cell 1991; 66:1095.

14. Delli Bovi,P.; Curatova,A.; Kern, F., et al. - An oncogene isolated by transfection of Kaposi's sarcoma DNA encodes a growth factor that is a member of the FGF family. Cell 1987; 50: 729.

15. Vlodavsky,I.; Folkman,J.; Sullivan,R., et al. - Endothelial cell-derived bFGF synthesis nd deposition into subendothelial extracellular matrix. Proc. Natl. Acad. Sci. USA 1987; 84:2292.

16. Gross,J.L.;Moscatelli,D. & Rifkin, D.B. - Increased capillary endothelial cell protease activity in response to angiogenic stimuli in vitro. Proc. Natl. Acad. Sci. USA 1983; 80:2623.

17. Sato,R. & Rifkin, D.B. - Autocrine activity of basic fibroblast gorwth factor: regulation of endothelial cell movement, plasminogen activator synthesis and SNA synthesis. J.Cell Biol. 1988; 107:1199.

18. Laiho, M.& Kesji-Oja,J. - Growth factors in the regulation of pericellular proteolysis: a review. Cancer Res. 1989; 49: 2533.

19. Maragoudakis, M.E.; Panoutsakopoulou,M. & Sarmonika,M. - Rate of basement membrane biosynthesis as an index to angiogenesis. Tissue Cell 1988; 20: 531.

20. Rogister, B.; Leprince,P.; Pettmann, B. & Labourdette, G. - Brain basic fibroblast growth factor stimulates the release of plasminogen activators by newborn rat cultured astroglial cells. Neuroscience Letters 1988; 91:321.

21. Torelli,S.; Dell'Era,P.; Ennas,M.G. et al. - Basic fibroblast gorwth factor in neuronal cultures of human fetal brain. J. Neurosci.Res. 1990;

22. Presta, M.; Ennas, M.G.; Torelli, S. et al. - Synthesis of urokinase-type plasminogen activator and type-1 plaminogen activator inhibitor in neuronal cultures of human fetal brain: stimulation by phorbol ester. J. Neurochem. 1990; 55:1647.

23. Polverini, P.J. & Leibovich,S.J. - Induction of neovascularization in vivo and endothelial proliferation in vitro by tumor associated macrophages. Lab. Invest. 1984; 51:635.

24. Gately,S.; Tsanaclis,A.M.C.;Klagsbrun,M. & Brem, S. - Signal peptide-bFGF transfected cells acquire an invasive phenotype. Proc.Am.Ass.Cancer Res. 1992; 33: 62.

25.Lee, J.K.; Choi,B.; Sobel,R.A., et al. - Inhibition of growth and angiogenesis of human neurofibrosarcoma by heparin and hydorcortisone. J. Neurosurg. 1990; 73: 435.

26. Folkman, J.; Langer,R.; Linhardt, R.J., et al. - Angiogenesis inhibition and tumor regression caused by heparin or a heparin frgment in the presence of cortisone. Science 1983; 221:719.

27. Brem, S.; Zagzag,D.;Tsanaclis,A.M.C. et al. - Inhibition of angiogenesis and tumor growth in the brain: suppression of endothelial cell turnover by penicillamine and the depletion of copper, an angiogenic cofactor. Am. J. Pathol. 1990;137:1121.

28. Brem,S.; Tsanaclis,A.M.C.; Zagzag, D. - Anticopper treatment inhibits pseudopodial protrusion and the invasive spread of 9L gliosarcoma cells in the rat brain. Neurosurgery 1990; 26: 391.

29. Gagliardi,A.; Hadd,H. & Collins,D.C.- Inhibition of angiogenesis by suramin.Cancer Res. 1992; 5225. Baird,A.;Esch,F.;Mormède,P. et al. - Molecular characterization of fibroblast growth factor: distribution and biological activities in various tissues. Recent Prog.Horm.Res. 1986; 52: 5073.

30. Takano,S.; Gately,S.;Engerhardt,H.;Tsanaclis,A.M.C. et al. - Angiossuppression and antiproliferative action of suramin - agrowth factor antagonist. Growth Factors, Peptides and Receptors. T.W.Moody, ed., 1993.

31..Gospodarowicz,D. Abraham,J.A. & Schilling, J. - Isolation and characterization of a vascular endothelial cell mitogen produced by pituitary-derived folliculo stellate cells. Proc. Natl. Acad. Sci. USA 1989; 86:7311.

32. Shweiki,D.; Itin,A;Soffer,D. & Keshet, E. - Vascular endothelial growth factor induced by hypoxia may mediate hypoxia-initiated angiogenesis. Nature 1992; 359: 843.

33. Berkman, R.A.; Merrill,J.J.; Reinhold,W.C. et al. - Expression of the vascular permeability factor/vascular endothelial growth factor gene in central nervous system neoplasma. J. Clin. Invest. 1993; 91:153.

34. Plate, K.H.; Breir,G.; Weich,R.A. & Risau,W. - Vascular endothelial growth factor is a potential tumour angiogenesis factor in human gliomas in vivo. Nature 1992; 359:845.

ANGIOGENESIS AND FIBRIN DEGRADATION IN HUMAN BREAST CANCER

W D Thompson, J E H Wang, S J Wilson, and N Ganesalingam

Department of Pathology,
Medical School,
Aberdeen Royal Infirmary,
Aberdeen, AB9 2ZD

INTRODUCTION

Poorer prognosis and likelihood of metastasis in patients with breast cancer have now been claimed by several groups to be associated with increased vascularity in breast biopsies from these cases (1, 2, 3). Fibrin deposition and lysis is known to occur at the invasive tumour margin of breast cancer where angiogenesis also occurs. Fibrin degradation products (FDP) have been shown previously by us to be angiogenic on the standard test system, the chick chorioallantoic membrane (CAM) (4). Solid fibrin is degraded by plasmin into soluble combinations of the two major components termed D and E, and it is fragment E that is angiogenic (5). Therefore our aim was to develop a method for assessing individual tumour angiogenic activity, and to relate this to tumour vascularity and content of FDP (6).

The problems encountered in tissue handling and extract preparation will be described, and discussed in relation to earlier work on human atherosclerotic plaques, and current work on experimental healing skin wounds.

METHODS

Preparation of Tissue Extracts

Samples were taken from fresh surgical specimens of breast tissue remaining after diagnostic requirements were met. A method for quantitating the angiogenic activity in extracts of a series of 21 samples of human breast biopsies was applied to 13 cancers, 6 samples of adjacent fat, and 2 non-malignant biopsies. Each sample was homogenised in buffer containing EDTA and EACA to inhibit clotting and fibrinolysis (0.005M tris, 0.1M NaCl, 5mM EDTA, 25mM EACA) (7). After buffer exchange to Dulbecco's buffer using a minicolumn and filter sterilisation, extracts were applied onto the chick chorioallantoic membranes of groups of 10 eggs which were subsequently assayed for the level of DNA synthesis (4, 5) This is a rapid assay that we have found to parallel other quantitative assays such as counting mesenchymal layer blood vessels in histological cross sections (8, 9).

Fertile hens eggs were prepared by "dropping" the contents at day 4 of growth, providing a chamber beneath the upper side of the egg of about 7 ml volume above the thin growing CAM. Test and control substances were applied onto the entire dropped area of CAM at day 10 when growth of the CAM itself is largely complete. Generally, groups of 10 eggs were used for each substance, allowing for 1 or 2 losses. After 18h, the level of DNA synthesis was measured by 30 min methyl-tritiated thymidine incorporation into the entire exposed area of each CAM. Ice-cold saline was injected promptly onto the test surface of all the CAMs to halt metabolism rapidly . Each CAM area was then excised and processed for liquid scintillation counting With this type of tissue, thorough delipidation by CCl_4 was essential to avoid artefactual depression of the level of tritiated thymidine uptake and incorporation into the chick CAM.

Tissue and Tumour Vascularity

Vascularity in the tissue or at the tumour edge was quantitated on histological sections. Immunoperoxidase staining for Factor VIII associated antigen was found helpful in locating small blood vessels amongst inflammatory cells. Sections were projected by a Gillet & Sibert projecting microscope set at X100 magnification in a darkened room down onto a Summagraphics bitpad linked to a computer running a "Digit" quantitation programme, and the vessels counted in each calibrated slide field area of 0.6mm X 0.6mm on a bitpad area of 30cm X 30cm. Pilot work had shown that a minimum of 10 random fields on one or two sections gave a count of 100 to 200 events, and these fields were delineated in sequence as half tumour/ half adjacent breast tissue, covering the whole edge on each section.

Measurement and Immunoblotting of Fibrinogen and Fibrin Degradation Products in Extracts

Rocket immunoelectrophoresis was used to measure the levels of total "Fibrin/fibrinogen - related antigens" (FRA) in extracts. The levels of fibrin fragment E were measured by a double window rocket technique. Each sample was run first through an anti D antiserum window to form the first rocket composed of fibrinogen and D containing fragments including DED, DD and D. The second window can contain either anti E antiserum or just antifibrinogen antiserum, and here rockets containing exclusively E components will form.

Extracts and affinity column treated extracts were loaded onto 3-20% gradient SDS polyacrylamide gels at 5µg FRA and run at a constant current of 20 mA/gel. The gels were electroblotted onto Immobilon membrane (Millipore) (9). Bands were demonstrated by incubation with specific antisera followed by alkaline phosphatase-conjugated antiglobulin. Bands were identified by position relative to standard molecular weight markers, and on the basis of previous experience with specific antisera against fragment D, fragment E and fibrinopeptide A (to distinguish fibrinogen from fibrin derived fragments)

Table 1.

ANGIOGENIC ACTIVITY OF EXTRACTS OF BREAST AND BREAST TUMOUR EXTRACTS

ANGIOGENIC ACTIVITY OF EXTRACTS OF BREAST AND BREAST TUMOUR EXTRACTS

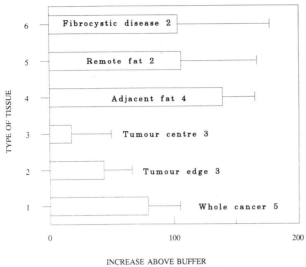

INCREASE ABOVE BUFFER
CONTROL GROUP mean +/- SEM

Table 2.

VASCULARITY OF DIFFERENT BREAST TISSUES

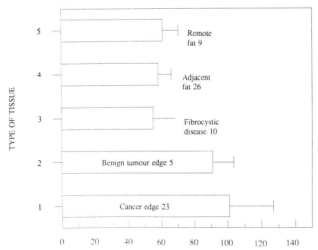

RESULTS

Effect of Extracts on the Chick CAM

Unexpectedly, virtually all samples from all types of breast tissue showed stimulation of DNA synthesis in the CAM (Table 1). There was no additional stimulation from samples of malignant breast tissue.

Breast Tissue and Tumour Vascularity

Vascularity is increased at the edge of malignant and benign tumours compared with normal breast fat and fibrocystic disease (Table 2).

Levels of Fibrin Degradation Products and Immunoblots

Correlations between the levels of vascularity ($r = 0.26$) and angiogenic activity ($r = 0.29$) and of total FRA are weak, but stronger correlation is seen between the levels of angiogenic activity and the levels of fragment E in all tissues ($r = 0.62$; $p < 0.005$) (Table 3). There was no apparent relationship between tissue type and levels of FRA or fragment E (Fig 1).

Immunoblotting showed that fibrin degradation products, including fragment E bands were present in all samples regardless of type (not shown).

DISCUSSION

Thorough delipidation of extracts turned out to be a critical step in avoiding artefactual suppression of the angiogenic response of the chick chorioallantoic membrane in terms of measurement of DNA synthesis. Thereafter, the surprising finding was that angiogenic activity was detectable in all tissue extracts, including normal fat (Table 1)and did not show any correspondence with the increased vascularity of malignant tumours (Table 2). In addition, fibrin fragment E was found in not only cancer extracts but in all other tissues including normal tissues where fibrin degradation would not be expected. Fragment E has been found previously to be the active angiogenic component of fibrin degradation products prepared in vitro (5).

Angiogenic activity in breast tissue extracts correlated weakly with vascularity, or content of total fibrinogen/ fibrin degradation products with regard to all tissues or subcategories (Table 3). Unexpectedly, the strongest correlation was seen comparing activity of all tissues with fibrin fragment E content (Fig 1). The delays involved in obtaining and chilling surgical breast specimens and samples allow clotting and lysis of plasma within the vascular tissue component, as evidenced by the presence of fibrin fragment E. It is proposed that much of the angiogenic activity of extracts may be attributable to the fragment E content, both present endogenously at the invasive tumour margin, and as an artefact due to delay in chilling warm surgical samples. A further contribution may come from fibrinogen clotting on the surface of the CAM

The problem of paradoxical angiogenic activity in normal tissue extracts has been encountered by us in the context of previous work on experimental skin wound healing (11). It was discovered then that this activity could be abolished by extract admixture with

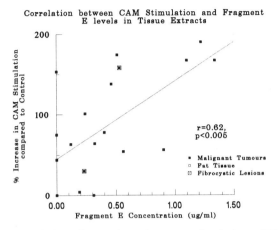

Fig 1   The strongest correlation observed was between stimulation of DNA synthesis in the chick chorioallantoic membrane, and fibrin fragment E levels in tissue extracts of all samples tested. There was, however, no relationship to the type of tissue sample.

Table 3.

SAMPLE TYPE	No	CORRELATION
Angiogenic Activity vs Vascularity (all tissues)	13 pairs	r = 0.26 NS
Angiogenic Activity vs Total FRA (all tissues)	21 pairs	r = 0.29 NS
Angiogenic Activity vs Frag E (all tissues)	20 pairs	r = 0.62 P < 0.01

serum, and it was suspected that the effect may have been attributable to the antithrombin activity of serum. Current work has been directed at more deliberate control of the two factors of fibrinogen persistence during extraction and of subsequent clotting and lysis, by subtraction of fibrinogen and more thorough prevention of fibrinolysis. Placing samples on ice as soon as possible after surgical removal is essential. Subsequent manipulation of temperature and buffer antifibrinolytic agents allows fibrinogen conversion and removal without fibrinolysis. Negative normal mouse skin extracts lacking in fibrinogen or FDP on immunoblots, and non stimulatory to the CAM, can now be prepared. Similar treatment of three day healing wound tissue demonstrates FDP and stimulatory activity to be present.

Such manipulation of the clotting and fibrinolytic mechanisms shows the relevance of these to the angiogenic pathway. Further than this, it is our intention to employ selective removal of FDP, and D and E containing fragments from extracts by affinity chromatography, to establish their relative contribution to extract activity. This has been achieved for extracts of human atherosclerotic plaques (12, 13), in which cell proliferation and angiogenesis also occur(14). Maximum stimulation of the CAM is achieved with extracts containing around 500 µg/ ml FRA, corresponding to the most active concentration of in vitro generated FDP. Most, if not all, of the activity from lesions of active type (gelatinous plaques) appears to be removed by an anti E column.

Thrombin activity is normally essential to the conversion of fibrinogen to fibrin, with cleavage of the relatively low Mr fibrinopeptides A and B. In our work on FDP prepared in vitro, the presence of platelets or fibronectin during clotting have not been relevant to the subsequent generation of FDP active on the CAM (5). Neither has crosslinking of fibrin been contributory. Fibrinogen itself, fibrinogen degradation products, and commercial fibrinogen fragments D and E are inactive. However treatment of fibrinogen fragment E to give fibrin E, as found in vivo, generates activity.

There is considerable current interest in the possible role of thrombin in the cell proliferation in atherosclerosis (15, 16) and angiogenesis. Although thrombin may bind and stimulate cell proliferation in serum free culture conditions, the biological relevance of this effect is uncertain in view of the ubiquitous presence of thrombin inhibitors in vivo. However there is one situation where thrombin is protected, and that is when bound to fibrin itself. Cells gaining access, or becoming surrounded by clot, may well become exposed to the direct stimulatory effect of thrombin.

Where might FDP find a place amongst the many factors now claimed to be angiogenic. Due to the limitations of the test systems currently available, it seems to us that too many factors have been labelled primarily "angiogenic" (17) . This is not to say that many such factors could not be involved in a putative pathway towards angiogenesis. However, we have observed that serum application to the chick CAM does not stimulate angiogenesis, in marked contrast to its long known, essential role in supporting cell growth in culture. Here our findings are in accord with the conclusions of Schwartz who, in the context of atherosclerosis research, has noted the emergence of many cell culture growth factors, apparently with an integral role in supporting smooth muscle cell proliferation, but doubted their role in initiating atherogenesis (18). A point worth making is, that the effect of many growth factors in vivo is now increasingly realised to be dependant not on the mere presence of the factor(s), but on the state of responsiveness on the target cells (19). Hence the lack of response to serum by a largely mature, normal chick CAM. This fits well with our view that FDP derived from the provisional matrix of fibrin act as a common pathway, bridging a variety of pathological events, and initiating cell responsiveness to autocrine, paracrine and matrix associated growth and angiogenic factors.

ACKNOWLEDGEMENTS

This work was supported by the research funds of Grampian Health Board and by grants from the MRC, Wellcome Trust and British Heart Foundation. We wish to acknowledge the technical assistance of Ms Allyson Reid .

REFERENCES

1: Weidner N, Semple JP, Welch WR Folkman J. Tumour angiogenesis and metastasis - correlation in invasive breast carcinoma. New Eng J Med 1991; 324: 1-8

2: Bosari S, et al. Microvessel quantitation and prognosis in invasive breast carcinoma. Hum Pathol 1992; 23: 755-761.

3: Horak ER et al. Angiogenesis, assessed by platelet/ endothelial cell adhesion molecule antibodies, as indicator of node metastases and survival in breast cancer. Lancet 1992; 340: 1120-1124.

4: Thompson WD, Campbell R, Evans T. Fibrin degradation and angiogenesis: quantitative analysis of the angiogenic response in the chick chorioallantoic membrane. J Pathol 1985; 145: 27-37.

5: Thompson WD, Smith EB, Stirk CM, Marshall FI, Stout AJ, Kocchar A. Angiogenic activity of fibrin degradation products is located in fibrin fragment E. J Pathol 1992; 168: 47-53.

6: Wang JEH, Wilson SJ, Thompson WD. Fibrin degradation and angiogenesis in human breast cancer. J Pathol 1993; 169 S, 178 (Abs).

7: Smith EB, Keen GA, Grant A. Fate of fibrinogen in human arterial intima. Arteriosclerosis 1990; 10: 263-275.

8: Thompson WD, Brown FI. Quantitation of histamine-induced angiogenesis in the chick chorioallantoic membrane: mode of action of histamine is indirect. Int J Microcirc 1987;6:343-357.

9: Clinton M, Duncan JI, Long WF, Williamson FB, Thompson WD. Effect of mast cell activator, Compound 48/80, and heparin on angiogenesis in the chick chorioallantoic membrane. Int J Microcirc. 1988; 7: 315-326.

10: Laemmli UK. Cleavage of structural proteins during the assembly of the head of bacteriophage T. Nature 1970; 227: 680-685.

11: Thompson WD, Harvey JA, Kazmi MA, Stout AJ. Fibrinolysis and angiogenesis in wound healing. J Pathol 1991; 165: 311-318.

12: Thompson WD, McGuigan CJ, Snyder C, Keen GA, Smith EB. Mitogenic activity in human atherosclerotic lesions. Atherosclerosis 1987 66:85-93.

13: Thompson WD, Smith EB, Stirk CM, Kochhar A. Atherosclerotic plaque growth: presence of stimulatory fibrin degradation products. Blood Coagulation and Fibrinolysis 1990; 1: 489-493

14: Kamat BR, Galli SJ, Barger AC, Lainey LL, Silverman KJ. Neovascularization and coronary atherosclerotic plaque: cinematic localization and quantitative histologic analysis. Hum Pathol 1987; 18: 1036-1042.

15: Schwartz SM. Serum-derived growth factor is thrombin? J Clin Invest 1993; 91: 4.

16: Sarembok IJ, Gertz SD, Gimple LW, Owen RM, Powers ER, Roberts WC. Effectiveness of recombinant desuphatohirudin in reducing restenosis after balloon angioplasty of atherosclerotic femoral arteries in rabbits. Circulation 1991; 84: 232-243

17: Thompson WD, Harvey JA, Kazmi MA, Stout AJ. Fibrinolysis and angiogenesis in wound healing. J Pathol 1991; 165: 311-318.

18: Schwartz SM, Heimark RL, Majesky MW. Developmental mechanisms underlying pathology of arteries. Physiol Rev 1990; 70: 1177-1209.

19: Flaumenhaft R, Rifkin DB. The extracellular regulation of growth factor action. Molec Biol Cell 1992; 3: 1057-1065.

THROMBIN-MEDIATED EVENTS IMPLICATED IN POST-THROMBOTIC RECOVERY

John W. Fenton II[1] and Frederick A. Ofosu[2]

[1]New York State Department of Health
Wadsworth Center for Laboratories and Research
Albany, New York, 12201;
Department of Physiology and Cell Biology and Department of
Biochemistry and Molecular Biology, Albany Medical College of Union
University, New York, 12208, USA; and

[2]Canadian Red Cross Society Blood Services
Department of Pathology
McMaster University
Hamilton, Ontario, Canada L8N 3Z5

UNIQUENESS OF THROMBIN

Thrombin (EC 3.4.21.5) is the activation product of prothrombin, where the α-thrombin form is the direct product of blood coagulation and, for the most part, possesses all activities ascribed to thrombin.[1-8] Although its gene may be expressed in central nervous tissue,[9] prothrombin is a hepatic synthesized glycoprotein, which, like coagulation factors VII, XI, and X, as well as proteins C and S, requires vitamin K-dependent γ-carboxylation of certain glutamic acids (e.g., 10 for prothrombin) for post-translational completion.[10] This requirement is the target of vitamin K antagonists, such as coumarins used as rodenticides or "blood thinner" in human medicine.[11,12] Unlike other proenzymes in blood coagulation, prothrombin circulates in blood at higher concentrations and its activation fragment does not remain attached to the activated enzyme. Consequently, all of thrombin specificity must be derived from the enzyme moiety, whereas the activation fragment partakes in the regulation of prothrombin activation.[2-8] Unlike other serine proteinases in blood coagulation, thrombin has no counterpart in the horseshoe crab,[13] and the mammalian prothrombin genes appear to be evolutionarily more evolved than those of related serine proteinases.[14] This suggests rapid recent evolution of thrombin structure and functions, such as its narrow active-site groove and exosites.[7,15]

In contrast to other blood serine proteinases, thrombin efficiently activates factors V and VIII providing cofactors and consequently amplifying prothrombin activation. Stated another way, thrombin regulates its own generation. Heparin is a recyclable cofactor for antithrombin III and accelerates the inhibition of thrombin by antithrombin III. Prevention of thrombin-mediated activation of factor V and VIII is the predominant mode of intervention of heparin and low molecular weight heparins, as well as thrombin-directed antithrombotics, such as hirudin and hirulogs.[16-20] Because

Angiogenesis: Molecular Biology, Clinical Aspects
Edited by M.E. Maragoudakis *et al*, Plenum Press, New York 1994

of blood flow-dilution and proteinase inhibitors in blood (e.g., antithrombin III, α_2-macroglobulin, heparin cofactor II) and irregardless of the specific target of intervention, the overall purpose of most antithrombotics is to control the rate of thrombin generation, since α-thrombin must be continuously generated in blood in order to maintain thrombotic processes.[8]

The most important distinction between thrombin and other serine proteinases in the thrombotic and thrombolytic pathways is that thrombin functions transcend all levels (i.e., plasma, blood cell, and vasculature) and phases of hemostasis (i.e., injury to healing), where these functions are central to normal and pathophysiologic events (Fig. 1). Hemostasis is a

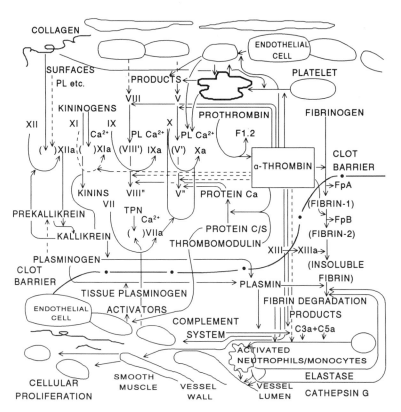

Figure 1. Bioregulatory functions of α-thrombin depicted in hemostasis. Roman numerals are used for coagulation factors where "a" denotes the activated serine proteinases. Primes and double primes signify activated and inactivated forms of the nonenzymic factors V and VIII, respectively. Other abbreviations include: F 1.2, the prothrombin activation fragment; FpA, fibrinopeptide A; FpB, fibrinopeptide B; PL, phospholipid; and TPN, thromboplastin. The dash-dot line indicates the clot barrier. The upper left-hand-facing corner shows pre-thrombotic while the lower right-hand-facing corner portrays post-thrombotic events. Thrombin reactions with various inhibitors (e.g., antithrombin III) and thrombin transformations are not shown. Redrawn from Figure 1 in Fenton[4] with updating corrections.

necessary adaptation of higher animals for the containment of circulating blood within blood vessels in order to supply nutrients and remove wastes from dependent tissues.[3,8] The participation of thrombin in thrombotic events is self-evident (e.g., platelet activation, fibrinogen conversion into clottable fibrin, factor XIII activation, and fibrin crosslinking). In contrast, the contributions of thrombin to post-thrombotic events are for the most part implicated from indirect experimental evidence (e.g., tissue culture models), while a few can be demonstrated in physiological models (e.g., thrombin-induced pulmonary edema).[21]

Human disability and death attributed to disorders in hemostasis are of considerable social and economic concern. These disorders include coronary thrombosis, stroke, disseminated intra-vascular coagulation, various embolic disorders, etc. On the other hand, the participation of thrombin is implicated in edema,[21] inflammation,[22,23] chemotaxis,[24,25] bone resorption,[26,27] cell proliferation,[28] angiogenesis,[29,30] atherosclerosis,[32,32] infectivity of certain viruses,[33] certain cancers,[34-36] as well as Alzheimer's disease and other disorders of nerve tissue.[37-41]

THROMBOTIC AND THROMBOLYTIC CYCLES

Blood coagulation is initiated by two pathways culminating in thrombin generation (Fig. 1). Because of blood-flow dilution, blood-borne proteinase inhibitors, and other consumptive mechanisms (see above), α-thrombin must be continuously generated at threshold levels (e.g., 1 to 5 nM) in order to sustain its thrombotic functions (e.g., stimulate vasculature, activate platelets, convert fibrinogen into clottable fibrin). While picomolar concentrations of α-thrombin cause vascular endothelial cells to produce a smooth muscle cell-relaxing substance (e.g., nitric oxide), nanomolar concentrations are needed for contracture of vasculature smooth muscles.[42] Furthermore, low concentrations of α-thrombin are well known to initiate platelet activation and aggregation prior to the onset of fibrin clotting. These two observations suggest thrombin concentration dependencies. However, the majority of thrombin activities appear to occur in a narrow nanomolar concentration range, suggesting that just about everything that thrombin does should more or less occur at once if there were not other considerations.

One very important factor in regulating thrombin functions is partitioning. Human plasma contains sufficient prothrombin to generate up to 1.37 μM α-thrombin.[42] During the activation of whole blood or recalcification by contact activation under nonflowing conditions only ~10% of the potential concentration of α-thrombin is ever achieved, despite the consumption of >80% of the proenzyme.[43] This consumption of α-thrombin cannot by explained simply by inactivation by antithrombin III and heparin cofactor II in the absence of heparin. But rather, incorporation of α-thrombin into forming fibrin clots provides a more probable explanation.

During fibrin deposition onto surfaces, α-thrombin is actively incorporated via its exosite which is involved in fibrinogen recognition but is independent of its catalytic site. While unlabeled α-thrombin will compete with labeled α-thrombin for incorporation into fibrin, a large excess of unlabeled α-thrombin will not readily displace previously incorporated labeled α-thrombin.[44,45] Therefore, incorporation behaves as an irreversible process involving a displacement reaction where incorporated α-thrombin is partitioned within fibrin micelles. By fibrinolysis with plasmin or compression of clots, active α-thrombin can be recovered, suggesting that thrombi are thrombin storage reservoirs.[45] This may explain the problem of rethrombosis upon fibrinolytic intervention with immature thrombi.[46] Since large proteinase inhibitors in blood do not readily penetrate,[47] therein lies the need for thrombus-penetrating thrombin-directed antithrombotics (Fig. 2).

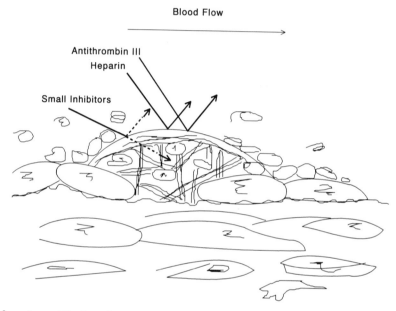

Figure 2. A small thrombus deflecting antithrombin III plus heparin is depicted while small inhibitors can partially penetrate. The thrombus shown consists of a fine-mesh fibrin canopy, coarse rod-structured internal fibrin, and includes erythrocytes, platelets, and neutrophils. The abluminal side contacts injured endothelium, protecting it from blood flowing within the vessel lumen.

Vascular and blood cell surfaces are also thought to be important in binding thrombin and modulating its concentrations and functions.[48] If cellular binding involves exosites on thrombin necessary for interacting with proteinase inhibitors, then cellular-bound thrombin would also serve as a thrombin reservoir and behave in the same manner as thrombus-entrapped thrombin. In pigs, the basal thrombin concentration (both cell- and thrombus-associated) has been estimated at 0.5 nM by hirudin complexing and shown to rise to 10 to 15 nM after administration of endotoxin.[49]

As thrombi form, they passively entrap or actively recruit blood cells, such as erythrocytes, platelets, and neutrophils. Neutrophil recruitment and activation are turning events in thrombin-induced pulmonary edema and lung injury.[21] In addition to various oxygen radicals, activated neutrophils release proteinases, which degrade fibrin and increase fluidity within thrombi.[50] Of these, neutrophil elastase and cathepsin G also proteolytically convert α-thrombin into ϵ- and ζ-thrombins, respectively. Such conversion does not significantly alter thrombin activities but reduces the thermal stability to approximately body temperature, thus allowing these thrombin forms to become enzymatically inactivated by denaturation.[51] On the other hand, denaturation is necessary for the expression of the Arg-Gly-Asp ligand in thrombin.[7] This ligand binds at the vitronectin receptor and causes cell spreading, as implicated in initial phases of wound healing.[52,53] This is clearly an example of a nonenzymatic activity of thrombin as is thrombin-induced chemotaxis with monocytes[24] and neutrophils.[25] However, the vast majority of thrombin activities require the catalytically competent enzyme. This includes activation by thrombin of its 7-stranded transmembrane receptor on various cells.[54,55]

Assuming that only a portion of the thrombin within aging thrombi undergoes denaturation then with increased fluidity, thrombin should become available to stimulate thrombus-contacted cells. Of such possible cell types, the endothelial cell is among the most responsive, where thrombin initiates cell rounding, changes in adhesive protein ligands, increases monolayer permeability to proteins, and the release of various substances, such as endothelin, tissue plasminogen activator, and proteinase inhibitors.[21] In the sense that thrombin induces tissue plasminogen activator release,[56] thrombin indirectly initiates thrombolysis. Furthermore, the plasminogen activator inhibitor released by thrombin may serve primarily as a thrombin inhibitor in the maturing phase of thrombi.[57] That is, residual thrombin released upon thrombolysis must be consumed in order to prevent rethrombosis.

VASCULAR MECHANISMS

Low concentrations of α-thrombin in the picomolar range stimulate vascular endothelial cells to produce a vascular dilating substance, presumably nitric oxide which causes smooth muscle tissues to relax. Increasing thrombin concentrations into the range of those for blood clotting in contrast causes vascular smooth muscle to contract (see above). Thus, initial dilation should increase blood flow, whereas with thrombus formation vessel contraction should enhance containment of thrombi to injury sites.

The fact that thrombin further stimulates vascular endothelial cells to undergo rounding or shape change and to increase transport of proteins, such as albumin, across the endothelial cell monolayer barrier (see above) provides a mechanism for edema and tissue swelling. In pulmonary edema experimentally induced with α-thrombin, protein clearance occurs into the lymphatic system.[21] With cultured endothelial-cell monolayers, preliminary experiments suggest that thrombin stimulation may cause selective transport of vitamin K-dependent coagulation factors,[58] implying that they become more available to subendothelial tissues. This observation suggests that injured tissues might be an important site of thrombin generation. In contrast to circulating blood, as within thrombi, tissues would provide a protected environment in contrast to circulating blood. When blood levels of the prothrombin activation fragments (e.g., fragment 1.2 and fragment 2) are compared with those of thrombin proteinase inhibitor complexes (e.g., antithrombin III), large discrepancies can be found, indicating that significant amounts of thrombin may exist in thrombi or elsewhere.[59]

Subendothelial thrombin generation should have some important consequences. Thrombin stimulates bone resorption[26,27] and if bone fragments are within the wound, then thrombin should aid in their removal. Furthermore, thrombin is further a potent angiogenic stimulus[29,30] and should promote revascularization. Thrombin is also a very potent mitogen for a variety of cell types, including vascular smooth-muscle cells.[28] Thus, chronic exposure to thrombin is implicated in development of atherosclerosis.[32] Another "aging disease" where thrombin is implicated is Alzheimer's disease.[40]

How thrombin stimulates cells has recently been partially clarified by the cloning of the 7-stranded transmembrane receptor for thrombin.[54,55] This receptor belongs to a large receptor family[60] but, unlike other members, requires proteolytic cleavage of its external amino-terminus peptide to generate a "tethered ligand," which in turn activates the receptor.[54,55] Synthetic-peptide analogues of the first 6 to 14 amino acids of the tethered ligand substitute for thrombin with a variety of cell types but not with all cellular responses.[61] In such systems, a second thrombin-mediated pathway appears to function in some cellular systems, as a double lock.[62-64] Furthermore, species differences have been found with thrombin receptor-activating peptides for platelets[65] and smooth-muscle cells.[66]

SUMMARY

Thrombin has several distinguishing properties that account for its several functions throughout hemostasis. Control of prothrombin activation or thrombin generation is the primary target of most antithrombotics, and thrombi-penetrating thrombin-directed antithrombotics can prevent rethrombosis after immature thrombi are lysed. During thrombus formation, α-thrombin is actively incorporated, and thrombin functions change with thrombus maturation. Thrombin increases protein transport across endothelial cell monolayers, and tissue generated thrombin may be important in the healing processes.

The scars of healing are but the manifestations of thrombin functions.

ACKNOWLEDGEMENTS

The author thanks Debra VonZwehl for her secretarial assistance and Diane V. Brezniak for computer-drawn illustrations. This work was supported in part by NIH Grant HL-13160-22 (JWF) and Heart and Stroke Foundation of Ontario (FAO).

REFERENCES

1. J.W. Fenton II, M.J. Fasco, A.B. Stackrow, D.L. Aronson, A.M. Young, and J.S. Finlayson, Human Thrombins. Production, evaluation, and properties of α-thrombin. J. Biol. Chem. 252:3587-3598 (1977).

2. J.W. Fenton II, B.H. Landis, D.A. Walz, and J.S. Finlayson, Human Thrombins, in: "Chemistry and Biology of Thrombin," R.L. Lundblad, J.W. Fenton II, and K.G. Mann, eds., pp. 43-70, Ann Arbor Science Publishers, Ann Arbor, MI (1977).

3. J.W. Fenton II, Thrombin specificity. <u>Ann. NY Acad. Sci.</u> 370:468-495, (1981).

4. J.W. Fenton II, Thrombin. <u>Ann. NY Acad. Sci.</u> 485:5-15, (1986).

5. J.W. Fenton II, Thrombin bioregulatory functions. <u>Adv. Clin. Ensymol.</u> 6:186-193, (1988).

6. J.W. Fenton II, Regulation of thrombin generation and functions. <u>Semin. Thromb. Hemost.</u> 14:234-240, (1988).

7. J.W. Fenton II, F.A. Ofosu, D.G. Moon, J.M. Maraganore, Thrombin structure and function: why thrombin is the primary target for antithrombotics. <u>Blood Coagul. Fibrinolys</u> 2:69-75, (1991).

8. J.W. Fenton II, F.A. Ofosu, D.V. Brezniak, and H.I. Hassouna, Understanding thrombin and hemostasis. <u>Hematol./Oncol. Clin. N. Am.</u>, in press, (1993).

9. M. Dihanich, M. Kaser, E. Reinhard, D.D. Cunningham, and D. Monard, Prothrombin mRNA is expressed by cells in the nervous system. <u>Neuron</u> 6:575-581, (1991).

10. S. Magnuson, T.E. Petersen, L. Sottrup-Jensen, and H. Claeys, Complete primary structure of prothrombin: isolation, structure, and reactivity of ten carboxylated glutamic acid residues and regulation of prothrombin activation by thrombin. <u>In</u>: Proteases and Biological Control. E. Reich, D.B. Rifkin, and E. Shaw, eds., Cold Spring Harbor Laboratory, Cold Spring Harbor, NY, pp 123-149 (1975).

11. R.A. O'Reily, P.M. Aggeler, Determinants of the response to oral anticoagulant drugs in man. <u>Pharmacol. Rev.</u> 22:35-96, (1970).

12. M.M. Bern, Considerations for lower doses of warfarin. <u>Hematol./Oncol. Clin. North Am.</u> 6:1105-1114, (1992).

13. S. Iwanaga, T. Miyata, F. Tokunaga, and T. Muta, Molecular mechanism of hemolymph clotting system in limulus. <u>Thromb. Res.</u> 68:1-32 (1992).

14. D.M. Irwin, K.A. Robertson, T.A. MacGillivray, Structure and evolution of the bovine prothrombin gene. <u>J. Mol. Biol.</u> 200:31-45 (1988).

15. M.T. Stubbs and W. Bode, A player of many parts: the spotlight falls on thrombin's structure, <u>Thromb. Res.</u> 69:1-58 (1993).

16. F.A. Ofosu, P. Sie, G.J. Modi, F. Fernandez, M.R. Buchanan, M.A. Blajchman, B. Boneu, and J. Hirsh, The inhibition of thrombin-dependent positive-feedback reactions is critical to the expression of anticoagulant effects of heparin. <u>Biochem. J.</u> 243:579-588 (1987).

17. S. Begiun, T. Lindhout, H.C. Hemker, The mode of action of heparin in plasma. <u>Thromb. Haemost.</u> 60:457-462 (1988).

18. F.A. Ofosu, J. Hirsh, C.T. Esmon, G.J. Modi, L.M. Smith, N. Anvari, M.R. Buchanan, J.W. Fenton II, and M.A. Blajchman, Unfractionated heparin inhibits thrombin-catalyzed amplification reactions of coagulation more efficiently than those catalyzed by factor Xa. <u>Biochem. J.</u> 257:143-150 (1989).

19. X.J. Yang, M.A. Blajchman, S. Graven, L.M. Smith, N. Anvari, F.A. Ofosu, Activation of factor V during intrinsic and extrinsic coagulation. Inhibition by heparin, hirudin and $\underline{D}$-Phe-Pro-Arg-CH$_2$Cl. Biochem. J. 272:399-406 (1990).

20. F.A. Ofosu, J.W. Fenton II, J. Maraganore, M.A. Blajchman, X. Yang, L. Smith, N. Anvari, M.R. Buchanan, J. Hirsh, Inhibition of the amplification reactions of blood coagulation by site-specific inhibitors of α-thrombin. Biochem. J. 283:893-897 (1992).

21. A.B. Malik, J.W. Fenton II, Thrombin-mediated increase in vascular endothelial permeability. Semin. Thromb. Hemost. 18:193-199 (1992).

22. S.M. Prescott, A.R. Seegert, G.A. Zimmerman, T.M. McIntyre, J.M. Maraganore, Hirudin-based peptides block the inflammatory effects of thrombin on endothelial cells. J. Biol. Chem. 265:9614-9616 (1990).

23. A. Tordi, J.W. Fenton II, T.T. Andersen, E.W. Gelfand, Functional thrombin receptors on human T-lymphoblastoid cells. J. Immunol., in press, 1993.

24. R. Bar Shavit, A. Kahn, G.D. Wilner, J.W. Fenton II, Unique exosite region in thrombin stimulates monocyte chemotaxis. Science 220:728-731 (1983).

25. R. Bizios, L. Lai, J.W. Fenton II, A.B. Malik, Thrombin-induced chemotaxis and aggregation of neutrophils. J. Cell. Physiol. 128:485-490 (1987).

26. G.T. Gustafson, V. Lerner, Thrombin, a stimulator of bone resorption. Biosci. Rep. 3:255-261 (1983).

27. D.N. Tatakis, C. Dolce, R. Dziak, J.W. Fenton II, Thrombin effects on osteoblastic cells. II. structure-function relationships. Biochem. Biophys. Res. Commun. 174:181-188 (1991).

28. D.H. Carney, G.S. Herbosa, J. Stiernberg, J.S. Bergmann, E.A. Gordon, D. Scott, J.W. Fenton II, Semin. Thromb. Hemost. 12:231-240 (1986).

29. D.A. Walz, J.W. Fenton II, and P.H. Johnson, Human thrombin induces angiogenesis. Thromb. Haemost. 65:1252 (1991).

30. N.E. Tsopanoglou, E. Pipili-Synetos, and M.E. Maragoudakis, Thrombin promotes angiogenesis by a mechanism independent of fibrin formation. Am. J. Physiol., in press, 1993.

31. N.A. Nelken, S.J. Sorfer, J. O'Keefe, T-KH, Vu, I.F. Charo, and S.R. Coughlin, Thrombin receptor expression in normal and atherosclerotic human arteries. J. Clin. Invest. 90:1614-1621 (1992).

32. C.A. McNamara, I.J. Sarembock, L.W. Gimple, J.W. Fenton II, S.R. Coughlin, G.K. Owens, Thrombin stimulates proliferation of cultured rat aortic smooth muscle cells by a proteolytically activated receptor. J. Clin. Invest. 91:94-98 (1993).

33. E.J. Dubovi, J.D. Geratz, and R.R. Tidwell, Enhancement of respirator syncytial virus-induced cytopathology by trypsin, thrombin, and plasmin. Infect. Immunol. 40:351-358 (1983).

34. M.Z. Wojtukiewicz, D.G. Tang, K.K. Nelson, D.A. Walz, C.A. Diglio, and K.V. Honn, Thrombin enhances tumor cell adhesive and metastatic properties via increased $\alpha_{IIb}\beta_3$ expression on the cell surface. Thromb. Res. **68**:233-245 (1992).

35. D.L. Morris, J.B. Ward Jr., P. Nechay, E.B. Whorton Jr., J.W. Fenton II, and D.H. Carney, Highly purified human α-thrombin promotes morphological transformation of BALB/c 3T3 cells. Carcinogenesis **13**:67-73 (1992).

36. L.R. Zacharski, M.Z. Wojtukiewicz, V. Costantini, D.L. Ornstein, V.A. Memoli, Pathways of coagulation/fibrinolysis activation in malignancy. Semin. Thromb. Hemost. **18**:104-116 (1992).

37. R.M. Snider, M. McKinney, C. Forray, E. Richelson, Neurotransmitter receptors mediate cyclic GMP formation by involvement of arachidonic acid and lipoxygenase. Proc. Natl. Acad. Sci. USA **81**:3905-3909 (1984).

38. G. Rovelli, S.R. Stone, K.T. Preisaner, D. Monard, Specific interaction of vitronectin with the cell-secreted protease inhibitor glia-derived nexin and its thrombin complex. Eur. J. Biochem. **192**:797-803 (1990).

39. H.S. Suidan, S.R. Stone, B.A. Hemmings, D. Monard, Thrombin causes neurite retraction in neuronal cells through activation of cell surface receptors. Neuron **8**:363-375 (1992).

40. J. Marx, A new link in the brain's defenses. Science **256**:1278-1280 (1992).

41. D. Monard, Thrombin and its inhibitors. Science **257**:145-146 (1992).

42. D.A. Walz, G.F. Anderson, R.E. Ciaglowski, M. Aiken, J.W. Fenton II, Thrombin-elicited contractile responses of aortic smooth muscle. Proc. Soc. Expl. Biol. Med. **180**:518-526 (1985).

43. D.L. Aronson, L. Stevan, A.P. Ball, B.R. Franza Jr., and J.S. Finlayson, Generation of the combined prothrombin activation peptide (F1.2) during the clotting of blood and plasma. J. Clin. Invest. **60**:1410-1418 (1977).

44. C.Y. Liu, H.L. Nossel, K.L. Kaplan, The binding of thrombin by fibrin. J. Biol. Chem. **254**:10421-10425 (1979).

45. G.D. Wilner, M.P. Danitz, S.M. Mudd, K.H. Hsieh, and J.W. Fenton II, Selective immobilization of α-thrombin by surface-bound fibrin. J. Lab. Clin. Med. **97**:403-411 (1981).

46. A.M. Lincoff, J.J. Popma, S.G. Ellis, T.A. Hacker, and E.J. Topol, Abrupt vessel closure complicating coronary angioplasty, clinical angiographic therapeutic. J. Am. Coll. Cardol. **19**:926-933 (1992).

47. J.J. Weitz, M. Hudoba, D. Massel, J. Maraganore, and J. Hirsch, Clot-bound thrombin is protected from inactivation by heparin-antithrombin III but us susceptible to inactivation by antithrombin III-independent inhibitors. J. Clin. Invest. **86**:385-391 (1990).

48. R. Bar-Shavit, A. Edlor, and I. Vlodavsky, Binding of thrombin to subendothelial extracellular matrix. J. Clin. Invest. **84**:1096-1104.

49. P. Zoldhelyi, J.H. Chesebro, and W.G. Owen, Hirudin as a molecular probe for thrombin in vitro and during systemic coagulation in the pig. Proc. Natl. Acad. Sci. USA 90:1819-1823 (1993).

50. C.W. Francis, and V.J. Marder, Degradation of cross-linked fibrin by human leukocyte proteases. J. Lab. Clin. Med. 107:342-352 (1986).

51. D.V. Brezniak, M.S. Brower, J.I. Witting, D.A. Walz, and J.W. Fenton II, Human α- to ζ-thrombin cleavage occurs with neutrophil cathepsin G or chymotrypsin while fibrinogen clotting activity is retained. Biochemistry 29:3536-3542 (1990).

52. R. Bar-Shavit, V. Sabbah, M.G. Lampugnani, P.C. Marchisio, J.W. Fenton II, I. Vlodavsky, and E. Dejane, An Arg-Gly-Asp sequence within thrombin promotes endothelial cell adhesion. J. Cell. Biol. 112:335-344 (1991).

53. D.H. Carney, R. Mann, W.R. Redin, S.D. Pernia, D. Berry, J.P. Heggers, P.G. Hayward, M.C. Robson, J. Christie, C. Annabli, J.W. Fenton II, and K.C. Glenn, Enhancement of incisional wound healing and neovascularization in normal rats by thrombin and synthetic thrombin receptor-activating peptides. J. Clin. Invest. 89:1469-1477 (1992).

54. U.B. Rasmussen, V. Vouret-Craviari, S. Jallot, Y. Schlesinger, G. Pages, A. Parirani, J.-P Lecocq, J. Pouyssegur, and E. Van Obberghen-Schilling, DNA cloning and expression of a hamster α-thrombin receptor coupled to Ca²⁺ mobilization. FEBS Lett. 288:123-128 (1991).

55. T.-K.H. Vu, D.T. Hung, V.I. Wheaton, and S.R. Coughlin, Molecular cloning of a functional thrombin receptor reveals a novel proteolytic mechanism of receptor activation. Cell 64:1057-1068 (1991).

56. E.G. Levin, D.M. Stern, P.P. Nawroth, R.A. Marlar, D.S. Fair, J.W. Fenton II, and L.A. Harker, Specificity of the thrombin-induced release of tissue plasminogen activator from cultured human endothelial cells. Thromb. Haemost. 56:115-119 (1986).

57. K.T. Preissner, and D. Jenne, Structure of vitronectin and its biological role in haemostasis. Thromb. Haemost. 66:123-132 (1991).

58. H. Lum, A.B. Malik, and J.W. Fenton II, unpublished data (1992).

59. L. Liu, M.A. Blajchman, L. Dewar, J.W. Fenton II, M. Andrew, M. Delorme, J. Ginsberg, and F.A. Ofosu, unpublished data (1993).

60. M.J. Berridge, Inositol trisphosphate and calcium signalling. Nature 361:315-325 (1993).

61. V. Vouret-Craviari, E. Van Obberghen-Schilling, E. Van Obberghen, and J. Pouyssegur, Differential activation of p44mapk (ERK1) by α-thrombin and thrombin-receptor peptide. Biochem. J. 289:209-214 (1993).

62. C. Kanthou, G. Parry, E. Wijelath, V.V. Kakkar, and C. Demoliou-Mason, Thrombin-induced proliferation and expression of growth factor-A chain gene in human vascular smooth muscle cells. FEBS Lett. 314:143-148 (1992).

63. Y. Sugama, C. Tiruppathi, K. Janakidevi, T.T. Andersen, J.W. Fenton II, and A.B. Malik, Thrombin-induced expression of endothelial P-selectin and ICAM-1: a mechanism for stabilizing neutrophil adhesion. J. Cell. Biol. 119:935-944 (1992).

64. H. Lum, T.T. Andersen, A. Siflinger-Birnboim, C. Tiruppathi, M.S. Goligorsky, J.W. Fenton II, and A.B. Malik, Thrombin receptor peptide inhibits thrombin-induced increase in endothelial permeability by receptor desensitization. J. Cell. Biol. 120:1491-1499 (1993).

65. J.L. Catalfamo, T.T. Andersen, and J.W. Fenton II, unpublished data (1992).

66. C.A. McNamara, I.J. Sarembock, L.W. Gimple, J.W. Fenton II, S.R. Coughlin, and G.K. Owens, unpublished data (1993).

ANTI-TUMOR EFFECTS OF GBS TOXIN ARE CAUSED BY INDUCTION OF A TARGETED INFLAMMATORY REACTION

Carl G. Hellerqvist[1], Gary B. Thurman[1], Bruce A. Russell[1], David L. Page[2], Gerald E. York[1], Yue-Fen Wang[1], Carlos Castillo[1] and Hakan W. Sundell[3]

[1]Department of Biochemistry
[2]Department of Pathology
[3]Department of Pediatric Neonatology
Vanderbilt University School of Medicine
Nashville, Tennessee 37232-0146, U.S.A.

INTRODUCTION

Group B Streptococcus is a major pathogen in hospital nurseries in the United States affecting 10,000 neonates each year with the mortality rate of approximately 15%. GBS pneumonia, often called early onset disease, presents with signs of sepsis, granulocytopenia, and respiratory distress and is characterized by pulmonary hypertension and increased vascular permeability and proteinaceous pulmonary edema. After treatment with antibiotic, the neonate is cured of the infection but the symptoms of early onset disease persist which suggests the involvement of an extracellular toxin similar to gram-negative endotoxin shock. These observations led us to analyze both bacteria and media components which could be responsible for the induction of respiratory distress in the neonate.

Our work was aided by a sheep model developed by Dr. Brigham at Vanderbilt University. Sheep, or lambs, are susceptible to infusion of live and heat-killed GBS bacteria and we demonstrated that a polysaccharide exotoxin of approximate molecular weight of 200,000 - 300,000, when infused in the sheep at picomole quantities, would induce respiratory distress mimicking the pathophysiology seen in human neonates which succumbed to GBS infection[1].

Using the sheep model, we demonstrated that GBS Toxin would bind to lung tissue[2] and that within the lung tissue, using immunohistochemistry with antibodies to GBS Toxin, the lung vascular endothelium was a target for GBS Toxin[3]. Our in vivo model, the lamb, thus suggested that GBS Toxin would bind to endothelium in the lung vasculature and induce an inflammatory reaction which would lead to respiratory distress[4]. If sufficient dose was given (100μg/kg), a lethal reaction could be seen.

Whereas the adult sheep is sensitive to GBS Toxin, there is no pathogenicity associated with Group B Streptococcus in uncompromised human adults. The Center for Disease Control does not consider GBS a human pathogen. Thirty to forty percent of all women in the United States are colonized in the birth canal with Group B Streptococcus which is the source for the bacteria which causes most early onset disease in neonates. Full-term neonates infected with GBS after four days of age will not develop early onset disease but may develop meningitis[5]. This suggests that GBS Toxin induces respiratory distress by binding to lung vascular structures that are only present at the early development phase of the lung capillary endothelium. It is worthy of note that autopsies on neonates who died of GBS early onset disease indicate that no other organ than the lung was affected.

RATIONALE

Based on these observations, we made the assumption that GBS Toxin binds only to developing neovasculature, and that, if this was the case, the developing endothelium in human tumors would also be susceptible to GBS Toxin. Immunohistochemical analysis using rabbit antibodies raised against highly purified GBS Toxin confirmed that GBS Toxin would bind to vascular endothelium within human carcinomas. Tumors originating from breast, lung, colon, and thyroid were tested. The neovasculature showed staining with GBS Toxin and no other cell types within the tumor were specifically stained.

ANTI-TUMOR EFFECT OF GBS TOXIN

In order to explore the potential of GBS Toxin as an agent to induce inflammation in developing tumor vasculature, we tested our hypothesis using mice as a model. Mice, as humans, show the same age-dependent susceptibility to GBS infection. Neonatal rats or mice are susceptible to GBS within twelve hours after birth and afterwards become totally nonsusceptible to GBS. In our hands, GBS Toxin can be infused intravenously in adult mice at 40mg/kg with no apparent toxicity or discomfort to the animals[1].

NUDE MOUSE MODEL

In order to test our hypothesis[6], nude mice were implanted with a human large cell adenocarcinoma capable of tumorigenic growth. Ten days after injection, the average tumor volume was 100 ± 12 cubic millimeters and the animals were divided into three groups with each group having an equivalent distribution of tumor sizes. GBS Toxin for inoculation was dissolved in PBS in concentrations to give 2µg/kg and 20µg/kg (corresponding to .25 and 2.5 picomole per 0.1ml injection, respectively). Dextran at 20µg/kg (7 picomole per injection) was used as control. A fourth group of mice without tumors was injected with 20µg/kg GBS Toxin at the same regimen as the other three groups and served as a control to determine any adverse effect of GBS Toxin on normal nude mice. The mice were injected intravenously through the tail vein with 100µl of PBS containing GBS Toxin or Dextran every other day for three weeks.

The result of this treatment showed that the group of mice receiving 20µg/kg of GBS Toxin had a statistically significant reduced tumor volume (42%) by Day 17 when compared to the Dextran-treated control ($p < 0.05$). A lower dose of GBS Toxin, 2µg/kg, caused less effect, 15% reduction.

HISTOLOGY

Combined gross and microscopic evaluation of the xenograft from these animals showed hemorrhage only in two of the seven tumors from the Dextran-treated control group. In contrast, six of the seven tumors from the group receiving the 20µg/kg dose of GBS Toxin had hemorrhage. The morphological appearance of the hemorrhage in the Dextran-treated control group was significantly different than that of the groups treated with GBS Toxin. Hemorrhage observed in the tumors of the groups treated with GBS Toxin were not only larger and more frequent, they were also found in the actively growing peripheral region of the tumor where new vasculature was being developed.

Further histological examination showed that the tumor cells from the groups treated with GBS Toxin showed more evidence of intracellular debris, greater variability of nuclear size, and much more evident cytoplasmic vacuolization, well-accepted signs of cell damage.

TOXICOLOGY

Histological examination of lung, liver, kidney, spleen and heart tissues at the end of the experiment from all the nude mice revealed no apparent toxicity of GBS Toxin at either dose. Use of human xenografts and nude mice established that GBS Toxin could induce neovascularitis in developing tumors, thus, demonstrating susceptibility of mouse neovasculature to GBS Toxin.

IMMUNOCOMPETENT MOUSE MODEL

Immunocompetent BALB/c mice were implanted with a Madison Lung Tumor, an adenocarcinoma of the lung. Mice were implanted with 10^5 tumor cells subcutaneously in the right ventral area and the tumors were allowed to progress to an average size of 200cmm. Mice were subdivided into three groups; one group receiving Dextran at 60µg/kg, one group receiving GBS Toxin at 20µg/kg and one at 60µg/kg of GBS Toxin. Mice were treated by i.v. injection three times per week. In this experiment, tumors from mice treated with 60µg/kg grew at a 50% slower rate, statistically significant when compared with the growth rate of the Dextran-treated control group (p<0.04). GBS Toxin at 20µg/kg reduced the rate of tumor growth by 29%[7].

HISTOLOGY

Histological examination of the lung and tumor tissue obtained corroborated and extended the findings of the nude mouse tumor model by showing that GBS Toxin treatment led to endothelial cell death, loss of capillary integrity, hemorrhage and thrombosis in the tumor tissue. In the immunocompetent animals, the inflammatory response caused by the GBS Toxin treatment was evident in the tumor with the development of wide borders of granulation tissue at the tumor connector tissue interface. The tumors in the animals treated with GBS Toxin were 50% of the volume of the control, however, histological analysis confirmed that a large portion of this tumor volume was occupied by inflammatory cells and thrombosis. Thrombi were not evident in the control animals, which had no sign of inflammatory reaction towards their tumors.

MECHANISM OF ACTION

To investigate the molecular mechanism of action of GBS Toxin in the mouse models, we treated tumor-bearing and normal BALB/c mice with GBS Toxin or Dextran and measured cytokine levels in their serum. IL-1α was measured using the Genzyme InterTest-

1αX™ Mouse IL-1α Elisa Kit. TNFα levels were measured using the Promega CellTiter 96™ Cytotoxicity Assay and L-929 cells. Our preliminary data indicates that TNFα levels peak at 1 hr. whereas IL-1α levels were still elevated at 2 hrs. post-infusion of GBS Toxin in the tumor-bearing mice. Mice without tumors show no elevation in IL-1α or TNFα over baseline levels as a result of injection of up to 480μg/kg of GBS Toxin. When tumor-bearing mice were tested with different dosages of GBS Toxin we found a dose response with the peak at a dosage of 60 to 120μg/kg. We interpret this to indicate that GBS Toxin, a bacterial polysaccharide composed of repeating critical epitopes, binds and cross-links critical endothelial receptors. Maximum cross-linking of the receptors by GBS Toxin would be dose-dependent in the observed manner. An optimum concentration of GBS Toxin is the concentration which will maximize cross-linking of the critical receptors. It is hypothesized that receptor cross-linking will cause a chemokine response from the endothelial cell similar to what would occur if a bacteria would cross-link receptors on a target cell. In both instances the target cell would respond by chemokine release which results in cytokine responses by inflammatory cells.

In order to further understand the mechanism of action of GBS Toxin, we have established that GBS Toxin binds mouse C3. Elisa plates were coated with GAM-C3 IgG and blocked. Various dilutions of mouse serum, normal or heat-inactivated, were incubated with biotinylated-GBS Toxin overnight and then added to the plates. Bound biotinylated-GBS Toxin was measured by adding streptavidin-β galactosidase and substrate and measuring optical density at 405nM. We found that a serial dilution of mouse serum gave a peak absorbance of biotinylated-GBS Toxin bound via the mouse anti-C3 capturing antibody. The heat-inactivated serum shows no binding of biotinylated-toxin. This indicates that a possible mechanism of action of GBS Toxin could be that GBS Toxin binds developing endothelium in the tumor. C3 then binds to the bound GBS Toxin, thus effectively opsonizing the endothelial cells for phagocytotic action by inflammatory cells. The involvement of the inflammatory cells is indicated by the IL-1α and TNFα release in response to GBS Toxin in tumor-bearing animals but not in normal mice.

LONGEVITY STUDIES

We established an optimum dose of 60μg/kg of GBS Toxin in our murine model and proceeded to investigate the effect of long-term treatment with GBS Toxin on tumor-bearing mice. Madison Lung Tumor cells were injected subcutaneously in the mouse and four days later i.v. treatment with GBS Toxin at different dosages, 15, 30, 60 and 120μg/kg, was initiated and maintained every other day for eleven weeks. The actual survival curves show that all the Dextran treated mice died by sixty-seven days giving a median survival time of forty-three days. The mice treated with GBS Toxin gave the following median survival times: 15μg/kg - 71 days; 30μg/kg - 59 days; 60μg/kg - >170 days; 120μg/kg - 57 days. Five of six mice treated with 60μg/kg GBS Toxin developed small tumors initially that regressed and did not recur. These five tumor-free mice were taken off treatment after eleven weeks and monitored for an additional three months. No tumor recurrences have been observed. One mouse in the 60μg/kg group developed a tumor that progressed to eventually kill the animal. Postmortem examination of dextran treated mice whose tumors progressed indicated that the mice usually died of one of two different causes: massive lung metastases that essentially destroyed the pulmonary capability of the mouse, or tumor invasion through the peritoneum or thoracic cavity leading to infections or loss of thoracic integrity.

CONCLUSIONS

The ability of GBS to induce fatal inflammation targeted to the lung in the human neonate led us to speculate that GBS Toxin could induce inflammation in the neovasculature recruited by human tumors. Our observation that GBS Toxin binds to human tumor vasculature and reduces growth rates of human tumors in nude mice corroborated this notion.

Our studies indicate a mechanism of action where GBS Toxin infused i.v. will react with structural elements present only on developing pathologic neovasculature. The repeating structural nature of GBS Toxin causes cross-linking of these unique endothelial receptors. A combination of these cross-links and C3 binding to GBS Toxin leads to an inflammatory reaction targeted to the tumor vasculature which leads to a breakdown of the vascular integrity, causing retarded tumor growth.

GBS Toxin has potential for use in human tumor therapy and can be effective in very low non-toxic doses. The effect of the action is targeted, naturally, directly to the tumor site. Although GBS Toxin is not directly cytotoxic to tumor cells, its ability to induce tumor neovascularitis, disruption of the tumor vasculature and its inducement of inflammatory reactions only within the tumors indicate great potential for use in cancer treatment. Clinical Phase I trials are being initiated.

REFERENCES

1. C.G. Hellerqvist, J. Rojas, R.S. Green, S. Sell, H. Sundell and M.T. Stahlman. "Studies on Group B β-Hemolytic Streptococcus. I. Isolation and Partial Characterization of an Extracellular Toxin," Pediatric Res. 15: 892 (1981).

2. C.G. Hellerqvist, H. Sundell and P. Gettins. "Molecular Basis for Group B β-Hemolytic Streptococcal Disease," Proc. Natl. Acad. Sci., U.S.A. 84: 51(1987).

3. B. Russell, R. Pappas, J. Brandt, H. Sundell, C.G. Hellerqvist. "Use of Polyclonal Rabbit Anti-GBS Toxin IgG for Localization of a Possible Toxin Binding Site in Sheep Lung Tissue," Annual Meeting of the Society for Complex Carbohydrates Glycoconj. J. 5: 298 (1988).

4. H.W. Sundell, W. Fish, K. Sandberg, K.E. Edberg, R.S. Pappas and C.G. Hellerqvist. "Lung Injury by Streptococcal Toxins," Vanderbilt University Press (1989).

5. C.J. Baker, F. Barnett, R.C. Gordon, and H.D. Yone. "Suppurative Meningitis due to Streptococci of Lancefield Group B: A Study of 33 Infants," J. Pediatr. 82:724 (1973).

6. C.G. Hellerqvist. "Therapeutic Agent and Method of Inhibiting Vascularization of Tumors," U.S.Patent 5,010,062 - April 23, 1991 - World Patent Pending.

7. C.G. Hellerqvist, G.B. Thurman, D.L. Page, Yue-Fen Wang, B.A. Russell, C.A. Montgomery and H.W. Sundell. "Anti-tumor Effects of GBS Toxin: A Polysaccharide Exotoxin from Group B β-Hemolytic Streptococcus," CancerResearch and Clinical Oncology (1993) (In Press).

BIOLOGICAL DATA AND DOCKING EXPERIMENTS OF bFGF-SULFONATED

DISTAMYCIN A DERIVATIVE COMPLEX

Nicola Mongelli and Maria Grandi

Farmitalia Carlo Erba
Research Center
Oncology Deptartment
Milano and Nerviano (Italy)

SUMMARY

A series of sulfonated derivatives of Distamycin A have been synthesized with the objective of identifying novel compounds able to complex bFGF, which is involved in tumor angiogenesis, and consequently to block the angiogenic process.
These new compounds have been characterized for their ability of inhibiting bFGF binding, *in vivo* bFGF induced angiogenesis and neovascularization of the chorioallantoic membrane and antitumor activity on M5076 murine reticulosarcoma.
A set of molecules was docked to the known three-dimensional structure of the growth factor and a possible feature of the bFGF-ligand complex was deduced.
Sulfonated Distamycin A derivatives act as bidental ligands interacting with both the heparin and receptor binding sites of bFGF in a 1:1 ratio. The consistency of computational and biological results supports this approach for explaning the interaction between bFGF and sulfonated Distamycin A derivatives.

INTRODUCTION

Basic fibroblast growth factor (bFGF) is a peptide which, since 1983 has been purified from different tissues, as shown by Benharroch[1]. Its amino acid sequence has also been defined and recently the three-dimentional structure has been elucidated by Zhu[2] through x-ray crystallografic analysis. Two heparin binding sites and two cell attachment sites have been described on bFGF by Eriksson[3] and the binding to heparin has been widely exploited for the isolation of this growth factor. bFGF has been detected in a wide variety of normal and malignant tissues where it has multiple physiological activities including mitogenic and angiogenic activities as well as a role in differentiation and development.

Angiogenesis: Molecular Biology, Clinical Aspects
Edited by M.E. Maragoudakis *et al*, Plenum Press, New York 1994

271

bFGF could play a major role in tumor progression, not only by acting as a mitogen for the tumor cells, but also by stimulating neoangiogenesis. In light of these considerations, the search for compounds able to inhibit bFGF is a promising approach to the development for new antitumor agents.

With the aim of identifying novel compounds able to complex bFGF a series of sulfonated Distamycin A derivatives has been recently synthesized by Mongelli[4]. The general chemical structure of these compounds is presented in Table 1. The compounds have been characterized for their ability to inhibit bFGF binding, *in vivo* bFGF induced angiogenesis, neovascularization of the chorioallantoic membrane (CAM) and solid tumor growth as shown by Ciomei[5].

It has been demonstrated that the bFGF binding inhibiting activity of these compounds is due to the formation of a complex with bFGF itself and not with the corresponding bFGF receptor as shown by Ciomei[5].

In order to investigate a possible model for the interaction the growth factor and our molecules, we utilized the 3-D structure as shown by Zhu[2] of bFGF as receptor for the binding of these compounds.

RESULTS

The biological activity of the tested sulfonated Distamycin A derivatives is summarized in Table 1. Equivalent values in terms of inhibition of bFGF binding *in vitro* and *in vivo* and inhibition of vascularization on the CAM assay were observed with the three molecules bearing respectively four-four and six pyrrolic rings, FCE 26644-FCE 27164 and FCE 27111; the molecule bearing only 2 pyrrolic rings FCE 26951 was conversely significantly less effective for all tested parameters.

As regards the inhibition of tumor growth on the M5076 murine reticulosarcoma model, only the fist two molecules were shown to possess antitumor activity.

Neither x-ray coordinates nor high field NMR spectroscopy data (no Overhauser effects were detected) were available to be used as input geometries for the computer assisted molecular modeling studies of the compounds. Consequently, optimization of the drawn structures by using the program MODEL (version KS 2.99) as shown by Steliou[6] was followed by extensive conformational analysis using as a method for conformational searching the torsionale Monte Carlo search in BatchMin (version 3.1) associated with the program MacroModel (versione 3.0) as shown by Still[7]. It was assumed a trans conformation for all of the amide bonds, on the basis of the very high energetical barrier (20 kca/mol) calculated with the program MOPAC[8] for the interconversion from cis to trans conformation, the latter being 4 kcal/mol more stable. The other torsional angles were rotated with 60° (for pyrrole C_4-N) and 180° (for pyrrole C_2-CO) increments. Due to the high hydrophilicity of the molecules, this conformational analysis was performed using a continum water model as implemented in BatchMin.

The FCE 26644 output structure appears as a highly symmetrical, quasi-planar arc of a circle. All the methyl groups are located in the outer part of the molecule, while four of the six NH groups lie in the concave face. A quite low (4 kcal/mol) torsional barrier around the naphthalene C_7-N bond was also observed.

Table 1. Chemical Structures and Biological Data of Selected FCE Compounds

compound	R	n	bFGF binding ID_{50} (µM) [1]	bFGF-induced angiogenesis inhib.(%) [2]	% AUC inhibition [3]	CAM vascularization inhib.(%) [4]
FCE 26644	H	2	142 ± 18	100	92	100
FCE 27164	SO_3K	2	116 ± 10	93	88	80
FCE 27111	H	3	163 ± 16	89	27	62
FCE 26951	H	1	550 ± 92	not determined	0	30

[1] ID_{50}: dose inhibiting by 50% the binding of bFGF to its receptor determined on BALB 3T3 cells after 4h incubation in the presence of 0.2 nM ^{125}I-bFGF and scalar concentraitons of each compound.

[2] Inhibition of in vivo angiogenesis induced by bFGF gelfoan implanted s.c. in C3H/He mice treated 24 h later with 200 mg/kg i.v. of each compound. The degree of angiogenesis was measured by counting the number of vessels in control and treated mice aftere 15 days implantation.

[3] Inhibition of M5076 solid tumor growth - results are reported as % inhibition of the area under the curve (% AUC). C_{57} mice were injected i.m. with 1.10^5 cells on day 0. - Treatment was performed on day 1 with 200 mg/kg of each compound administered i.v.

[4] Inhibition of angiogenesis in the chorioallantoic membrane assay (CAM). Results are reported as % of CAMs presenting an avascular zone with at least 2 mm in diameter after 48h treatment with 350 nmol of each compoundincorporated into a methylcellulose pellet.

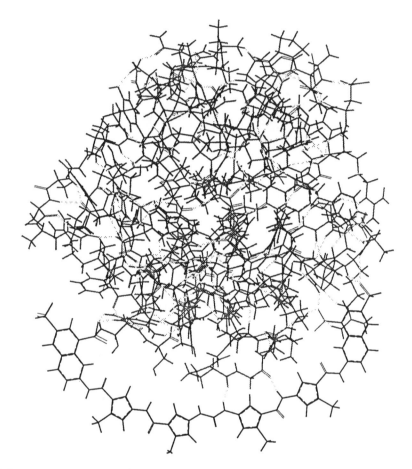

Figure 1. 3-D representation of BFGF structure derived from x-ray crystallografic studies (2) docked with FCE 26644. Dotted lines indicate hydrogen bonds.

The distance between the sulfonate groups at the 1 position of the two naphthalene rings in this conformation is 24-26 Å, which fits well with the calculated distance (25 Å) between the heparin and receptor binding sites on bFGF as shown by Eriksson[3].

Furthermore, manual docking experiments[9] performed on the bFGF-FCE 26644 complex (Fig. 1) showed a very good interaction between Arg^{121}, Lys^{126} e Asn^{28} of the heparin binding site on one hand and Arg^{108} of the receptor binding site, on the other hand, with the sulfonate groups of the drug. It is interesting to note that in such a representation the stability of this complex is further increased by additional interactions (hydrogen bonds) between the NH groups in the concave face of the drug and amino acid carbonyls of the bFGF. Moreover, once the complex structure was optimized, the electrostatic potential maps were calculated[9] and a great complementary was found between the positive surfaces corresponding to the above cited amino acids and the negative ones on the sulfonate groups.

Examination of the Van der Waals volumes calculated on the complexes between bFGF, FCE 26644, FCE 27111 and FCE 26951 shows a perfect complementarity in the case of FCE 26644. A quite good complementarity is still observed for compound FCE 27111 while FCE 26951 is too short for giving profitable interaction without steric repulsion. Finally, no affinity was shown by compounds having the same naphthyl moieties at the same relative distance as in FCE 26644, but differing in the nature of the spacers (γ-aminobutyric acid)[10]. A possible explanation of the lack of affinity in the latter case could be found in the greater conformational freedom of these molecule, which might prefer an extended conformation and find difficult the binding to a single bFGF molecule. In view of both the biological and computational results we can conclude that this series of molecules interact with the growth factor in a previously never described manner, by the attachment of their naphthyl- sulfonic moieties to both the heparin and the receptor binding sites of bFGF, thus preventing the interaction of the growth factor with the corresponding receptor.

ACKNOWLEDGMENTS

This work was supported by I.M.I in behalf of the Ministery for University and Scientific and Technological Research.
The biological assays were performed by M.Ciomei, F.Sola, M.Farao and M.Mariani, Farmitalia Carlo Erba, Research Center, Experimental Oncology.
We thanks prof. B.Bolla and prof. F.Corelli of Siena University for the all computational work.

REFERENCES

1. D.Benharroch and D.Birnhaum, Irs.J.Med.Sci. **26,** 212 (1990).

2. X.Zhu et al., Science **251,** 90 (1990).

3. A.E.Eriksson et al., Proc.Natl.Acad.Sci. **88,** 3441 (1991).

4. N.Mongelli et al., U.K. Patent n. 91902204.6 (1991).

5. M.Ciomei et al., in preparation.

6. K.Steliou, Program MODEL (Version KS 2.99) and BAKMDL, University of Montreal, Canada.

7. W.C.Still, Program MacroModel (Version 3.0) and BatchMin (Version 3.1), Department of Chemistry, Columbia University, New York, N.Y., USA.

8. MOPAC (Molecular orbita package), QCPE Creative Arts Bldg. 181, Indiana University, Bloomington, IN, USA.

9. Sybyl, Tripos Assiactes, St. Louis, MO, USA.

10. A full account on the chemistry of FCE compounds will be reported elesewhere.

CLINICAL OCULAR ANGIOGENESIS

David BenEzra

Immuno Ophthalmology and
Ocular Angiogenesis Laboratory
Hadassah University Hospital
Jerusalem, Israel

INTRODUCTION

Neovascularization - angiogenesis of the ocular tissues and their adnexa is a widespread phenomenon and one of the most common underlying causes for severe loss of vision. During embryogenesis of the eye, the growth of vessels within tissues which later become avascular is a continuous and smooth normal process finalized only around the seventh month of gestation. Involution of the embryonic vessels is closely associated with the evolution of the adult vessel system within the eye. These processes are tightly regulated and are most probably influenced by the sequential appearance of specific inhibitors and stimulators. Any interference with the timely production (or release) of these factors results in the persistence of embryonic vessels, the lack of proper adult vessel growth and inadequate development of the ocular structures. Within the adult eye, on the other hand, the growth of new blood vessels is an integral process of tissue damage with attempts at repair.

New vessel growth within the adult eye structures may be triggered by trauma, chronic inflammatory processes, changes in the metabolism of ocular tissues or the abnormal accumulation of "waste products".

Following astute clinical observations and meticulous experimental investigations, Michaelson suggested in the 1940's that the process of new blood vessel growth (angiogenesis) is influenced by the emergence or accumulation of a specific chemical factor. He coined this postulated product the "X-Factor of Neovascularization" (1,2).

In this paper, I will review some of the clinical ophthalmic diseases in which neovascularization-angiogenesis is an important underlying pathological process. Arbitrarily, the various clinical manifestations and diseases will be divided in the context of their associations with embryological events, perinatal events or events taking place within the adult eye.

Angiogenesis: Molecular Biology, Clinical Aspects
Edited by M.E. Maragoudakis *et al*, Plenum Press, New York 1994

DISEASES ASSOCIATED WITH EMBRYOLOGICAL EVENTS

Persistence of the "normal" embryological blood vessels and the lack of proper development of the adult system is by far the most common underlying pathology in this group of diseases. In rare cases, angiogenesis of ocular tissues after "normal" involution of the embryonic vessels is also observed in neonates.

Angiogenesis of the cornea is observed in "central corneal ulcers of von Hippel" and the various disease entities associated with "anterior chamber cleavage syndrome". Neovascularization of the iris or persistence of the embryonic system is also seen. It is generally accompanied by persistent pupillary membranes and vasculosa lentis (presence of vessels surrounding the lens). Most common neonatal "angiogenic diseases", however, are the diseases associated with the persistence of the embryonic hyaloid system and primary vitreous. This latter group of diseases is responsible for 10 to 15% of the causes of blindness (mostly in one eye) occurring in neonates.

DISEASES ASSOCIATED WITH PREMATURITY

Along with the involution of the embryonic hyaloid vessels between the second and third months of gestation, a steady progressing vascularization of the retina by the permanent retinal vessel system is taking place. This process is probably regulated by the increased metabolic needs of the thickening retinal structures and the extent of its avascular zones (2). The maturation of retinal vessels occurs earlier in the nasal retina, reaching its final stages in the temporal retina after the seventh month of gestation. Disturbances in the "normal" metabolism of the retinal cells during the premature stages of the retinal vessels may induce an uncontrolled and poorly regulated growth of these vessels. This process is known as retrolental fibroplasia (RLF) or retinopathy of prematurity (ROP). It is still a common cause of bilateral loss of vision in premature neonates and has been associated with the uncontrolled use of high concentrations of oxygen (3,4). The exact role of oxygen in the development of ROP is not fully understood. It derives most probably from the relative hyperoxia/hypoxia and the influence on the metabolism of the retinal cells within the avascular (mostly temporal) zone in these premature babies (5).

DISEASES IN THE ADULT EYE

Neovascularization of the adult ocular tissues is detrimental to normal visual functions. The new blood vessels interfere physically with the avascular ocular structures and tend to bleed. The recurrent bleeding from these new vessels is the most important underlying factor and common cause of visual loss in humans. Ocular diseases derived from angiogenic processes may be observed in and around the orbit and within all the structures of the globe. A description of some prevalent diseases classified according to the most important underlying cause (and anatomical site) is given below.

Extraocular tumors

Hemangioma, the most common (generally benign) tumor in childhood, occurs in high incidence in and around the orbit. Although most of the hemangiomas tend to involute, giant tumors may induce amblyopia and blindness (Figures 1, 2).

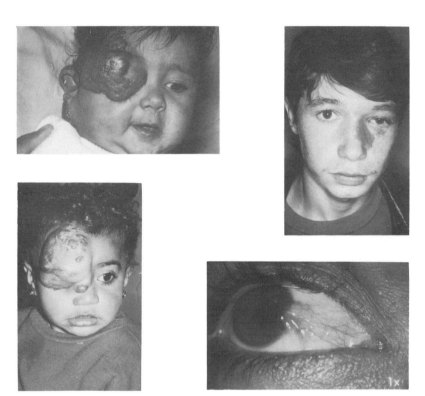

Figure 1. (L) Baby girl with giant hemangioma of the neck and forehead including the upper eyelid.
Figure 2. (R) Baby girl with giant hemangioma extending from the forehead and involving the upper eyelid.
Figure 3. (L) Young boy with Sturge Weber syndrome and hemifacial hemangioma and choroidal hemangioma inducing intractable glaucoma of the left eye.
Figure 4. (R) Proliferation of new blood vessels associated with a rapidly growing pterygium.

Hemangiomas in the Sturge Weber syndrome involve also the choroid and are the cause of intractable glaucoma and loss of vision in the affected eyes (Figure 3). Carcinomas of the eyelids or conjunctiva are rare and the accompanying angiogenic process of little significance in these cases. Pterygium, however, is a benign overgrowth of conjunctival tissues whose progress is closely associated with the degree of angiogenic activity in the progressing tissue (Figure 4).

Intraocular tumors

Retinoblastoma and malignant melanoma are the most common intraocular tumors. In retinoblastoma, the formation of new blood vessels does not appear to have any significance on the growth and/or invasiveness of this tumor. In malignant melanoma, the role of vascularization as a predictable factor of malignancy and invasiveness is under evaluation. Benign masses of the retina as in angiomatosis retinae or Coat's disease are closely associated with the formation of new blood vessels within, on or around the pathological masses. In some of these cases, angiogenic phenomena are also observed at a site distant from the lesion.

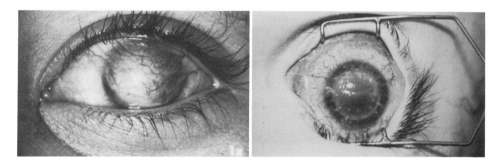

Figure 5. (L) Angiogenesis of the cornea associated with chronic recurrent herpetic keratitis.
Figure 6. (R) Neovascularization of corneal graft during an immune rejection process.

Inflammatory reactions

Chronic inflammation following infection, trauma or immune phenomena is an important trigger of angiogenesis of ocular tissues. Probably a process involved in repair and healing, the growth of new blood vessels within the cornea is frequently observed in cases of recurrent herpetic keratitis (Figure 5) and rejection of corneal graft (Figure 6). Chronic uveitis and vasculitis retinae induce neovascularization of the intraocular tissues. The angiogenic process is often observed on and around the optic nerve in the eyes of patients suffering from Behcet's disease (Figure 7) or in the periphery of the retina in eyes with chronic intermediate uveitis.

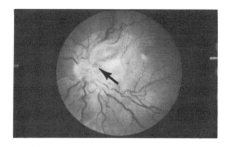

Figure 7. Formation of new blood vessels on the optic nerve of the left eye of a patient suffering from Behcet's disease.

Metabolic perturbations

This group includes the most widespread angiogenic ocular diseases, being one of the major underlying causes of blindness in adults. Disturbances in the normal metabolism of the retinal tissues as occur in hyperglycemia and diabetes trigger the growth of new blood vessels. One of the earliest events in this process is the formation of microaneurysms of the retinal capillaries (Figure 8). Perturbation of the normal metabolism of the retinal cells following central retinal vein occlusion is a common trigger for rubeosis iridis and intractable neovascular glaucoma. Lastly, the formation of subretinal neovascular membrane under the macula as it frequently occurs in age-related macular degeneration (ARMD) is one of the leading causes of blindness in the elderly (Figure 9). In this disease, the continuous "wear and tear" of the macular photoreceptors and the accumulation of abnormal metabolites are most probably the triggering factor(s) for the angiogenic processes. As the life expectancy is increasing steadily, it is anticipated that ARMD will be more prevalent and a major cause of severe visual disability in the elderly population.

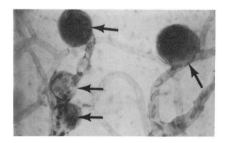 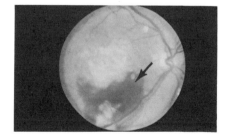

Figure 8. (L) Microaneurysm of the retinal capillaries as observed in a flat retina digest preparation of a patient who suffered from background diabetic retinopathy.
Figure 9. (R) Total loss of vision in the right eye of a 65-year-old lady with age-related macular degeneration. Note the profuse bleeding from the newly formed subretinal neovascular membrane.

CONCLUDING REMARKS

It is undeniable that the initial angiogenic triggering processes in the various clinical diseases described in this chapter are different. However, close follow-up and observation of the clinical manifestations lends support to the postulation that the final pathways and modulating events are identical and common in all diseases except (possibly) in hemangiomas. Furthermore, it is most probable that similar pathways are also responsible for the formation of new blood vessels in non-ocular organs. Thus, it is possible that the pathways leading to angiogenesis and growth of new blood vessels in all tissues are basically governed by similar processes. These processes may be affected by various chemical mediators during the early stages but are finally determined by the presence of one (or a few) pivotal factor(s) reaching a local critical concentration at the site of the angiogenic event (6).

ABSTRACT

Angiogenesis within the ocular tissues is a frequent ophthalmic manifestation and the underlying cause of numerous blinding diseases. An attempt to describe these diseases in association with embryological events, prematurity and specific processes within the adult eye has been made. Analysis of the various clinical events taking place during the development of the new blood vessels in these diseases leads to the assumption of a possible final common pathway. Initially, angiogenic triggers may be different and specific for each disease. However, the final pathways and crucial steps are most probably regulated and modulated by similar mediators.

REFERENCES

1. I.C. Michaelson, The mode of development of the retinal vessels and some observatrions on its significance in certain retinal diseases, *Trans Ophthalmol Soc UK.* 68:137, 1948.
2. I.C. Michaelson. "Retinal Circulation in Man and Animals,", Charles C. Thomas, Publisher, Springfield, Illinois (1954).
3. N. Ashton, B. Ward, G. Serpell, Effect of oxygen on developing retinal vessels with particular reference to the problem of retrolental fibroplasia, *Br J Ophthalmol.* 38:397, 1954.
4. V.E. Kinsey, Retrolental fibroplasia. Cooperative study of retrolental fibroplasia and the use of oxygen. *Arch Ophthalmol.* 56:481, 1956.
5. I.C. Michaelson, D. BenEzra, D. Berson, Possible metabolic mechanisms modulating blood vessel development in the inner eye and their significance for vascular pathology in the definite. *Met Ped Syst Ophthalmol.* 6:1, 1982.
6. D. BenEzra, Neovascularization: A unitarian phenomenon, *Doc Ophthalmol.* 25:125, 1981.

IN VIVO ASSAYS FOR ANGIOGENESIS

Wanda Auerbach and Robert Auerbach

Center for Developmental Biology
University of Wisconsin
1117 West Johnson Street
Madison, WI 53706 USA

INTRODUCTION

Angiogenesis, the formation of new blood vessels, is an important element of a large number of normal and disease processes. The most dramatic physiologically normal angiogenesis occurs during the development of the placental and embryonic vasculature. The most striking disease-associated angiogenesis is seen in the growth of solid tumors. But angiogenesis also accompanies normal processes such as the cyclical changes of the endometrium and the hair follicle, and the reparative events involved in the healing of wounds. The range of angiogenesis-associated disease processes is even broader: Angiogenesis is seen in sites of immunological activity ranging from autoimmune diseases to inflammation; it is almost universally associated with ocular diseases and injury, it is prominent in psoriasis and scleroderma, and it is a hallmark of granulomatous lesions. Studies of angiogenesis and its inhibition are therefore an integral part of numerous research programs both in academic institutions and in pharmaceutical firms interested in the development of angiogenic or anti-angiogenic strategies.

In recent years, much emphasis has been placed on the development of in vitro tests of angiogenesis (see Auerbach et al, 1991; Auerbach and Auerbach, 1994 for review). The establishment of culture techniques for the isolation and growth of endothelial cells has permitted the screening of reagents for their effect on endothelial cell proliferation, migration, and reorganization into three-dimensional patent vessels. However, while these in vitro systems are valuable, they are not always translatable into the in vivo situation. For example, TGFß is a potent inhibitor of endothelial cell proliferation and migration in vitro, yet it is angiogenic in vivo. Conversely, prostaglandin E is a powerful inducer of angiogenesis in vivo but is virtually inert when tested in vitro.

Inasmuch as the ultimate goal for the development of angiogenic and anti-angiogenic strategies is to influence blood vessel formation in the intact individual, there are now concerted efforts to provide animal models for more quantitative analysis of in vivo angiogenesis. It is these that will be discussed in this paper.

MORPHOLOGICAL STUDIES

The careful description of blood vessel formation as visualized by low power microscopy is still a valid means of characterizing angiogenesis. The grading of a vascular reaction based on low power microscopic examination serves as an excellent primary means of screening for induction or inhibition of a neovascular reaction. Subsequent analysis using high power resolution microscopy provides a reliable diagnosis of angiogenesis, by virtue of its ability to identify morphological characteristics such as the absence or reestablishment of tight junctional complexes, of pericyte-depleted endothelial cell projections, and of characteristic microangiopathies.

The use of fluorescent dyes to obtain better visualization of the vasculature has long been employed in the clinic, especially in the examination of ocular neovascularization. Low molecular weight fluorochrome-complexed tracers or fluorogenic dyes, moreover, have provided an additional method for detecting new blood vessels, based on the property of these vessels to be particularly "leaky", thus permitting tracer dyes to diffuse into perivascular sites.

What has now become a major new tool in morphological analysis, however, is the application of computer-assisted imaging technology to the delineation and measurement of both residual and newly-formed blood vessels. Image analysis can provide measurements of total vascular length, correlate fluorochrome content with vessel size, identify leakiness by measurements of fluorochrome-generated emission at perivascular sites, and calculate the relative number of divarication sites and vessel tortuosity. Each of these measures can then be correlated with source and type of angiogenic stimulus or inhibitor. Because of the speed with which such measurements can now be made, and because they may in many instances not require histological preparations but rather be carried out on whole mounts or in intact animals there is no longer the formidable limitations on sample numbers that previously limited in vivo tests for angiogenesis.

IMAGE ANALYSIS OF NEOVASCULARIZATION IN THE EYE

Perhaps the best example of modernization of a classic assay for angiogenesis is the study of corneal neovascularization. Initially described for rabbit corneas (Gimbrone et al, 1974), and now expanded to include the corneas of guinea pigs, rats and mice (cf. Muthukkaruppan and Auerbach, 1979), the assay involves the placement of an angiogenesis inducer (tumor tissue, activated lymphocytes or growth factors) into a corneal pocket in order to evoke vascular outgrowth from the peripherally located limbal vasculature (Fig.1).

This assay has seemed the most critical in vivo one, inasmuch as it measures only new blood vessels, since the cornea is initially avascular. In the past, angiogenesis has been described qualitatively (strong, weak, absent) or by measuring the length of the most advanced vascular sprout at daily intervals. More recently, flat mount preparations of corneas were used to obtain more quantitative assessment using imaging techniques and vector analysis to determine the directional properties of newly-penetrating blood vessels (Proia et al, 1988). Lately, the use of fluorescence to provide better imaging properties has led to more precise measurement of neovascularization, as described above, and the availability of sensitive video cameras and image recording will permit sequential recording of the neovascular process in individual animals without the need to resort to the dissection and fixation processes previously needed for obtaining imaging data from flat mount preparations.

Figure 1. Mouse corneal implants. Hydron-coated polyvinyl sponges containing bFGF (left) or bFGF + RNasin ribonuclease inhibitor (right) were placed 1mm from the edge of the limbal vasculature. Note extensive penetration of blood vessels induced by bFGF and the inhibitory effect of RNasin on this reaction (from Polakowski et al, 1993).

Quantitative assessment of neovascularization in the posterior segment of the eye can also now be obtained using similar techniques. Intra-vitreal injection of test cells, angiogenesis-inducing growth factors or anti-angiogenic reagents can be followed by characterization of the vascular response in the retina. Using fluorochrome-coupled high molecular weight dextran, for example, the entire retinal vasculature can be photographed and subjected to image analysis technology (D'Amato et al, 1993). By varying the route of injection of the tracer (right ventricle vs. intravenous), the penetration of the dextran can be modulated to reflect afferent vs. efferent vasculature. In addition to providing composite images of the vascular system, even low-power images can give some indication of the levels within the retina in which the blood vessels are located. In our own laboratory we have combined such imaging with data obtained from confocal microscopy. Composites of the confocal images yield a complete picture of the retinal vasculature, while individual optical sections furnish detailed information on individual blood vessels. The intensity of fluorescence, moreover, converted into pseudocolor images based on the number of activated pixels can reflect both the loss of fluorescence due to leakiness and the areas into which the fluorogenic tracers have penetrated.

THE DISC ANGIOGENESIS ASSAY

A few years ago, Fajardo and his colleagues described an assay procedure in which a polyvinyl alcohol sponge disc is implanted subcutaneously into test animals (Fajardo et al, 1988). By placing test material into the center of such a disc and by limiting diffusion to the edge of the disc (by sealing the top and bottom) "angiogenesis" could be determined by the extent of penetration of cells from the edge towards the center of the disc. Although the assay initially measured total penetration of cells, thus including not only endothelial cells but pericytes, fibroblasts and macrophages, recent extension of the earlier studies provided convincing evidence of true angiogenesis based both on histological identification of endothelial cells (Polakowski et al, 1993) and vascular channels and by the measurement of hemoglobin content (Kowalski et al, 1992).

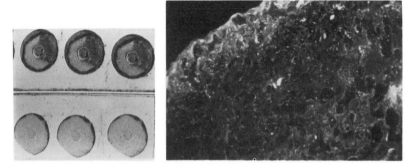

Figure 2. Angiogenesis induced by bFGF (top row) and inhibition of angiogenesis achieved by RNasin (bottom row) as measured in the disc angiogenesis assay system. Penetration of cells is seen in hematoxylin-stained histological sections through the sponge discs (left), and the computer-generated quantitation is shown by the white areas generated in pseudocolor by image analysis protocols (right).

A powerful modification of the Fajardo assay was developed in our own laboratory. Measurements were limited to the endothelial cell component of cellular penetration by using a fluorochrome-coupled lectin with binding affinity for endothelial cells. By using UV-illumination to examine sponge disc sections, it was possible to record only the penetration of endothelial cells, since fibroblasts and other cells accompanying the influx of those endothelial cells failed to label with the lectin. By use of low power (1X) objectives and sensitive video camera, it was possible to record the fluorescent images and provide numerical data for total fluorescence/preparation as a measure of angiogenesis. Although the method to date has been used only on sectioned discs, whole mounts can be used for both conventional and confocal analysis, thus eliminating one of the most time-consuming procedures of the disc angiogenesis system.

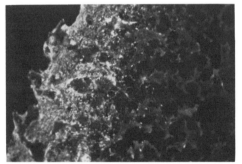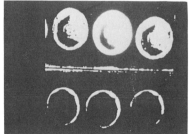

Figure 3. Sectioned discs labeled with a fluoresceinated lectin (^{FITC}DBA) which selectively stains endothelial cells. Control discs show typical penetration of endothelial cells (left), whereas the inclusion of RNasin ribonuclease inhibitor in the center of the disc markedly inhibits the entry of endothelial cells into the sponge implant (right). (from Polakowski et al, 1993)

The techniques developed for the disc angiogenesis assay have been modified to permit assessment of angiogenesis in the cornea (see above). Polyvinyl sponges loaded with test substances and then coated with slow release polymers have been introduced into the mouse cornea where they release the test factors and provide a target for the penetration of neovascular sprouts from the limbus.

Quantitation of angiogenesis by Xenon clearance analysis has been used in other types of sponge implant assays (Fan et al, 1993). Since the original description in 1987, several improvements have been made to simplify the assay procedures and important new data have been generated with this procedure (Hu et al, 1993). Since blood clearance depends primarily on diffusion from the microvasculature, the assay correlates well with angiogenesis.

FIBRIN CLOTS, SODIUM ALGINATE BEADS AND MATRIGEL PLUGS

Several types of matrices have been produced to provide a three-dimensional mesh that attracts and maintains new blood vessels. Although qualitative, vascularization of fibrin clots and gelatin occur under natural (base-line) conditions. By incorporating angiogenesis-inducing growth factors such as basic fibroblast growth factor (bFGF) or inhibitors such as antibodies to bFGF, useful information on both angiogenesis augmentation or inhibition can be generated.

Sodium alginate, a liquid which rapidly coalesces in the presence of added calcium, was used to suspend cells or reagents, dispersed into controlled-size beads by use of nebulizers, and implanted subcutaneously (Downs et al, 1992). Alginate beads maintained integrity for up to one month and the angiogenic response could be assessed by measurement of total hemoglobin content as well as by detection of systemically-administered, radioisotope-labelled erythrocytes. Although useful, the detection methods measure total blood content, and thus reflect predominantly the content of larger vessels and sinuses and possibly minimize the contribution of the newer microvascular projections which are the primary product of the angiogenic process.

A new addition to the matrix assay systems is Matrigel, a complex product generated by certain tumors that is rich in laminin but also contains a large number of growth modulating factors such as bFGF and TGFß and lower amounts of other matrix molecules such as collagen. Matrigel is liquid when cool but forms a strong coagulum at 37°C. Cells or test substances can be incorporated into liquid Matrigel, which is then injected subcutaneously where it rapidly solidifies. Angiogenesis can then be measured by recovering the material at specified times after inoculation, and determining hemoglobin content (Passaniti et al, 1992; see also studies of Grant, Kleinman and their colleagues in this volume). A more precise analysis requires examination of histological preparations. The method has already generated important information on both angiogenic and anti-angiogenic factors. The potential application of fluorescence imaging techniques (see above) to the Matrigel system should prove most fruitful both by providing yet better quantitation and by eliminating the need for histology.

CHICK EMBRYO ASSAYS

In a sense, the chick embryo, both in the shell or when explanted as a whole embryo, can be considered an in vivo assay, in that it measures the response of the whole animal. However, tests of angiogenesis in the chick have relied entirely on measurement of responses induced on the extra-embryonic membranes, i.e. the response of the yolk sac or the chorio-allantoic membrane. Therefore we will not discuss this assay in the present paper. The principal methodology is, however, discussed in an earlier review (Auerbach et al, 1991), and its application to the study of anti-angiogenesis has also been recently reviewed (Auerbach and Auerbach, 1994).

OTHER IN VIVO TECHNIQUES

We have previously reviewed other methods of in vivo analysis, including the use of diffusion chambers, the enumeration of blood vessels in the skin, the study of omental and other mesenteric vascularization, and the detection of neovascularization of the peritoneal lining. In all of these assays, the major limitations have involved the difficulty of obtaining quantitation. For example, the intradermal angiogenesis assay, originally described as a means of quantitating lymphocyte-induced angiogenesis, requires careful dissection, followed by immediate assessment of three-dimensionally distributed vascular divarication sites. All of these assays can be made markedly more useful if they are combined with easy methods of making preparations permanent and easy quantitation of the vascular response. The use of high molecular weight markers, of endothelium-specific lectins and antibodies, and of low molecular weight tracers for leakage combined with video-recording and imaging techniques will permit all of these methods to be more widely utilized.

CONCLUDING REMARKS

Perhaps the most important assays for angiogenesis will be those that measure directly the neovascular response within the system of interest. For example, tumor-associated angiogenesis might best be measured in tumors, the inflammatory angiogenic response in sites of inflammation, the estrus-associated angiogenic reaction in the endometrium and the vascularization of the placenta in the placenta. We and others already have established that there is extensive endothelial cell heterogeneity, reflected both in cell surface expression of organ-specific antigens and cell adhesion molecules, and in functional attributes ranging from selective transport mechanisms and specific enzyme-mediated biochemical reactions to selective synthesis of different mRNAs and variation in subsequent translational and post-translational events. Therefore it seems rational to propose that all angiogenesis is not alike, and that the angiogenic processes will vary depending on the organ site as well as on the nature of the angiogenesis-inducing factors or disease processes. Angiogenesis in vivo, moreover, will in large part depend on physiological factors that may vary for different individuals, both because of underlying individual genetic variability and because of modulations due to cyclical fluctuation of systemic factors such as hormones. To study this "differential angiogenesis" will be nearly impossible in vitro but should be possible using modifications of in vivo assays for angiogenesis such as the ones we have described in this review.

REFERENCES

Auerbach, R., Auerbach, W., and Polakowski, I., 1991, Assays for angiogenesis: a review, *Pharmacol. Ther.* 51: 1-11.

Auerbach, W., and Auerbach, R., 1994, Angiogenesis inhibition: A review, *Pharmacol. Therapeut.* (in press).

D'Amato, R., Wesolowski, E., and Hodgson-Smith, L.E., 1993, Microscopic visualization of the retina by angiography with high-molecular-weight fluorescein-labeled dextrans in the mouse. *Microvasc. Res.* 46:135-142.

Downs, E.C., Robertson, N.E., Riss, T.L., and Plunkett, M.L., 1992, Calcium alginate beads as a slow-release system for delivering angiogenic molecules in vivo and in vitro, *J. Cell. Physiol.* 152:422-429.

Fajardo, L.F., Kowalski, J., Kwan, H.H., Prionas, S.D. and Allison, A.C., 1988, The disc angiogenesis system, *Lab. Invest.* 58:718-724.

Fan, T.-P.D., Hu, D.-E., Guard, S., Gresham, G.A., and Watling, K.J., 1993, Stimulation of angiogenesis by substance P and interleukin-1 in the rat and its inhibition by NK_1 or interleukin-1 receptor antagonists, *Br. J. Pharmacol.* 110:43-49.

Gimbrone, M.A. Jr., Leapman, S.B., Cotran, R.S. and Folkman, J., 1974, Tumor growth and neovascularization: An experimental model using the rabbit cornea. *J. Natl. Cancer Inst.* 52:413-427.

Hu, D.E., Hori, Y., and Fan, T.-P.D., 1993, Interleukin-8 stimulates angiogenesis in rats. *Inflammation*, 17:135-143.

Kowalski, J., Kwan, H.H., Prionas, S.D., Allison, A.C. and Fajardo, L.F., 1992, Characterization and applications of the disc angiogenesis system, *Exp. Mol. Pathol.* 56:1-19.

Malcherek, P. and Franzén, L., 1991, A new model for the study of angiogenesis in connective tissue repair, *Microvasc. Res.* 42:217-223.

Muthukkaruppan, VR., and Auerbach, R., 1979, Angiogenesis in the mouse cornea. *Science* 205:1416-1418.

Passaniti, A., Taylor, R.M., Pili, R., Guo, Y., Long, P.V., Haney, J.V., Pauly, R.R., Grant, D.S. and Martin, G. R., 1992, A simple, quantitative method for assessing angiogenesis and antiangiogenic agents using reconstituted basement membrane, heparin, and fibroblast growth factor, *Lab. Invest.* 67:519-528.

Polakowski, I., Lewis, M.K., Muthukkaruppan, VR., Erdman, B., Kubai, L. and Auerbach, R., 1993, A ribonuclease inhibitor expresses anti-angiogenic properties and leads to reduced tumor growth in mice, *Am. J. Pathol.* 143:507-517.

Proia A.D., Chandler, D.B., Haynes, W.L., Smith, C.F., Suvarenamani, C., Erkel, F.H., and Klintworth, G.K., 1988, Quantitation of corneal neovascularization using computerized image analysis. *Lab. Invest.* 58:473-479.

Sidky, Y.A. and Auerbach, R., 1975, Lymphocyte-induced angiogenesis: A quantitative and sensitive assay of the graft-vs.-host reaction, *J. Exp. Med.* 141:1084-1100.

Sidky, Y.A. and Borden, E.C., 1987, Inhibition of angiogenesis by interferons: effects on tumor- and lymphocyte-induced vascular responses, *Cancer Res.* 47:5155-5161.

QUANTITATIVE ANALYSIS OF EXTRACELLULAR MATRIX FORMATION *IN VIVO* AND *IN VITRO*

Papadimitriou,E.[*][#], Unsworth,B.R.[*], Maragoudakis,M.E.[#] and Lelkes,P.I.[@]

[*]Marquette University, Dept of Biology, Milwaukee, WI; [#]Dept of Pharmacology, University of Patras, Greece; [@]University of WI Medical School, Milwaukee Clinical Campus, Milwaukee, WI

INTRODUCTION

Angiogenesis, the formation of new blood vessels from pre-existing ones, is a fundamental process, essential in both health situations, such as reproduction, development and wound repair, as well as in diseases, such as tumor growth, diabetes and arthritis. In the first case, angiogenesis is highly regulated, i.e. turned on for brief periods of time and then completely inhibited. In the latter case, angiogenesis is persistent and unregulated.

The formation and maturation of new blood vessels is associated with the sequential deposition of extracellular matrix (ECM) proteins, such as fibronectin (FN), laminin (LM), collagen-I and collagen-IV (Murphy and Carlson, 1978; Ausprunk *et al.*, 1991; Risau and Lemmon, 1988). The chorioallantoic membrane (CAM) of the chicken embryo is a well characterized and widely used model to study *in vivo* neovascularization (Stockley, 1980; Missirlis *et al.*, 1990). CAM is an extraembryonic membrane, which is formed at day four of embryo development, by the fusion of the chorion with the allantois. In this double layer of mesoderm an extensive capillary network develops. It has been shown that the number of vessels in the CAM increases rapidly between days 9 and 14, after which time it remains practically constant until hatching (Maragoudakis *et al.*, 1988a). Previous, descriptive, immunohistochemical studies have determined the sequence of appearance and the distribution of FN, LM and collagen-IV during formation of new vessels in the CAM (Ausprunk *et al.*, 1991). These studies showed that in the developing CAM, the distribution of the ECM proteins changed concomitant with formation and maturation of new blood vessels, suggesting a role of these components in regulating the growth and patterning of blood vessels *in vivo*. Recently, basement membrane collagen biosynthesis has been suggested as a quantitative index of angiogenesis (Maragoudakis *et al.*, 1988a).

In vivo, endothelial cells serve as the principal cellular component of the neovascular response, migrating into the interstitium, proliferating, reorganizing and anastomosing to form new capillary networks (Ausprunk and Folkman, 1977). With the development of isolation and culture techniques, it has become possible to study endothelial cells *in vitro* (Folkman and Haudenschild, 1980). This has allowed for a systematic approach in which many variables can be controlled, the ultimate goal being to mimic as closely as possible the *in vivo* condition. It has been shown that the same ECM proteins that are believed to regulate formation of new blood vessels *in vivo* (Maragoudakis *et al.*, 1988b; Madri *et al.*, 1991; Schor and Schor, 1983; Ingber and Folkman, 1988), are also implicated in endothelial cell proliferation, migration and differantiation into tube-like structures in culture (Madri *et al.*, 1983; Madri and Pratt, 1986; Form *et al.*, 1986; Herbst *et al.*, 1988; Kubota *et al.*, 1988; Carley *et al.*, 1988; Madri *et al.*, 1989).

Angiogenesis: Molecular Biology, Clinical Aspects
Edited by M.E. Maragoudakis *et al*, Plenum Press, New York 1994

As an *in vitro* model, we used endothelial cells from the rat adrenal medulla (RAME cells). The main purpose of this study was to establish the time course and extent of expression of collagenous and non-collagenous ECM proteins in cultured RAME cells and to compare this situation with an *in vivo* model, viz. the developing chick CAM. Ideally, a comparison between *in vivo* and *in vitro* models of neovascularization should employ CAM endothelial cells. However, we were unsuccessful in isolating and culturing endothelial cells from the chick CAM. Therefore, we chose RAME cells as an appropriate *in vitro* model, with the rationale that the adrenal medulla in mammals undergoes significant postnatal development (Coupland, 1965) and is a site of vasculogenesis (Lelkes and Unsworth, 1992).

Chromaffin cells in the adrenal medulla contain, and secrete extracellularly, very high amounts of ascorbate (Levine and Morita, 1985). The role of this secreted ascorbate is not known, but it has been postulated to be necessary for maintaining the vascular integrity and structure (Levine and Morita, 1985). Ascorbate is known to promote *in vitro* angiogenesis through enhancement of collagen synthesis (Nicosia *et al.*, 1991; Grant *et al.*, 1991). Therefore, we examined the effects of ascorbate on the levels of the ECM proteins FN, LM, collagen-IV and collagen-I, in cultured RAME cells.

METHODS

For the *in vivo* studies, CAMs at various days post fertilization were removed from the eggs and the proteins were extracted. Equal amounts of total protein were loaded on SDS-PAGE gels and Western blotting techniques were subsequently applied, probing the nitrocellulose with polyclonal antibodies against FN, LM and collagen-I. Collagen-IV was not examined in the developing chick CAM in the present study, because in our hands, all tested antibodies (n=6) either cross-reacted non-specifically with other proteins or did not react in Western blots. The protein levels were evaluated using quantitative image analysis. The distribution of collagen-I during the development of the chick CAM was determined using an immunocytochemical assay, performed according to (Reed *et al.*, 1992).

For the *in vitro* studies, RAME cells were isolated and cultured for one to 10 days. The cells were fixed with methanol at each time point. Enzyme-linked immunoassay (ELISA) techniques were used in order to quantitate the relative expression of the ECM proteins FN, LM, collagen-I and collagen-IV produced by these cells. The results were normalized per number of cells, using a novel fluorescent assay (Papadimitriou and Lelkes, 1993). A limitation of the ELISA technique was the inability to differentiate between intracellular and extracellulrly deposited proteins. Therefore, the localization of selected ECM molecules during culture was visualized by immunofluorescence, using polyclonal antibodies against FN, LM, collagen-I and collagen-IV. Detailed methodology has been described previously (Papadimitriou *et al.*, 1993).

RESULTS AND DISCUSSION

In Vivo

Trichrome staining of parrafin sections of the CAM showed that early in development, there was a small number of vessels, surrounded by an amorphous acellular matrix (Figure 1a). Throughout development, the tissue became more ordered and the number of vessels increased (Figure 1b). The relative abundance and distribution of the ECM proteins in the developing CAM changed, concomitant with formation and maturation of new blood vessels. All the ECM proteins studied were present from the onset of CAM development, however in quantitatively different amounts.

FN was expressed in a transient fashion. At the onset of CAM development, FN was present at about 40% of the day seven maximum. Subsequently, FN levels in the CAM gradually decreased, reaching at day 19 approximately 65% of the day seven maximum (Figure 2). Previous descriptive studies (Ausprunk *et al.*, 1991) have shown that this ECM material is one of the earliest components to be deposited in the CAM vascular basal laminae and diminishes with time. This decrease in FN is in agreement with other *in vivo* findings, where a FN-rich matrix is assosiated with migrating and proliferating embryonic endothelial cells, but not with tubule formation (Risau and Lemmon, 1988). Since mature capillaries are seated on a LM-rich basement membrane, which is largely devoid of FN (Risau and Lemmon, 1988), FN does not seem to be involved in the maintainance of a mature basal lamina.

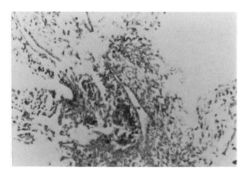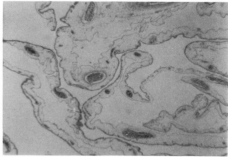

Figure 1: Trichrome staining of parrafin sections of the chick CAM at day five (a) and day 17 (b) of embryo development.

In contrast to the transient expression of FN, the amount of LM gradually increased until day 19 of gestation (Figure 3). At day five, the earliest day studied, the amount of LM was about 15% of that at day 19. Previous immunocytochemical studies (Ausprunk *et al.*, 1991) have shown that LM is not present in the blood vessel walls until two days after FN can be detected at that location and it appears in the perivascular matrix at the same time that a basal lamina can first be detected ultrastructurally. The presence of LM during all stages of vessel development in the CAM suggests that it might have a role in both the proliferation and migration of cells during early formation of the capillary tubes, as well as the later maturation of the primitive vessels into mature arterioles, venules and capillaries.

Collagen-I was a major protein component of the CAM. Both α_1 and α_2 chains were present at day five of embryogenesis, but represented less than 10% of the corresponding amounts at day 19, while ß chains appeared only after day seven. The levels of expression of each of the above chains increased gradually as the CAM developed (Figure 4). Immunocytochemical studies using paraffin sections of the chick CAM showed that immunoreactivity to collagen-I was evident from day five of embryo development and gradually increased in quantity. Collagen-I seemed to be ubiquitously expressed throughout the whole tissue during all stages of development. By day 13, the intensity of staining increased in the vicinity of the basal lamina of all blood vessels (Figure 5). It seems that collagen-I forms the core of this amorphous, mostly acellular tissue. In addition, this molecule may provide a permissive substrate for spreading and migration of endothelial cells during tubular organization (Schor *et al.*, 1983). Since angiogenesis in the growing chick CAM can be inhibited by preventing collagen synthesis (Maragoudakis *et al.*, 1988b; Lee *et al.*, 1992), collagen-I in the chick CAM may be important for supporting endothelial cell migration and growth, as well as for stabilizing newly formed vessels.

As mentioned above, we were unable to biochemically detect the expression of collagen-IV in the chick CAM. However, it was shown in previous, descriptive, immunocytochemical studies (Ausprunk *et al.*, 1991), that collagen-IV was the last ECM protein of those tested to appear in the matrix surrounding the CAM vessels, being absent during the earlier stages of vessel development, when the endothelial cells are proliferating rapidly. These data suggested that collagen-IV could be involved in stabilizing the progressive differentiation of endothelial and smooth muscle cells, occuring at later stages of vessel development in the chick CAM (Ausprunk *et al.*, 1974).

Fibronectin

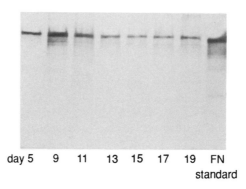

day 5   9   11   13   15   17   19   FN
standard

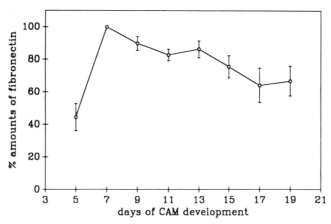

Figure 2: Fibronectin expression during development of the chick CAM. At day five, fibronectin was already present in significant amounts peaking at day seven of development. Data were quantitated by computer assisted image analysis. Data points are mean ± SEM from at least seven independent experiments.

The cellular origin of the selected ECM proteins in the *in vivo* CAM system was not ascertained in this study. Endothelial cells and smooth muscle cells seem the most likely source for LM, while FN and collagen-I could also be produced by connective tissue cells. Both endothelial cells and smooth muscle cells have been shown to produce FN (Macarak *et al.*, 1978; Jaffe and Mosher, 1978; Birdwell *et al.*, 1978; Yamada and Olden, 1978) and LM (Herbst *et al.*, 1988; Hedin *et al.*, 1988) *in vitro*. FN has also been shown to be produced by cultured fibroblasts (McDonald *et al.*, 1982; Pesciotta Peters *et al.*, 1990). Collagen-I has been shown to be present *in vitro* in cultures of both smooth muscle cells and connective tissue cells (Kleinman *et al.*, 1981), as well as in cultures of endothelial cells (Iruela-Arispe *et al.*, 1991).

Laminin

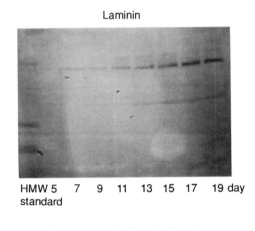

HMW 5   7   9   11   13   15   17   19 day
standard

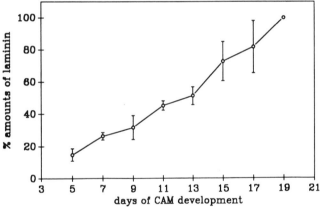

Figure 3: Laminin expression during development of the chick CAM. Laminin amounts gradually increased from day 5 to day 19 of development. Data were quantitated by computer assisted image analysis. Data points are mean ± SEM from four independent experiments.

In Vitro

Based on previously described isolation techniques (Mizrachi *et al.*, 1989), a population of RAME cells was obtained that was morphologically homogeneous with a cobblestone-like appearance. The endothelial cell phenotype was confirmed by demonstrating contact inhibition and positive staining for vWF and ac-LDL (data not shown). We plated the cells at such a density that they reached confluence three days after seeding.

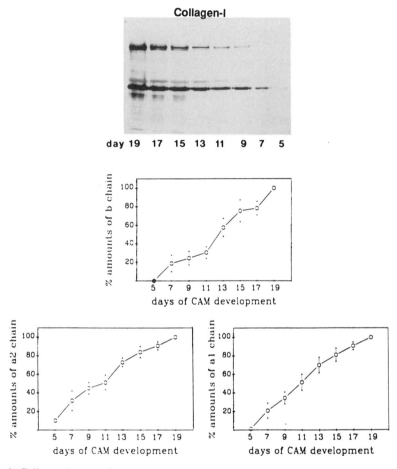

Figure 4: Collagen-I expression during development of the chick CAM. Both α_1 and α_2 chains of collagen-I appeared at day five of embryo development while ß chain appeared only after day seven. All the chains increased gradually during CAM development. Data were quantitated by computer assisted image analysis. Data points are mean ± SEM from at least five independent experiments.

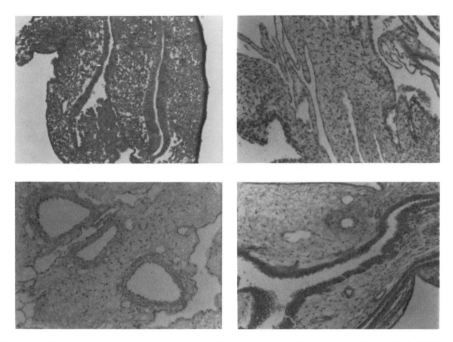

Figure 5: Immunoperoxidase staining for collagen-I on parrafin sections of the chick CAM, at day five (a), nine (b), 13 (c) and 19 (d) of embryo development.

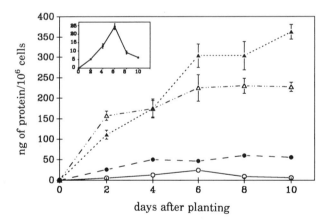

Figure 6: Amounts of fibronectin (○), laminin (●), collagen-I (▲) and collagen-IV (△) deposited into the ECM of RAME cells. Data points are mean ± SEM from at least three different experiments each performed in triplicates. Where no error bars are apparent, the SEM falls within the diameter of the symbol. Insert: Enlargement of the scale to emphasize the transient increase in fibronectin expression.

Collagen-IV, the most abundant ECM protein produced by these cells, was detectable before cells were confluent and its levels increased steadily to reach a plateau six days after plating. LM was also detectable before cells reached confluence and reached a plateau level two days earlier than collagen-IV. The maximal concentration of LM was approximately ten times less than that of collagen-IV.

Collagen-I was detectable in subconfluent concentrations of cells as well. Its level increased steadily, to reach a plateau eight days after plating.

Figure 6 shows the time course of the relative expression of FN, LM, collagen-IV and collagen-I, produced by RAME cells. The amounts of these proteins were quantitated by enzyme-linked immunoassays, using specific polyclonal antibodies.

The maximal level of FN was less than that of LM, collagen-IV, or collagen-I. FN was expressed two days after the cells were plated and increased with time in culture until six days after plating, when its concentration was similar to the maximal levels of LM. After that time, the amounts of FN decreased by about 80%, reaching the same levels as in early cultures.

The localization of the same ECM molecules during culture was visualized by immunofluorescent techniques.

FN was localized mainly intracellularly at the onset of the culture, but a few extracellular fibrils were also present, prior to the cells achieving confluence. At confluence, the extracellular fibrillar meshwork of FN was extensive, while the intracellular immunofluorescent signal decreased. The extracellular fibrils appeared to form a continuous meshwork underneath the cells. After four days of confluency, both the extracellular fibrillar meshwork and intracellular FN appeared to decrease (Figure 7a, b and c). This apparent loss of FN was also reflected in the quantitative data (Figures 6 and 9).

The transient nature of FN expression *in vitro* is a noteworthy similarity between the *in vitro* and *in vivo* systems used in the present study. FN may be necessary to mediate initial adhesion of the cells to the culture dish and promote their migration to form a continuous monolayer (Kowalczyk *et al.*, 1990; Macarak *et al.*, 1978; Yamada and Olden, 1978). However, in later stages after attaining confluence, when the cells begin to deposit sufficient LM, collagen-IV and collagen-I, they may no longer need FN and degradation would occur. A down regulation of FN matrix assembly has also been suggested in cultures of bovine pulmonary artery endothelial cells, in order to explain the depolarization of FN secretion, which was observed after several days of confluency (Kowalczyk *et al.*, 1990).

The techniques used for the *in vitro* studies cannot provide direct information regarding the cellular and/or serum origin of FN fibrils. It has been shown that in culture, both plasma FN and that produced by the cells, bind to the tissue culture plastic to mediate the attachment (Kleinman *et al.*, 1981; Hayman and Ruoslahti, 1979). Also, in cultures of human fibroblasts, plasma and cellular FNs were shown to be present in the same matrix fibrils and also in the same locations on the cell surface (Pesciotta Peters *et al.*, 1990). According to that study, the contribution of plasma FN to the ECM becomes significant at a concentration of $250\mu g/ml$ in the culture medium, while it is insignificant at a concentration of $20\mu g/ml$ in the medium. Since the medium used in the present study contained approximately $45\mu g/ml$ FN, FN fibrils seen in our *in vitro* cultures were apparently predominantly of cellular origin.

LM was present intracellularly immediately after the cells were plated and did not appear extracellularly until the cells became confluent. After cell contact was established, there was gradual appearance of extracellular fibrils, which organized into a continuous fibrillar meshwork during culture. The intracellular levels of LM gradually decreased to very low levels (Figure 7d, e and f). This redistribution of LM was not accompanied by quantitative alterations in total LM (Figure 6).

Similarly to LM, collagen-IV was initially confined intracellularly and could be located extracellularly only after confluency of the cells was established. With progressing culture, collagen-IV formed a diffuse, continuous, nonfibrillar meshwork. Collagen-IV was observed intracellularly during the entire study (Figure 7g, h and i).

The above data provide evidence for the importance of cell-cell contact for the extracellular deposition of LM and collagen-IV. Each of these proteins is produced by RAME cells from the onset of culture. However, they are not deposited extracellularly until cell contact has been established. This is in agreement with our *in vivo* results and with previous *in vivo* and *in vitro*

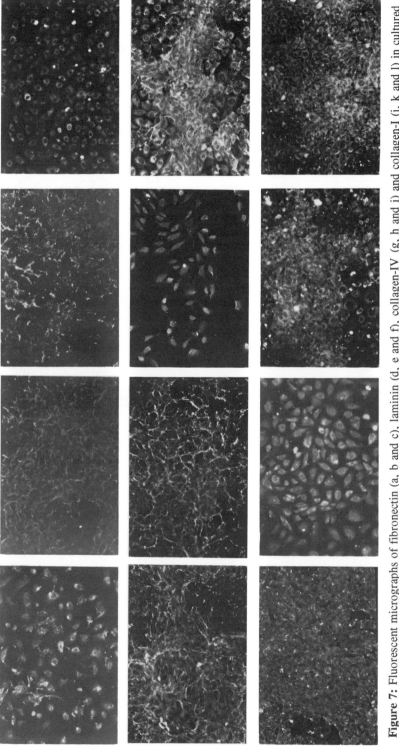

Figure 7: Fluorescent micrographs of fibronectin (a, b and c), laminin (d, e and f), collagen-IV (g, h and i) and collagen-I (j, k and l) in cultured RAME cells. Two days after plating (a, d, g and j), six days after plating (b, e, h and k) and ten days (c, f, i and l) after plating of the cells. Note that at two days after plating only fibronectin was localized extracellylarly. The intracelular levels of fibronectin and laminin decreased at ten days after plating, while those of collagen-I and collagen-IV remained the same.

299

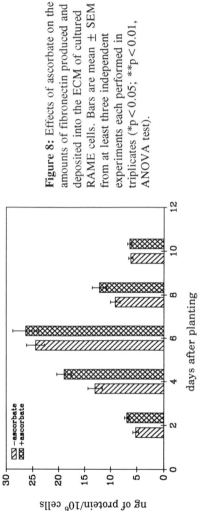

Figure 8: Effects of ascorbate on the amounts of fibronectin produced and deposited into the ECM of cultured RAME cells. Bars are mean ± SEM from at least three independent experiments each performed in triplicates (*p<0.05; **p<0.01, ANOVA test).

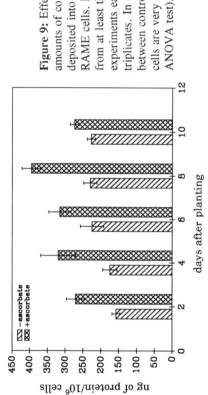

Figure 9: Effect of ascorbate on the amounts of collagen-IV produced and deposited into the ECM of cultured RAME cells. Bars are mean ± SEM from at least three independent experiments each performed in triplicates. In each case, the differences between control and ascorbate-treated cells are very significant (p<0.01, ANOVA test).

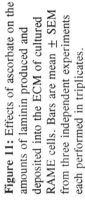

Figure 10: Effects of ascorbate on the amounts of collagen-I produced and deposited into the ECM of cultured RAME cells. Bars are mean ± SEM from three independent experiments each performed in triplicates.

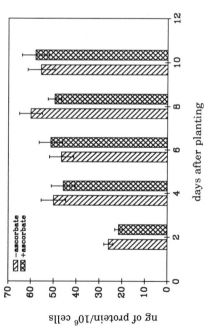

Figure 11: Effects of ascorbate on the amounts of laminin produced and deposited into the ECM of cultured RAME cells. Bars are mean ± SEM from three independent experiments each performed in triplicates.

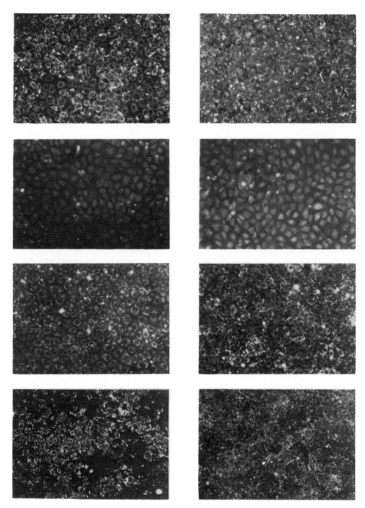

Figure 12: Fluorescent micrographs of fibronectin (g and h), laminin (e and f), collagen-I (c and d) and collagen-IV (a and b) in cultured RAME cells, four days after plating, with (b, d, f and h) or without (a, c, e and g) ascorbate.

studies, which showed that LM and collagen-IV are essential for endothelial cell differentiation to form tube-like structures, which occurs at a later stage of vessel development, when the cells are in contact with each other (Ausprunk *et al.*, 1991; Risau and Lemmon, 1988; Kubota *et al.*, 1988). During formation of new vessels, collagen-IV appears in the basal lamina later than LM (Ausprunk *et al.*, 1991). This observation agrees with our *in vitro* finding, that laminin reached a plateau level two days earlier than collagen-IV.

Collagen-I was localized only intracellularly in the beginning of the culture and it appeared extracellularly after the cells became confluent. With time in culture, collagen-I formed a continuous, extensive meshwork. Similar to collagen-IV, collagen-I was observed intracellularly during the entire study (Figure 7i, k and l).

Ascorbate is known to promote collagen biosynthesis and to enhance *in vitro* angiogenesis (Grant *et al.*, 1991; Nicosia *et al.*, 1991). We examined whether ascorbate would specifically enhance collagen-IV and collagen-I levels and whether it had any effect on other, non-collagenous ECM proteins, such as LM and FN. Ascorbate, at a concentration of 20μg/ml, was found to differentially affect the expression of the ECM proteins tested. As expected, collagen-IV levels were significantly increased by ascorbate at all the periods of culture (Figures 8 and 12a and b).

Surprisingly however, it had no effect on the expression of collagen-I (Figures 10 and 12c and d). To the best of our knowledge, this study is the first to detail ascorbate-mediated promotion of the synthesis of specific collagen types. In examining the effects of ascorbate on other, non-collagenous ECM proteins, we obtained contrasting results. Ascorbate had no effect on LM, based on the quantitative data (Figure 11). However, the immunofluorescent studies showed an increase in the extracellular amounts of LM by ascorbate (Figure 12e and f). This effect might be due to a redistribution by ascorbate of intracellular and extracellular LM, favouring an increase in the extracellularly deposited LM. FN synthesis and deposition were transiently increased by ascorbate (Figure 9). The transient nature of this effect might be due to a shut-off of the FN gene after several days in culture, which cannot be reversed by ascorbate. Alternatively, there might be a shift from secreted FN molecules to cell-associated insoluble molecules with the addition of ascorbate at the beginning of the culture, and as long as the cells require FN in their matrix. Further experiments are required to elucidate this point, but they are beyond the scope of this study.

CONCLUSIONS

There is a great similarity between the *in vitro* and the *in vivo* systems used in the present study. The composition of the ECM is changing both throughout the formation and the maturation of new blood vessels in the developing CAM, or throughout the culture period of RAME cells. In both systems, FN is one of the earliest components to be deposited in the ECM, while its amounts diminish after the establishment and maturation of new vessels *in vivo*, or after several days of confluency *in vitro*. Secretion and deposition of both LM and collagen-IV in the matrix of the cells requires cell-cell contact. Collagen-I is expressed early in both systems and its levels increase gradually. It may be important for supporting endothelial cell migration and growth (Schor *et al.*, 1983), as well as stabilizing later differentiation of cells.

ACKNOWLEDGEMENTS

We would like to thank Dr. D. Malone for the ac-LDL, Dr. R. Montgomery for the polyclonal vWF antibody, Dr. D. Amrani for the human FN and the anti-FN antibody, Dr. D. Grant for the anti-collagen-IV antibody and Dr. C. Little for the chicken anti-collagen-I antibody. We also thank Ms. D. Chick for help with the cell cultures and Dr. M. Samet for his help in digitizing the gel images.

REFERENCES

Ausprunk, D.H., Knighton, D.R., and Folkman, J. (1974) Differentiation of vascular endothelium in the chick chorioallantois: A structural and autoradiographic study. Dev. Biol., *38*:237-248.

Ausprunk, D.H., Dethlefsen, S.M., and Higgins, E.R. (1991) Distribution of fibronectin, laminin and type IV collagen during development of blood vessels in the chick chorioallantoic membrane. In: The development of the vascular system. R.N. Feinberg, G.K. Sherer, and R. Auerbach, eds. Karger, Basel, pp. 93-107.

Ausprunk, D.H. and Folkman, J. (1977) Migration and proliferation of endothelial cells in preformed and newly formed blood vessels during tumor angiogenesis. Microvascular Research, *14*:53-65.

Birdwell, C.R., Gospodarowicz, D., and Nicolson, G.L. (1978) Identification, localization, and role of fibronectin in cultured bovine endothelial cells. Proc. Natl. Acad. Sci. USA, *75(7)*:3273-3277.

Carley, W.W., Milici, A.J., and Madri, J.A. (1988) Extracellular matrix specificity for the differentiation of capillary endothelial cells. Exp. Cell Res., *178*:426-434.

Coupland, R.E. (1965) The natural history of the chromaffin cell. Little, Brown and Co., Boston, pp. 1-279.

Folkman, J. and Haudenschild, C. (1980) Angiogenesis in vitro. Nature, *288*:551-556.

Form, D.M., Pratt, B.M., and Madri, J.A. (1986) Endothelial cell proliferation during angiogenesis In vitro modulation by basement membrane components. Lab. Invest., *55(5)*:521-530.

Grant, D.S., Lelkes, P.I., Fukuda, K., and Kleinman, H.K. (1991) Intracellular mechanisms involved in basement membrane induced blood vessel differentiation in vitro. In Vitro Cell. Dev. Biol., *27A*:327-336.

Hayman, E.G. and Ruoslahti, E. (1979) Contribution of fetal bovine serum fibronectin and indogenous rat cell fibronectin in extracellular matrix. J Cell Biol., *83*:255-259.

Hedin, U., Bottger, B.A., Forsberg, E., Johansson, S., and Thyberg, J. (1988) Diverse effects of fibronectin and laminin on phenotypic properties of cultured arterial smooth muscle cells. J Cell Biol., *107*:307-319.

Herbst, T.J., McCarthy, J.B., Tsilibary, E.C., and Furcht, L.T. (1988) Differential effects of laminin, intact type IV collagen, and specific domains of type IV collagen on endothelial cell adhesion and migration. J Cell Biol., *106*:1365-1371.

Ingber, D. and Folkman, J. (1988) Inhibition of angiogenesis through modulation of collagen metabolism. Lab. Invest., *59(1)*:44-51.

Iruela-Arispe, M.L., Hasselaar, P., and Sage, H. (1991) differential expression of extracellular proteins is correlated with angiogenesis in vitro. Lab. Invest., *64(2)*:174-186.

Jaffe, E.A. and Mosher, D.F. (1978) Synthesis of fibronectin by cultured human endothelial cells. J. Exp. Med., 1779-1791.

Kleinman, H.K., Klebe, R.J., and Martin, G.R. (1981) Role of collagenous matrices in the adhesion and growth of cells. J Cell Biol., *88*:473-485.

Kowalczyk, A.P., Tulloh, R.H., and McKeown-Longo, P.J. (1990) Polarized fibronectin secretion and localized matrix assembly sites correlate with subendothelial matrix formation. Blood, *75*:2335-2342.

Kubota, Y., Kleinman, H.K., Martin, G.R., and Lawley, T.J. (1988) Role of laminin and basement membrane in the morphological differentiation of human endothelial cells into capillary-like structures. J Cell Biol., *107*:1589-1598.

Lelkes, P.I. and Unsworth, B.R. (1992) Role of heterotypic interactions between endothelial cells and parenchymal cells in organospecific differentiation: a possible trigger of vasculogenesis. In: Angiogenesis in health and disease. M.E. Maragoudakis, P. Fullino, and P.I. Lelkes, eds. Plenum Press, New York,NY, pp. 27-43.

Levine, M. and Morita, K. (1985) Ascorbic acid in endocrine systems. In: Vitamins and hormones advances in research and applications volume 42. G.D. Aurbach and D.b. McCormick, eds. Acadmic Press,Inc., Orlando,FL, pp. 2-53.

Macarak, E.J., Kirby, E., Kirk, T., and Kefalides, N.A. (1978) Synthesis of cold-insoluble globulin by cultured calf endothelial cells. Proc. Natl. Acad. Sci. USA, *75(6)*:2621-2625.

Madri, J.A., Williams, S.K., Wyatt, T., and Mezzio, C. (1983) Capillary endothelial cell cultures: phenotypic modulation by matrix components. J Cell Biol., *97*:153-165.

Madri, J.A., Pratt, B.M., and Yannariello-Brown, J. (1989) Endothelial cell--extracellular matrix interactions. In: Endothelial cell biology in health and disease. N. Simionescu and M. Simionescu, eds. Plenum Press, New York, pp. 167-186.

Madri, J.A., Bell, L., Marx, M., Merwin, J.R., Basson, C., and Prinz, C. (1991) Effects of soluble factors and extracellular matrix components on vascular cell behavior in vitro and in vivo: models of de-endothelialization and repair. J. Cell. Biochem., *45*:123-130.

Madri, J.A. and Pratt, B.M. (1986) Endothelial cell--matrix interactions: In vitro models of angiogenesis. J. Histochem. Cytochem., *34(1)*:85-91.

Maragoudakis, M.E., Panoutsacopoulou, M., and Sarmonika, M. (1988a) Rate of basement membrane biosynthesis as an index to angiogenesis. Tissue and Cell, *20*:531-539.

Maragoudakis, M.E., Sarmonika, M., and Panoutsacopoulou, M. (1988b) Inhibition of basement membrane biosynthesis prevents angiogenesis. J. Pharmacol. Exp. Ther., *244*:729-733.

McDonald, J.A., Kelley, D.G., and Broekelmann, T.J. (1982) Role of fibronectin in collagen deposition: Fab' to the gelatin-binding domain of fibronectin inhibits both fibronectin and collagen organization in fibroblast extracellular matrix. J Cell Biol., *92*:485-492.

Missirlis, E., Karakiulakis, G., and Maragoudakis, M.E. (1990) Angiogenesis is associated with collagenous protein synthesis and degradation in the chick chorioallantoic membrane. Tissue and Cell, *22(4)*:419-426.

Mizrachi, Y., Lelkes, P.I., Ornberg, R.L., Goping, G., and Pollard, H.B. (1989) Specific adhesion between pheochromocytoma (PC12) cells and adrenal medullary endothelial cells in co-culture. Cell Tissue Res., *256*:365-372.

Murphy, M.E. and Carlson, E.C. (1978) An ultrastructural study of developing extracellular matrix in vitelline blood vessels of the early chick embryo. Am. J. Anat., *151*:345-376.

Nicosia, R.F., Belser, P., Bonanno, E., and Diven, J. (1991) Regulation of angiogenesis in vitro by collagen metabolism. In Vitro Cell. Dev. Biol., *27A*:961-966.

Papadimitriou, E., Unsworth, B.R., Maragoudakis, M.E., and Lelkes, P.I. (1993) Time-course and quantification of extracellular matrix maturation in the chick chorioallantoic membrane and in cultured endothelial cells. Endothelium, *1*:207-219.

Papadimitriou, E. and Lelkes, P.I. (1993) Measurement of cell numbers in microtiter culture plates using the fluorescent dye Hoechst 33258. J. Immunol. Method., *162*:41-45.

Pesciotta Peters, D.M., Portz, L.M., Fullenwider, J., and Mosher, D.F. (1990) Co-assembly of plasma and cellular fibronectins into fibrils in human fibroblast cultures. J. Cell Biol., *111*:249-256.

Reed, J.A., Manahan, L.J., Chang-Soo, P., and Brigati, D.J. (1992) Complete one-hour immunocytochemistry based on capillary action. BioTechniques, *13*:434-443.

Risau, W. and Lemmon, V. (1988) Changes in the vascular extracellular matrix during embryonic vasculogenesis and angiogenesis. Dev. Biol., *125*:441-450.

Schor, A.M., Schor, S.L., and Allen, T.D. (1983) Effects of culture conditions on the proliferation, morphology and migration of bovine aortic endothelial cells. J. Cell Sci., *62*:267-285.

Schor, A.M. and Schor, S.L. (1983) Tumor angiogenesis. J. Path., *141*:385-413.(Abstract)

Stockley, A.T. (1980) The chorioallantoic membrane of the embryonic chick as an assay for angiogenic factors. Br. J. Dermatol., *102*:738-742.(Abstract)

Yamada, K.M. and Olden, K. (1978) Fibronectins--adhesive glycoproteins of cell surface and blood. Nature, *275*:179-184.

MICROVASCULAR MORPHOMETRY

K. Narayan and C. Garcia

Peter MacCallum Cancer Institute
Melbourne, Vic, Australia

INTRODUCTION

Julius Cohnheim (1889) was one of the first to describe the microvasculature of tissues. However, it was not until sixty years later that it could be quantified (Chalkley, et al 1949). Popular methods to visualise microvasculature had been colloidal carbon injection (Hilmas and Gillette 1974) and perfusion with a mixture of India Ink and gelatin (Jee and Arnold 1960). Perfused specimens were either cleared in glycerol and made transparent or embedded in wax and sectioned to be examined under the microscope. Recently colloidal carbon and gelatin has been replaced by a mixture of fixative, copper pigment and latex (Reinhold et al 1983) or photographic emulsion (Reempts et al 1983). Once the microvasculature is "labelled", it is photographed and the images so obtained are analysed either with a point counting grid according to the Chalkley method (Hilmas and Gillette 1974) or the area or perimeter of the stained vessel cross sections measured using an image analysing computer system (Wilkinson et al 1981). Most of the studies have emphasised the importance of perfusing the microvasculature at near physiological pressures. The introduction of polymerising casting media which withstand electron bombardment in the scanning electron microscope has opened up an elegant avenue to study microvascular patterns (Murakami 1971). It soon became evident that complete filling of microvasculature does not occur at physiological pressures (Lametschwandtner et al 1990). It would thus be difficult to measure a modest reduction in microvascular volume using perfusion at a physiological pressure. If the purpose of a study is to detect an anatomical reduction in the microvasculature then complete filling of the microcirculation would be essential. Microcorrosion casts can demonstrate the extent of filling of the microcirculation in a given tissue. It is, however, difficult to quantitate three dimensional vascular casts. The histology of cast specimens on the other hand can be quantified. The use of vascular casts in microvascular morphometry has been described (Narayan et al 1991). However, in this technique it is not possible to distinguish between vascular spaces, air bubbles or spaces due to loss of tissue as all these appear clear in the histology of a mercox (CL-2B-S Blue, Vilene Hospital, Japan) cast specimen. The present paper describes a technique to measure only mercox filled vascular spaces in a histological section. An attempt has also been made to correlate the perfusion pressure and its effects on capillary filling, with the resulting variation in microvascular morphometry. Murine brains, bone marrow, liver and kidney were used in this study.

METHODS

Animals: Thirty-four twelve week old male CBA mice, approximately 25 grams in weight were used.

Microvascular casting

Mice were deeply anaesthetised using Penthrane (Methoxyflurane, Abbot Laboratories, USA). The ascending aorta in the case of brain, thoracic aorta in the case of liver, upper abdominal aorta in the case of kidney and lower abdominal aorta, just below the origin of renal blood vessels in the case of bone marrow were exposed and cannulated. Following cannulation, the right atrium was incised to allow blood and subsequently clear fluid to drain out of the circulation. Blood was cleared from the circulation by perfusing iso-osmotic fluid which consisted of a 6.0% solution of Polyvinyl pyrolidone (BDH, England) dissolved in normal saline. Ten thousand units of heparin were added to one litre of clearing fluid. Perfusions were carried out at 37°C. Various perfusion pressures were maintained for different organs (Table 1). These perfusion pressures were measured in the perfusion bottle. Once the blood was cleared, mercox resin (MERCOX CL-2B5, Vilene, Japan) mixed with methyl methacrylate in a 3:1 ratio with accelerator (Benzoyl Peroxide) added was perfused. The perfused animal was kept at room temperature (18°C for 10 - 15 minutes) to allow for the polymerisation of the resin. Four mice had their brains perfusion-fixed with 10% formalin only and no mercox was perfused. These when compared with the mercox specimens provided the normal histological controls. Twelve mice had their brains perfused, four were used for obtaining femur samples and six were used for both kidney and liver samples.

TISSUE SAMPLING

Bone Marrow

The femora were dissected free. The right femora were fixed by immersion in 10% formalin for 24 hrs for histology and subsequent morphometry. Following fixation, they were placed in 10% EDTA solution for three weeks for decalcification. Decalcified femora were cut into 2 mm long pieces and the upper and lower ends containing active bone marrow were embedded in plastic (see below). Left femora from each animal were placed directly in 10% EDTA for 3 weeks and decalcified for specimen corrosion and subsequent scanning electron microscopy.

Brain, Liver and Kidney

Formalin perfused brains were embedded in plastic for ordinary histology. Four brains were cut into 5mm pieces and corroded in 40% KOH. The others were cut into 2-3mm pieces and embedded in plastic for morphometry.

Cast kidneys were freed of perinephric fat and connective tissue. Each kidney was bisected coronally. One half of the right and left kidney from each animal was fixed in 10% formalin and subsequently embedded for histology. The remaining halves were corroded for scanning electron microscopy. Likewise, liver was cut in pieces 5 - 7 mm maximum dimension and corroded for scanning electron microscopy and pieces 2 - 3 mm in thickness were fixed in formalin and embedded in plastic for morphometry.

Scanning electron microscopy

Tissues were corroded in either 40% or 20% potassium hydroxide for 2 - 4 days at room temperature. Corroded specimens were sonicated and washed in several changes of distilled water to remove tissue debris and then air dried at room temperature. The dried specimens were mounted onto aluminium stubs using electro-conductive glue (Carbcement, Bio-Rad, USA) sputtered with gold, and examined on a Cambridge scanning electron microscope at 5 - 15 KV.

Histology

Mercox perfused and formalin fixed tissue as well as formalin fixed tissue only, were dehydrated through an ascending series of water and methyl methacrylate mixtures prior to embedding in pure methacrylate plastic (Polaron, Australia). Hardened tissue blocks were trimmed and 3 μm sections were cut on a 2030 series Reichert-Jung autocut microtome. Sections were floated on distilled water and transferred to glass slides which were dried at 40°C for 24 hours. These sections were stained with haematoxylin and eosin.

Mercox specific staining

Gordon and Sweet reticulin stain (Gordon and Sweets 1936) was found to stain Mercox. Stained slides were air dried on a hot plate at 60°C to remove wrinkles and mounted in DPX medium. Stained slides showed pale brown Mercox and the rest of the tissue appeared colourless and transparent.

Morphometry

a)   Cast histological specimen

Uniformly perfused regions as shown by the stained mercox were photographed on a Leitz Biomed microscope using x 10 objective and x 6 eye-piece aligned to an Ikegami colour camera and Sony V02630 video cassette recorder on Umatic video tape. Sixteen frames of the field to be analysed were integrated by computer (Image - 1/AT system, Universal Imaging Corporation West Chester, Pennsylvania, USA) to obtain a noise reduced, black and white image of the field on the video monitor. The image analysis system was equipped with the facility to select 255 shades of grey. An appropriate grey filter was then selected to include only the grey tones of the stained mercox and all other shades were excluded, thereby registering only the mercox field vasculature. This was measured as the number of objects and the combined surface area given as a percentage of the microscopic field. The field size during each measurement was 200 to 300 μm^2. Fifty regions at each perfusion pressure were analysed. In the case of the formalin fixed brains which did not contain mercox, H&E stained slides were analysed. The clear spaces within the section were regarded as the blood vessels and their surface area calculated. These served as controls when compared to mercox filled specimens.

b)   Measurement of the vascular calibre from microvascular corrosion casts under the scanning electron microscope

A Cambridge scanning electron microscope was equipped with separation lines (electronic micrometer) that could be super- imposed on the image of the specimen. These lines were aligned on either side of the capillary vascular cast and the distance between the lines was read off the electronic micrometer data panel in the TV monitor. All normal capillaries had uniform diameters throughout the length of the vascular cast. A single reading from one capillary loop was recorded. Glomerular tufts were measured with two readings at right angles and their mean value was taken as the measurement. Capillaries and glomeruli were selected randomly. Fifty measurements were made from each specimen.

Results and Discussion

Brain

In the case of the brain, mercox filled specimens showed twice as many vessels in the similar area as in the control. The mean size of the individual vascular area in mercox was nearly twice that in the control brain slides, signifying that mercox prevents vascular collapse in histological sections. In mercox filled specimens, mercox occupied $10\pm2\%$ of the entire surface area whereas in the control specimen vascular area was only $2.0\pm0.4\%$.
In the case of the kidney, the diameter of capillaries and glomeruli were measured. The cortical microvascular area was measured. For morphometry, cortical regions were selected so as to include a minimum of two glomeruli in the field to be measured. In the case of the liver only sinusoidal capillaries were measured, and for morphometry the central vein or other larger vessels were excluded, thereby only the pure sinusoidal cross section area was measured. In the case of the bone marrow only the microvascular cross section at the sub epiphyseal region was measured. Results are shown in Table 1.

Kidney:

Plain histology of the kidney does not reveal its vasculature (fig 1a). However vascular cast preparation did show its rich vascularity. The vascular cast consisted of arteries, afferent and efferent arteries, glomeruli, venules and veins (fig. 1b). The relationship between parenchyma and blood vessels was shown in the histology of a cast specimen (Fig. 1c). The extent of filling of the microvasculature could be best seen at the surface of the kidney. At 150 mmHg the entire surface of the kidney was covered by capillaries and even the most peripherally situated glomeruli could not be seen from the surface. The capillaries overlying the glomeruli were completely filled, leaving no gap through which underlying structures could be seen. Occasional glomeruli could, however, be seen in specimens perfused at 100 mmHg and glomeruli could be easily seen from the surface of a kidney perfused at 70 mmHg. The vascular cast showed several regions of low vascularity at the cut surface in a specimen filled at 70 mmHg although no obvious differences appeared in the specimen perfused at 100 and 150 mmHg. The mean diameter of glomeruli was measured at 70.92 ± 12.05 µm at 100 mmHg. Image Analysis: the vascular cross section area measured from histology sections (fig 1d) at different perfusion pressures is shown in Table 1.

Liver:

Almost all of the central veins were cast even at 70 mm pressure. A filigree of surrounding capillary sinusoids was also filled which in places appeared continuous with the sinusoids of the adjacent lobules. Large empty spaces denoting incomplete filling of sinusoids were quite evident. The capillary sinusoidal diameter at 70 mmHg was 6.9µm (± 1.3µm) (Table 1). At 100 mmHg pressure there were fewer empty spaces and more continuous filling was observed with a capillary diameter of 7.14 µm (± 1.8). At 150 mmHg pressure continuous filling was obtained (Fig 2a) and a sinusoidal capillary diameter of 7.24µm ± (1.9 µm) was recorded.

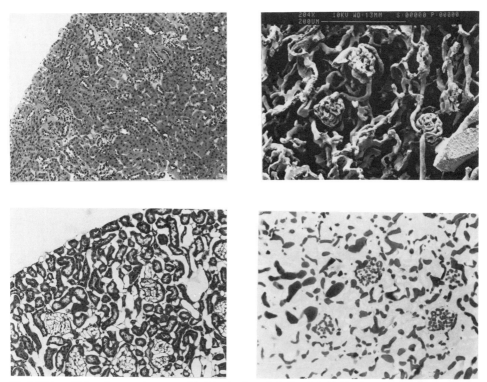

Fig.1 Kidney (light microscopic mag. x 10 objective 8 ocular). Top left (a) histology specimen stained with H & E. Top right (b) Micovascular corrosion cast showing glomeruli. Mag: As one of the data panel. Bottom left (c) histology of a cast specimen, stained with H & E showing clear glomular tuft and vessels surrounding tubules. Bottom right (d) histology of a cast specimen, stained with reticulin stain showing mercox within the vascular network. This was scanned for morphometry.

Table 1

Organ	Perfusion Pressure mmHg	Capillary Diameter	Microvasculature area as % of the parenchyma in the perfused region.
Liver	70	6.90μ ± 1.3	36 ± 8
Liver	100	7.14μ ± 1.8	37 ± 7
Liver	150	7.24μ ± 4	34 ± 4
		Glomerular Diameter	
Kidney Cortex	70	70.92μ ± 11.8	23 ± 5
Kidney Cortex	100	79.92μ ± 12.02	27 ± 6
Kidney Cortex	150	76.73μ ± 16.8	26 ± 5
Bone marrow	150	5.78μ ± 0.15	26 ± 0.6
Brain	150	-	10 ± 2

<u>Image Analysis:</u> of reticulin stained perfused regions of the histological sections (fig 2b) showed 36, 37 and 34% capillary cross-section area per unit parenchymal region (Table 1).

<u>Bone marrow:</u>

Femora were perfused only at 150 mmHg . The diameters of the sinusoids showed great variation and were difficult to measure. Only reticulin stained histological sections were measured and these showed a variation in the microvascular density throughout the length of the femur. However, when the sections passing through sub-trochanteric regions were measured, vascular density was uniform measuring around 26.1.

Many previous investigators have used the endothelial nuclear count per unit volume (Calvo et al 1986 and Ljubimova et al 1991) as an index of microvascularity. The endothelial cell count does not necessarily correlate with the microvascular surface area. Narayan and Cliff (1982) have shown that in irradiated vasculature, the number of cells per unit of capillary area decreases dramatically but the remaining endothelial cells spread out to cover, on

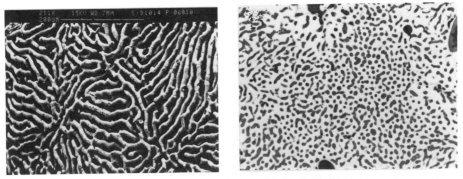

Fig. 2 Liver, Left hand side (a) microcorrosion cast of liver. Right hand side (b) histology of cast specimen, stained with reticular stain, showing sinusoids (mag x 10 objective and x 8 ocular).

average, three times the area covered by non-irradiated endothelial cells. Thus fewer endothelial cells can maintain a relatively larger capillary volume. In an irradiated specimen, therefore, a diminishing number of endothelial cells does not accurately reflect with a smaller microvascular volume. The present technique provides a reliable method of measuring microvascular volume in selected regions of the tissue. Corrosion casting of the same tissue provides a further means to assess the extent of the vascular filling. Microvascular density within the actual perfused region is almost non-pressure-dependent. At higher perfusion pressures a larger area is filled. This is particularly true of tissues which have uniform microvascular density like liver, kidney and muscle. Any process that may affect the entire microvasculature uniformly (eg. ionizing radiation) can therefore be effectively studied by this technique even though the tissue is not completely perfused. Physiological microvascular volume is under neuro-hormonal control as well as being affected by the tissue's metabolism and cannot be measured by this method. This technique is purely an anatomical one. This technique may also not be used in measuring the total microvascular volume in a given tissue, as complete filling cannot be guaranteed.

REFERENCES

Calvo W, Hopewell J W, Reinhold H S, van den Berg A P and Yeung T K (1987). Dose-dependent and time-dependent changes in the choroid plexus of the irradiated rat brain. Br. J. Radiol, 60: 1109 - 1117

Chalkley H W, Cornfield J and Park H (1949). A method for estimating volume-surface ratios. Science 110: 295 - 297

Cohnheim J (1989). Lectures in General Pathology, translated by McKee A D, from the second German edition, Vol I London, New Sydenham Society

Gordon H and Sweets H H (1936). Gordon and Sweets method for reticular fibres. In Theory and Practice of Histological Techniques: Ed by Bancroft J D and Stevens A, 2nd Edition 1982 pp 142 - 143. Published by Churchill/Livingstone

Hillmas D E and Gillette E L (1974). Morphometric analysis of the microvasculature of tumors during growth and after x- irradiation. Cancer 33: 103 - 110

Jee W S S and Arnold J S (1960). India ink - gelatin vascular injection of skeletal tissues. Stain Technol 35: 59 - 65

Lametschwandtner A, Lametschwandtner U and Weiger T (1990). Scanning electron microscopy of vascular corrosion casts - technique and applications: updated review. Scanning microscopy 4: 889 - 941

Ljubimova N V, Levitman M K H, Plotnikova E D, and Eidus L K H (1991). Endothelial cell population dynamics in rat brain after local irradiation. Br. J. Radiol 64: 934 - 940

Murakami T (1971). Application of the scanning electron microscope to the study of the fine distribution of the blood vessels. Arch Histol Jap 32: 445 - 454

Narayan K and Cliff W J (1982). Morphology of irradiated microvasculature: a combined in vivo and electron microscopic study. Am J Pathol 106: 47 - 62

Narayan K, Strohmeier R and Swann K (1991). Use of vascular casts in microvascular morphometry. Progress in Microcirculation Research edited by M A Perry, D G Garlick 1: 69 - 70

Reempts J V, Haseldon X M and Borgers M (1983). A simple technique for the microscopic study of microvascular geometry and tissue perfusion, allowing simultaneous histopathologic evaluation. Microvasc Res 25: 300 - 306

Reinhold H S, Hopewell J W and van Rijsoost A (1983). A revision of the Spalteholz method for visualising blood vessels. Int J Microcirc: Clin Exp 2, 47 - 52

Wilkinson J H, Hopewell J W and Reinhold H S (1981). A quantitative study of age-related changes in the vascular architecture of the rat cerebral cortex. Neuropath Applied Neurobiol 7: 451 - 462

THE CORNEAL MODEL FOR THE STUDY OF ANGIOGENESIS

David BenEzra

Immuno Ophthalmology and
Ocular Angiogenesis Laboratory
Hadassah University Hospital
Jerusalem, Israel

INTRODUCTION

Many models for the study of the processes taking place during angiogenic reactions in vivo have been designed. Due to the complexity of the reactions leading to the growth of new vessels in vivo adequate interpretation of the observed processes is, in most cases, difficult. The phenomena in vivo are generally the end results of complex reactions, interactions and feedback mechanisms, most of which remain unknown. In order to enhance accuracy, the in vivo models have to be as simple as possible and amenable to direct visualization and follow-up. Furthermore, a thorough knowledge regarding the anatomy, physiology and histo-pathology of the organ used is mandatory.

Angiogenesis of the ocular tissues in adults is the most common underlying cause for visual impairment. Therefore, over the years, this phenomenon has attracted the attention of the most renowned ophthalmologists worldwide. Many ocular structures have been used as models for the study of neovascularization in vivo. However, experiments with the cornea started during the 1940's [1] have yielded extensive useful information. In this chapter, the experimental model using the rabbit cornea and polymer implants as it evolved in our laboratory during the last two decades will be described.

MATERIALS AND METHODS

Animals

The use of mice, rats or guinea pigs is feasible. The relatively small size of the eyes in these animals (mice and rats) and the presence of preexisting vessels within the cornea in some of the strains are serious disadvantages. Furthermore, the need for general anesthesia of these animals during each step is a drawback for their routine use in large-scale experiments. Therefore, during the last two decades we have seldom used the cornea of the above species for the study of neovascularization.

Angiogenesis: Molecular Biology, Clinical Aspects
Edited by M.E. Maragoudakis *et al*, Plenum Press, New York 1994

Rabbits have a larger cornea which has been found avascular in all strains examined so far. Rabbits are more docile and amenable to handling and experimentation. Only local anesthesia is needed for surgery and is occasionally needed during the daily examinations. Because of these characteristics the rabbit cornea has been our preferred model for the study of angiogenesis.

Stimuli

Corneal angiogenesis can be induced by various stimuli: trauma, cautery of the corneal tissue, or chemical burn. Generally, these types of stimuli are not amenable to meticulous titrations. Injection of cells and soluble factors within the stroma at a precise distance from the limbus provided more reproducible results (2). However, the rapid diffusion of the tested materials from the stroma affected the reproducibility of the experiments and hampered the proper evaluation of the true angiogenic potential of these compounds.

Implants

The use of polymers for the slow release of substances has opened newer avenues for more accurate investigations. After initial experimentations with various polymers, it was found that the copolymer ethylene-vinyl-acetate (40% vinyl-acetate by weight), Elvax-40, was the most suitable for routine use in the cornea (3). However, in order to avoid non specific reactions the Elvax-40 for implantation have to be carefully prepared as follows: After extensive washings of the polyvinyl-acetate copolymer beads in absolute alcohol for 20 days, a 10% casting stock solution of Elvax-40 (1.0 gram) in methylene chloride (10 ml) is prepared and tested for its biocompatibility. 200 μl of this solution are layered on a glass plate and the methylene chloride is allowed to evaporate. The resulting film of polymer is cut into pieces which are used as implants. Ten implants of this preparation are inserted intrastromally within the rabbit cornea in ten eyes. In each eye, the implant is positioned at 2.5 mm from the limbus and the corneal reaction is monitored for 14 days after implantation. If any of the ten implants induces the slightest clinical or histological reaction in the rabbit cornea, the casting stock solution is discarded. Further washing of the polyvinyl acetate beads is carried out for ten additional days and a new casting solution is prepared and retested as above. A casting stock solution is eligible for use only when all ten test implants remain absolutely inert. The same casting solution is then used throughout the entire study. For testing, a pre-determined volume of the Elvax-40 casting solution is mixed with a given amount of the compound to be tested and the polymer is allowed to dry under a laminar flow hood. After drying, the resulting polymer film sequestering the compounds to be tested is rolled and cut into pieces yielding implants of 0.25 to 1.0 mm according to the design of the study. Empty implants of Elvax 40 and/or those sequestering "nonsense" compounds are used as controls.

Surgical procedure of implantation

Under the operating microscope and aseptic conditions,

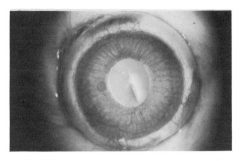
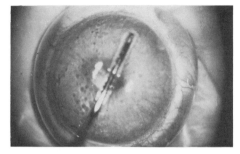

Figure 1. (L) Initial central mid thickness incision at the apex of the cornea.

Figure 2. (R) Insertion of cyclodialysis spatula and the formation of the stromal tunnel.

an incision 4 mm wide and half of the corneal thickness deep is made at the center (Figure 1). A radial midstromal tunnel is created, pushing a thin cyclodialysis spatula toward the limbus in the direction of the site with least engorgement of limbal vessels (Figure 2).

Implants are carefully positioned at 2.5 mm from the limbus using a surgical caliper (Figures 3 and 4).

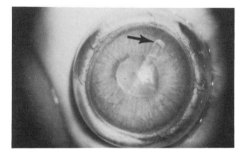
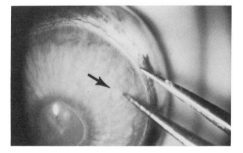

Figure 3. (L) Positioning the implant at the planned distance from the limbus.

Figure 4. (R) Checking the position of the implant and its distance from the limbus using a surgical caliper.

During the surgery, the cornea is irrigated with 0.5% chloramphenicol eye drops and local anesthetic (Novesine) as needed. On completion of the surgery the operated eyes are treated with one drop of Atropine 1% and Synthomycetin 5% eye ointment. Implants sequestering the tested materials and the controls are coded and implanted in a double-masked manner. Each test is repeated four to six times. Figure 5 illustrates the limited and controlled neovascular reaction which can be observed six days after implantation of an implant sequestering a compound with angiogenic ability. Figure 6 illustrates a similar implant sequestering a non-angiogenic compound six days after implantation.

The clinical reaction of the cornea including: edema, cellular infiltrate and angiogenesis are recorded daily. Edema and cellular infiltrates are scored arbitrarily on a scale from 0 (none), ± (barely detectable), and (+) mild to (+4) very severe. Angiogenesis, on the other hand, is monitored by meticulous measurements of the new vessel growth within the previously avascular cornea. Assessment of the angiogenic potential is made by measuring the surface of the growing neovascular bed. This parameter is obtained by daily recording, under the microscope, of the following: 1) the length of the leading vessel sprouting from the "original" limbal vessels and directed toward the implant within a previously avascular cornea; 2) the extent of the specific reactive "neovascular bed" at the limbus in front of the implant. As a rule, the neovascular tuft of vessels progresses toward the stimulating implant in a triangular fashion (4). Therefore, the "angiogenic surface" is obtained by the multiplication of the length of the leading vessel by

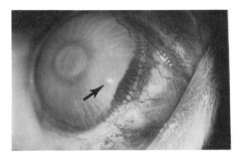

Figure 5. (L) Positive neovascular reaction induced by an implant sequestering a potential angiogenic factor six days after implantation.
Figure 6. (R) No reaction is observed with an implant sequestering a non angiogenic compound six days after implantation.

the extent of the reactive neovascular bed around the limbus divided by two (3). Daily reactions are also routinely recorded by photography via the microscope.

Histology

At various intervals, the eyes are enucleated and fixed in buffered formalin for 48 hours. Then, the corneas are dissected, a 5.0 x 6.0 mm section of corneo-sclera with the implant at its center is removed and subjected to routine histology (3).

RESULTS AND DISCUSSION

When all of the above are carried out meticulously, we have found that the corneal assay of in vivo angiogenesis may be a reliable and reproducible method of assessment only if the following criteria are satisfied:

1. The size of the implant must be identical in each cornea throughout the entire study.

2. The position of the implant(s) within the avascular cornea must be at an equal distance from the limbal vessels. We have found that in order to obtain the same degree of stimulation by a given angiogenic substance, the required concentration is 5-fold larger if the implant is positioned at a distance of 3.5 mm instead of 2.0 mm from the limbus and 10 times larger if the implant is positioned 1.0 mm further away. Moreover, less than 50% of the concentration of an angiogenic factor is needed in order to obtain the same degree of neovascularization if the implant is positioned at 2.0 mm instead of 2.5 mm from the limbus.

3. Positioning the implant(s) at a distance less than 2.0 mm from the limbus may induce a non specific angiogenic response. In fact, "anything" implanted in the rabbit cornea (or any other cornea of a living animal/human) will induce a non specific angiogenic response if the distance between the implant and the limbal vessels is less than 2.0 mm. Within this 2.0 mm limit, there probably exists a "critical zone" where stimulation of angiogenesis is (non specifically) triggered. This trigger follows the disturbance of the corneal structural integrity by the implantation technique and the release of "metabolites" having angiogenic potential. Because of the proximity to the limbal vessels, these metabolites reach the critical concentration needed for the induction of an angiogenic process.

4. The preparation of the implants and their insertion into the cornea have to be carried out under strict aseptic conditions. Subclinical infections and the release of endotoxins may stimulate a non specific angiogenic response.

5. In vivo corneal angiogenic assays have to be performed in a double-masked manner to avoid any possible biases.

6. Exact measurements of the new blood vessel growth must be performed. Arbitrary assessment of angiogenesis as present or absent (or absent - present) may be misleading and grossly inaccurate.

7. Daily monitoring of the blood vessel growth is necessary and their kinetics a very important source of information. If the growth of new blood vessels is not observed by the second day after implantation, the tested compound is, most probably, either not diffusing from the implant, or it is not a primary angiogenic stimulant. Furthermore, if sprouting of the new vessels is initially observed only on the 5th or 6th day after implantation, secondary phenomena related to the host cornea are, in most cases, responsible for this reaction with little (or no) correlation to the angiogenic ability of the tested substance.

8. Most importantly, all corneal implant assays (and angiogenic assessments) have to be performed under the high magnification of a microscope. The use of a surgical microscope and a thorough understanding of the corneal structure and anatomy are essential for the adequate performance of these studies. They are, of course, an

absolute must if one uses animals with smaller eyes and thinner corneas.

Due to the complexity of the in vivo angiogenic corneal assay, it may not be surprising that contradictory results may be obtained. If all above cautionary measures are taken, however, reliable and reproducible results are to be expected and controversies circumvened.

ABSTRACT

The use of the rabbit cornea and implants of Elvax-40 as a reliable and reproducible model for the study of angiogenic stimuli in vivo is described. After two decades of experience with this model, it was realized that the following factors are of primordial importance for the performance of a reliable and reproducible experiment: a) Positioning of the implant at less than 2.0 mm from the limbus induces non specific reactions. b) Implants within the same experiments must be of identical size. c) The position of all implants within the same experiments must be at the same exact distance from the limbus. d) Preparation of the implants and their insertion into the cornea must be carried out under aseptic conditions. e) All evaluations have to be carried out in a double-masked manner. f) Exact daily measurement of the surface of neovascularization is the most adequate practical parameter. g) Surgery for implantation and the daily recording of reactions must be performed under high magnification of a microscope or a slit lamp. h) Double-masked assessment of the results is mandatory in order to circumvene unavoidable biases.

REFERENCES

1.   I.C. Michaelson, The mode of development of the retinal vessels and some observatrions on its significance in certain retinal diseases, *Trans Ophthalmol Soc UK*. 68:137, 1948.
2.   D. BenEzra, Mediators of immunological reactions: Function as inducers of neovascularization. *Metab Ophthalmol*. 2:339, 1978.
3.   D. BenEzra, Neovasculogenic ability of prostaglandins, growth factors and synthetic chemoattractants. *Am J Ophthalmol*. 86:455, 1978.
4.   I.C. Michaelson, Effect of cortisone upon cornea vascularization produced experimentally. *Arch Ophthalmol*. 47:459, 1952.

FACTITIOUS ANGIOGENESIS: NOT SO FACTITIOUS ANYMORE? THE ROLE OF ANGIOGENIC PROCESSES IN THE ENDOTHELIALIZATION OF ARTIFICIAL CARDIOVASCULAR PROSTHESES

Peter I. Lelkes, Dawn M. Chick, Mark M. Samet, Viktor Nikolaychik, Gregory A. Thomas*, Robert L. Hammond*, Susuma Isoda* and Larry W. Stephenson*

Laboratory of Cell Biology, University of Wisconsin Medical School, Milwaukee Clinical Campus, Milwaukee, WI., (*) Division of Cardiothoracic Surgery, Wayne State University, Detroit, MI

INTRODUCTION

Due to genetic predisposition and, even more frequently, due to habitual mistreatment, (e.g., by malnutrition, smoking, etc.), the human cardiovascular system is particularly prone to massive failure, (occlusion of the blood vessels due to atherosclerosis or thromboembolism, cardiac insufficiency due to myocardial infarct, congestive heart failure), and repair, (angioplasty, cardiac transplantation). Both the anatomical repair of malfunctioning blood vessels and/or their replacement by grafting autologous blood-conduits are only partially successful: a large percentage (up to 30%) of all angioplasty as well as (coronary) bypass procedures eventually fail mainly due to re-stenosis. Similarly, although the percentage of 5-year survivors after cardiac transplantation has dramatically risen since the introduction of potent immunosuppressants, there are increasing signs of long-term adverse effects of these drugs on other organs. Furthermore, the rejection process *per se* is not abrogated, and often it manifests itself in the form of accelerated atherosclerosis in the blood vessels of the transplanted tissues (Mills *et al.*1992). Last but certainly not least, the availability of suitable donor hearts is extremely limited, so that, eventually, most of the potential cardiac transplant patients will never receive a donor heart and die while on the waiting list. Thus, there is an increasing world-wide demand for permanent cardiovascular prostheses, such as vascular grafts, cardiac assist devices or total artificial hearts. However, currently, the long-term use of such prostheses poses serious problems, such as neointimal hyperplasia and occlusion, sepsis, thromboembolism, and calcification. Most of these complications can be traced back to the inadequate hemocompatibility of the (polymeric) blood contacting surfaces in these devices. At the previous NATO ASI on angiogenesis we delineated the ongoing attempts by us and others to restore "nature's biocompatible blood container" through "*factitious angiogenesis*", i.e., by precoating the luminal surfaces of these prostheses with a functional, nonthrombogenic monolayer of autologous endothelial cells (ECs) prior to implantation (Lelkes *et al.*1992). In this chapter we discuss new evidence, indicating that endothelialization of the blood contacting surface in some cardiovascular prostheses may occur through "genuine" angiogenic processes.

ENDOTHELIALIZATION OF VASCULAR PROSTHESES

Decades of extensive research have been devoted to improving the hemocompatibility of biomaterials used for artificial vascular grafts. However, to date these studies have failed to produce the ultimate, non-thrombogenic blood contacting surface (Massia *et al.*1992). In recent years, ENDOTHELIALIZATION of vascular grafts, i.e., lining their luminal surfaces with autologous ECs, may have solved several major problems in cardiovascular surgery (Park *et al.*1990; Schneider *et al.*1990; Rupnick *et al.*1991; Lelkes *et al.*1991; Miwa *et al.*1992; Eberl *et al.*1992). Furthermore, these studies are paving the way for new areas, such as gene therapy (Wilson *et al.*1989; Callow, 1990; Nabel *et al.*1992). At this point the question is no longer **if** endothelialization is clinically feasible, but, rather, **how** can it be optimized. In the context of this book it is noteworthy that some of the recent approaches to successfully endothelialize artificial vascular grafts exploit fundamental aspects of angiogenesis.

Adhesive and trophic factors in the subendothelial cell matrix are critical for the establishment and maintenance of a non-thrombogenic EC lining, that will not be sheared off by the blood flow. Thus, initial endothelial cell attachment and spreading on artificial surfaces is significantly accelerated by precoating the biomaterials with extracellular matrix (ECM) proteins such as fibronectin (Seeger *et al.*1988). Admixing of growth factors of the FGF family at the time of seeding or using preformed complex extracellular matrices greatly enhances endothelial cell attachment and subsequent growth, and yields the most promising, shear resistant monolayers (Greisler *et al.*1987; Rosengart *et al.*1988).

Upon initial seeding, successful maintenance of an intact EC monolayer requires the continued, sequential production of extracellular matrix proteins and their deposition into the subendothelial space (Lelkes *et al.*1991; Papadimitriou *et al.*1993). Such a complex ECM can assure the retention of the EC monolayer on the biomaterials even after focal injury and detachment, and guide the migration of ECs to repair the denuded areas.

Endothelial cells can migrate over long distances (several centimeters) from the host vessel, across the site of anastomosis onto the luminal surface of the graft, and thus generate complete vascular healing (Niu *et al.*1990). This observation is true for most mammalian ECs, including monkeys but, alas, not in humans (Zilla *et al.*1987; Hoffman *et al.*1992). In humans, the lack of endothelial cell ingrowth across the site of anastomosis remains an interesting, yet puzzling enigma, which might attest to the uniqueness of human cells and their interactions with biomaterials (Niu *et al.*1992). Thus, for example, under standardized culture conditions (same low passage numbers, optimized culture media, etc.), aortic ECs, isolated from middle aged humans (HAEC), exhibit a significantly lower rate of proliferation, attain a lower cell density at confluence and experience cellular senescence more rapidly than their bovine (BAEC) counterparts. Similar observations have been made in the past using other human large vessel-derived ECs (Kent *et al.*1989). This trend is even more pronounced when the cells are grown on medical grade polyurethanes, rather than on the "gold-standard", tissue culture plastic (see Figure 1).

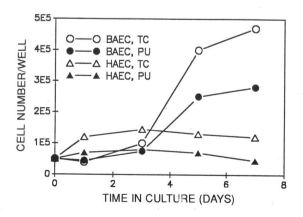

Figure 1: Comparison of the growth of human and bovine aortic endothelial cells (HAEC and BAEC) seeded under identical conditions onto tissue culture plastic (TC) and polyurethanes (PU).

Therefore, to date, most successful strategies for endothelializing vascular grafts rely on rapid establishment of a complete endothelial cell coverage of the luminal surface at, or prior to, the implantation of the graft. This goal is achieved by *"sodding"* (i.e., initial seeding at monolayer density of more than 1.5×10^6 cells/cm^2) the graft surface with abundant microvascular endothelial cells derived from human omentum/adipose tissue (Rupnick *et al.*1991). In order to improve initial cell attachment and enhance the shear resistance of the EC monolayer, numerous variations to this basic technique have been proposed, including precoating graft surfaces with adhesive proteins, trapping the cells in a preformed clot, or pressure seeding into the preclotted, porous grafts (Kaehler *et al.*1989; Greisler *et al.*1992; Lin *et al.*1993). Alternatively, rapid endothelialization (in a dog model) was achieved by impregnating (under pressure) the porous fabric of vascular grafts with mechanically minced fragments of autologous venous tissue: the rationale for these studies being that angiogenic processes would facilitate vascular healing (Noishiki *et al.*1992). Indeed, within a short time ECs (and smooth muscle cells) were found to migrate from those tissue fragments onto the luminal surface, leading to the formation of a well organized neointima, which within 2 weeks was completely covered with a contiguous, non-thrombogenic endothelial cell layer. An occasional contiguous EC lining which connects graft surface and the wall of the feeding capillaries attested to the origin of the luminal EC lining from the underlying tissue. Interestingly, in this study large vessel-derived ECs seemed to have taken over the entire repertoire of angiogenesis, including migration and formation of new capillaries underneath the luminal surface (e.g., vasa vasorum), suggesting phenotypical and functional plasticity of the endothelial cells.

In a parallel development, recent observations suggest that the porous fabric of prosthetic vascular grafts attracts tissue ingrowth and that this tissue will then become neovascularized (Menger *et al.*1990). Furthermore, given the proper pore size, the newly formed blood vessels underneath the vessel lumen will serve as the principal cellular source for the migration of endothelial cells onto the luminal surface of the artificial grafts, which in due time will become fully endothelialized (Golden *et al.*1990; Kohler *et al.*1992). Thus, in this case, microvessel-derived ECs differentiated phenotypically and (presumably) functionally to assume the role of large vessel ECs.

The implications from these studies are many-fold: In terms of clinical relevance, current work is focussing on how to optimize the rapid, initial attachment and spreading of "sodded" ECs, and how to attain an instantaneous, shear-resistant, non-thrombogenic EC monolayer. Also, given the involvement of angiogenic processes in the endothelialization of porous grafts, efforts are being made to promote tissue ingrowth into the prosthetic material by selectively incorporating angiogenic factors (e.g., growth factors) and thus enhance EC migration and neovascularization. Finally, from the vantage point of basic research it will be fascinating to explore the phenotypical plasticity of ECs and establish the cellular and molecular events that accompany the interconversion of large-vessel type ECs and microvascular ECs.

SKELETAL MUSCLE VENTRICLES

Skeletal muscle tissue can be trained by electrical stimulation to acquire biochemical and mechanical characteristics similar to those found in a cardiac muscle (Magovern, 1991). In particular, an electrically conditioned skeletal muscle can become fatigue-resistant (Acker *et al.*1986; Pette, 1991). Several possibilities exist to use conditioned skeletal muscle flaps as autologous blood pumps, in order to assist or replace the failing heart (Anderson *et al.*1991; Salmons *et al.*1992). Currently, an experimental procedure called "cardiomyoplasty" is being explored in several clinical centers world wide, in which a skeletal muscle is wrapped around the heart and then mechanically assists with the contractions. Another option is to wrap the *latissimus dorsi* muscle around a mandrel, which within a few weeks will be encapsulated by a fibrous pouch. Such a muscle pouch, also termed skeletal muscle ventricle (SMV), can then be connected to the circulation and used as an autologous cardiac contractile assist device (Acker *et al.*1986; Pochettino *et al.*1991). However, the potential of thrombus formation on the blood contacting surfaces, which is comprised of fibrous material, granulation tissue, adipocytes and also, perhaps, mesothelial cells, have so far restricted the long term clinical use of such pouches as ventricular replacements. Several attempts have been made in the past to enhance the thromboresistance of the SMV surface, e.g., by re-designing the shape of the SMV and/or by covering the mandrel with pericardium or with pleura (Anderson *et al*.1991). While these approaches may have reduced the overall incidence of thromboembolism, local thrombus formation and occasional thromboembolism remain a significant factor. Thus, as in the case of artificial grafts, the problem reduced itself to the issue of hemocompatibility of the blood-contacting (foreign) surfaces and, again, the optimal solution appears to be the establishment of a functional, non-thrombogenic endothelial cell lining.

As previously described, we have successfully tested the feasibility of harvesting autologous (subcutaneous fat-derived) ECs from mongrel dogs, and seeding the luminal surface of static (non-conditioned) SMVs (Lelkes *et al*.1990; Lelkes *et al*.1992). In this procedure, SMVs were constructed around a solid silicone mandrel. Approximately six weeks later, after a fibrous pouch had formed around this mandrel, the SMV was opened again and filled with a solution containing endothelial cells. Subsequently, the wound was closed and the animal was allowed to recover, and it was sacrificed two weeks later. The main findings from these initial, pioneering experiments were: **1.** after 8 weeks (without electrical conditioning) the luminal side of the untreated muscle pouches was covered with a thick network of mostly collagenous fibers, **2.** by contrast, the luminal side of the seeded pouches was coated with a continuous, smooth cellular lining, positive for EC markers, such as van Willebrand factor (vWF) (Lelkes *et al*.1993).

It was not clear from these experiments whether **1.** the ECs seen at the end of the experiments were actually the same (or progenies of) the ECs that were seeded, **2.** whether the seeded ECs would withstand the contractions induced by electrical conditioning and/or hemodynamic forces upon connection of the EC-lined SMVs into circulation, and **3.** whether endothelialization would, indeed, reduce the incidence of thromboembolism.   To address these issues we established a close collaboration between the laboratories in Milwaukee and in Detroit, combining our respective expertises in endothelial cell biology and the preclinical testing of SMVs. Thus, in adapting the type of mandrel used in Detroit, we are now seeding SMVs, similar to those that have previously been connected into the circulation and have successfully supported cardiac output in dogs. However, when examining this type of SMV, Nakajima et al (1992) reported a large incidence of local thrombus formation and some degree of distal thromboembolism.

Most dogs in our facilities are rather lean. Thus, it is difficult to obtain, without prolonged cell culture, large numbers of pure ECs from subcutaneous fat. This raises the problem of potential over growth of the culture by non-EC cells and/or senescence and de-differentiation of the ECs. Since the goal is to use minimally-passaged ECs for reintroduction, we are currently isolating jugular vein-derived ECs, which then are cultured for 1-2 passages to yield large numbers of cells at high purity. The quantities obtained are sufficient to seed an SMV with confluent densities (ca. 1.5×10^5 cells/cm^2 of SMV area).

Rather than having to go through an additional surgical procedure (for the open seeding routine), we developed a novel seeding technique, in which we introduce the cells in aliquots via percutaneous injection into the SMV-space around the mandrel. This modification reduces the risks associated with yet another surgical step and increases the potential, clinical feasibility of the SMV as an autologous blood pump.

We used a long-lasting, non-exchangeable fluorescent cell marker PKH26 (Horan *et al*.1990), to label the cells prior to seeding and identify the original cells (and or their progenies) at the time of termination of the experiment. This particular label has been successfully used in related *in vivo* applications for extended periods of time. We verified the usefulness of this technique by tracing the label in dog jugular vein endothelial cells through approximately 20 population doublings for more than 6 weeks in tissue culture (not shown).

With these modifications, we have successfully verified and extended our initial findings, that the fibrous SMV surface can be lined with EC (Gao *et al*.1990; Lelkes *et al*.1990; Lelkes *et al*.1992). At this time (September 1993), we have successfully completed several sets of experiments:

1. Canine SMVs were constructed bilaterally as previously described (Thomas *et al*.1993b). Four to six weeks later both SMVs in all but 2 dogs were seeded with fluorescently labeled allogeneic EC (i.e., ECs derived from generic dogs other than the recipients), using either the open or the percutaneous seeding technique. One dog was seeded with non-labeled ECs, while the SMVs in the last dog were left as untreated controls. Because of the use of heterologous ECs, the animals had to be sacrificed one week after the seeding and prior to the onset of any tissue rejection.

However, the purpose of this particular set of experiments was mainly to re-examine the feasibility of the approach and to compare both seeding techniques. Upon examination of the SMV by light and fluorescence microscopy we observed significant coverage of the SMV surfaces (30-70 %, depending on the location) with labeled endothelial cells (Figure 2).

Also, examination of the SMV surface by scanning (Figure 3) and transmission electron microscopy (Figure 4) confirmed, respectively, the EC nature of the cell lining on the surface (presence of Weibel-Palade bodies) and the extent of cellular coverage in the seeded SMVs.

micrograph
ing   with
s
rage of
50 μm

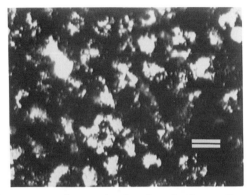

Figure 2: Fluorescence micrograph taken 1 week after seeding with PKH26-labeled EC reveals extensive cellular coverage of SMV surface. Bar = 50μm

Figure 3: Ultrastructure of an endothelialized SMV (transmission electron micrograph). An EC, characterized by Weibel Palade bodies (arrow) is seen on top of an extensive network of colleagenous fibers. Bar = 1μm

Taken together, these data suggest that the ECs lining the luminal surface were, at least in part, derived form the cells originally used for seeding. Importantly, there was no statistically significant difference in the seeding efficiency between the open and the percutaneous seeding technique. This finding is consequential for any future clinical usefulness of this approach. After construction and vascular healing, endothelial cell seeding of an SMV can thus be achieved by a rather simple subcutaneous injection, thus alleviating a hitherto necessary cumbersome and risky second surgical (open) procedure.

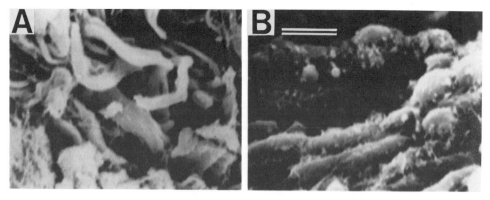

Figure 4: Ultrastructure of untreated SMVs (**Panel A**) and 1 week after endothelialization (**Panel B**). Note the prominent fibrous network and the absence of a cellular coverage in untreated SMVs contrasting the presence of cells after seeding. Bar = 20 μm

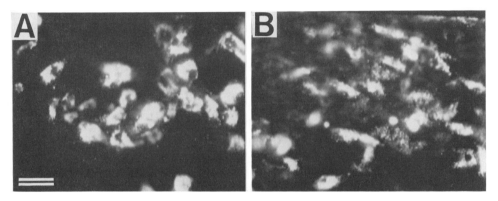

Figure 5: Light microscopic examination of electrically conditioned SMVs after endothelialization. En-face fluorescence micrographs of PKH26-staining of the SMV surface reveals minimal EC seeding in the top portion (**Panel A**) and extensive cellular coverage in the apex (**Panel B**). Bar = 50 μm

2. The second major innovative step was to seed endothelial cells onto the luminal surface of electrically conditioned, contracting SMVs (Thomas *et al.*1993a). *A priori*, it was not clear whether the ECs seeded onto the fibrous pouches would remain firmly anchored when the SMV was electrically stimulated and began to contract. For these experiments unilateral SMVs were constructed, as previously described (Acker *et al.*1987). A nerve cuff electrode was placed around the thoracodorsal nerve and connected to an implantable neurostimulator (Itrel, from Medtronic). After a 2 weeks, the stimulators were turned on and the muscle was conditioned with a pulse train of 2 V at 33 Hz. Three to four weeks later the animals were returned to the operating room, the

stimulators were switched off, and the SMVs were seeded with autologous endothelial cells by percutaneous injection (as above). During the first week after seeding, the electrical stimulation of the SMVs was discontinued in order to allow the cells to adhere, spread and form a confluent monolayer. Then, the stimulators were turned on again and the SMVs continuously paced for another 2 weeks, before the animals were sacrificed.

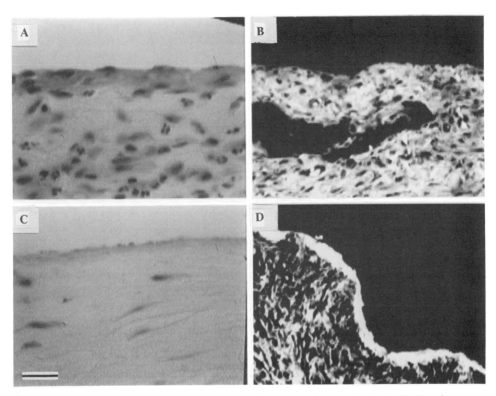

Figure 6: Routine histology of paraffin sections (H&E staining) under bright field and fluorescence illumination: Top Row: Seeded SMVs, Bottom Row Unseeded Controls, **Panels A, C**: Conventional bright field illumination, **Panels B, D**: Fluorescence (FITC) illumination **Bars: Panels A,C** = 16 μm, **Panels B,D** = 50 μm

To verify the' presence of endothelial cells on the luminal surface of the contracting SMVs, we used three independent markers: **1.** some of the SMVs had been seeded with EC prelabeled with PKH26 (as above), **2.** uptake of diI-labeled acetylated LDL, and **3.** staining for vWF. The latter two markers were tested on SMVs seeded with unlabeled EC.

PKH26 staining revealed the presence of ECs throughout the central and the apical portion of the SMV, while the upper part appeared less well seeded (Figure 5). However, using the other two EC markers, we observed a much more profound coverage of the apex and the center of the SMV with a contiguous EC monolayer (not shown). In addition, the extent of endothelialization was also evaluated by light and fluorescence microscopy after routine histological (Hematoxylin and Eosine, H & E) staining (Figure 6). Conventional bright field illumination shows a contiguous cellular coverage after endothelialization (**Panel A**) and the absence of a cellular lining in untreated SMVs (**Panel C**). Fluorescence visualization highlights the thick, acellular surface layer containing fibrous (elastic) ECM proteins in unseeded (**Panel D**), but not in seeded SMVs (**Panel B**).

Remarkably, in spite of (as might have been expected), or, probably, because of the electrical conditioning (see below) the extent of seeding throughout the apex and middle portion of the SMV was significantly higher (80-100%) than in the unconditioned SMVs. These data suggest that 3 weeks after EC seeding, and during 2 weeks of contractions, the SMV surface is covered to a large extent with an EC monolayer. This finding by itself is a break-through: to the best of our knowledge, this is the first time that a beating "artificial blood pump" has been successfully endothelialized. Current experiments are in progress to evaluate the ultimate goal of these procedures, namely to attain a non-thrombogenic SMV surface, once the endothelialized SMVs are connected into circulation.

Besides its clinical potential, our data raise some interesting angiogenesis-related issues, which will be examined in detail in the future. Our *in vitro* studies suggest that during the time course of this experiments (3 weeks), the PKH26 label is maintained in all progenies of the originally seeded cells. However, *in vivo* not all the ECs present on the SMV surface express PKH26. Therefore we hypothesize that not all these cells are derived from the initial EC population.

TABLE 1: QUANTITATION OF SUBLUMINAL BLOOD VESSEL FORMATION IN CANINE SMVs

SMV-type	Number of Vessels[*]	Distance of nearest blood vessel from surface (μm)
control, static (n=4)[@]	3 ± 2	330 ± 130
control, conditioned (n=1)	5 ± 1	240 ± 75
endothelialized, static (n=4)	$12 \pm 3^{\#}$	$120 \pm 60^{\&}$
endothelialized, conditioned (n=4)	$18 \pm 4^{\#,\$}$	$60 \pm 50^{\#}$

@:   n = number of animals

*:   3 different sections of n individual, H&E-stained SMVs were examined. All sections were examined at low power, photographed at an original magnification of 125 x, and for each section 5 micrographs were chosen randomly for further analysis. All vessels in a 250,000 μm^2 area (500 x 500 μm^2) adjacent to the SMC surface were evaluated with respect to the number of blood vessels present and the distance of the nearest vessel to the SMV surface.

\#:   p < 0.01 (endothelialized vs. controls)

&:   p < 0.02 (endothelialized vs. controls)

$:   p < 0.05 (conditioned vs. static)

One possible explanation for the discrepancy between the appearance of two kinds of EC labels, with PKH26 applied at the time of seeding and diI-ac-LDL and vWF being employed at the time of termination of the experiments, is that PKH26-negative ECs might have migrated to the SMV surface from underlying capillaries by angiogenic processes. Indeed, a closer examination of the area adjacent to the SMV surface indicates a higher degree of vascularization in the seeded SMVs (Thomas et al.1993a). A quantitative analysis of the first 500 μm underneath the pouches confirms this visual impression (Table 1). In the unseeded SMVs, the area underneath the luminal surface, whether static or after conditioning, is poorly vascularized. Only a few vessels, if any, are found in close proximity to the SMV surface. By contrast, upon endothelialization, both the number of capillaries and their proximity to the SMV surface increase significantly. It is noteworthy, that electrical conditioning leads to enhanced capillary density in SMVs.

These data suggest that "endothelial cells will beget endothelial cells". We hypothesize that initially seeded ECs might release angiogenic factors which facilitate the formation of new blood vessels in close proximity to the SMV surface and, in turn, recruit from these new vessels ECs to form the SMV surface lining. Thus, this process seems to be similar to that postulated in the case of endothelialization of porous vascular grafts (see above).

Of particular interest is the fact that the capillary density in electrically stimulated, contracting SMVs is significantly higher than in the non-stimulated ones. While this observation is generally in line with data presented by O. Hudlicka and coworkers on the angiogenic effects of mechanical stimulation in muscles (Hudlicka, 1992), the validity of our findings will have to be confirmed using more stringent controls.

The involvement of angiogenic processes in the endothelialization of contracting SMVs has important implications for any potential clinical use of this procedure. If "natural" angiogenesis is, indeed, a major process that may lead to successful endothelialization of the SMV's luminal surface, we should refocus our experimental approach of "factitious" angiogenesis. In future studies we will explore in detail the beneficial role of EC-specific angiogenic factors, such as VEGF, alone or as adjuvants to endothelial cell seeding.

ACKNOWLEDGEMENTS

The original results reported in this communication were obtained while the authors were supported, in part, by grants-in-aid from the Milwaukee Heart Research Foundation, Promeon/Medtronic Inc. and the American Heart Association, Wisconsin-Chapter (GIA # 92-65-26C)(P.I.L), and by grants from NIH [HL-34778 (L.W.S) and HL-8384 (G.A.T.)].

REFERENCES

Acker M.A., Mannion J.D. and Stephenson L.W. (1986) Methods of transforming skeletal muscle into a fatigue-resistant state: potential for cardiac assistance. In *Biomechanical cardiac assist: cardiomyoplasty and muscle-powered devices* (ed. R.C.-J. Chiu), Futura Pub.Co., pp. 19-28.Mount Kisco, N.Y.

Acker M.A., Anderson W.A., Hammond R.L., Chin A.J., Buchanan J.W., Morse C.C., Kelly A.M. and Stephenson L.W. (1987) Skeletal muscle ventricles in circulation: one to eleven weeks' experience. *J Thorac Cardiovasc Surg*, **94**, 163-174.

Anderson D.R., Pochettino A., Hammond R.L., Hohenhaus E., Spanta A.D., Bridges C.R.,Jr., Lavine S., Bhan R.D., Colson M. and Stephenson L.W. (1991) Autogenously lined skeletal muscle ventricles in circulation: up to nine months' experience. *J Thorac Cardiovasc Surg*, **101**, 661-670.

Callow A.D. (1990) The vascular endothelial cell as a vehicle for gene therapy. *J Vasc Surg*, **11**, 793-798.

Eberl T., Siedler S., Schumacher B., Zilla P., Schlaudraff K. and Fasol R. (1992) Experimental in vitro endothelialization of cardiac valve leaflets. *Ann Thorac Surg*, **53**, 487-492.

Gao H., Lelkes P.I., Edgerton J.R., Flemma R.J. and Christensen C.W. (1990) Endothelialized latissimus dorsi pouches as possible replacements for cardiac muscle. *Pharmacologist*, **32**, 123. (Abstract)

Golden M.A., Hanson S.R., Kirkman T.R., Schneider P.A. and Clowes A.W. (1990) Healing of polytetrafluoroethylene arterial grafts is influenced by graft porosity. *J Vasc Surg*, **11**, 838-845.

Greisler H.P., Klosak J.J., Dennis J.W., Karesh S.M., Ellinger J. and Kim D.U. (1987) Biomaterial pretreatment with ECGF to augment endothelial cell proliferation. *J Vasc Surg*, **5**, 393-402.

Greisler H.P., Cziperle D.J., Kim D.U., Garfield J.D., Petsidas D., Murchan P.M., Applegren E.O., Drohan W. and Burgess W.H. (1992) Enhanced endothelialization of expanded polytetrafluoroethylene grafts by fibroblast growth factor type 1 pretreatment. *Surgery*, **112(2)**, 244-255.

Hoffman D., Gong G., Liao K., Macaluso F., Nikolic S.D. and Frater R.W.M. (1992) Spontaneous host endothelial growth on bioprostheses. *Circulation*, **86[suppl II]**, II-75-II-79.

Horan P.K., Melnicoff M.J., Jensen B.D. and Slezak S.E. (1990) Fluorescent cell labeling for in vivo and in vitro cell tracking. *Meth in Cell Biol*, **33**, 469-490.

Hudlicka O. (1992) Role of mechanical factors in angiogenesis under physiological and pathophysiological circumstances. In *Angiogenesis in Health and Disease* (eds. M.E. Maragoudakis, P. Gullino and P.I. Lelkes), Plenum Publishing Co., pp. 207-215. New York, London

Kaehler J., Zilla P., Fasol R., Deutsch M. and Kadletz M. (1989) Precoating substrate and surface configuration determine adherence and spreading of seeded endothelial cells on polytetrafluoroethylene grafts. *J Vasc Surg*, **9**, 535-541.

Kent K.C., Shindo S., Ikemoto T. and Whittemore A.D. (1989) Species variation and the success of endothelial cell seeding. *J Vasc Surg*, **9**, 271-276.

Kohler T.R., Stratton J.R., Kirkman T.R., Johansen K.H., Zierler B.K. and Clowes A.W. (1992) Conventional versus high-porosity polytetrafluoroethylene grafts: clinical evaluation. *Surgery*, **112**, 901-907.

Lelkes P.I., Gao H., Flemma R.J., Edgerton J.R. and Christensen C.W. (1990) Endothelial cell seeding of latissimus dorsi pouches: possible use as skeletal muscle ventricles. *Clin Res*, **38**, 497. (Abstract)

Lelkes P.I. and Samet M.M. (1991) Endothelialization of the luminal sac in artificial cardiac prostheses: a challenge for both biologists and engineers. *J Biomech Eng*, **113**, 132-142.

Lelkes P.I., Samet M.M., Christensen C.W. and Amrani D.L. (1992) Factitious angiogenesis: endothelialization of artificial cardiovascular prostheses. In *Angiogenesis in health and disease* (eds. M.E. Maragoudakis, P. Gullino and P.I. Lelkes), Plenum Press, pp. 339-353. New York, NY

Lelkes P.I., Gao H., Edgerton J.R. and Christensen C.W. (1994) Endothelial cell seeding of latissimus dorsi muscle pouches. *J Surg Res*, in Press.

Lin H.B., Sun W., Mosher D.F., Garcia-Echeverria C., Schaufelberger K., Lelkes P.I. and Cooper S.L. (1994) Endothelial cell adhesion and growth on polyurethanes: effect of RGD-peptide incorporation. *J Biomed Mat Res*, in press.

Magovern G.J. (1991) Introduction to the history and development of skeletal muscle plasticity and its clinical application to cardiomyoplasty and skeletal muscle ventricle. *Sem Thorac Card Surg*, **3**, 95-97.

Massia S.P. and Hubbell J.A. (1992) Tissue engineering in the vascular graft. *Cytotechnology*, **10**, 189-204.

Menger M.D., Hammersen F., Walter P. and Messmer K. (1990) Neovascularization of prosthetic vascular grafts. quantitative analysis of angiogenesis and microhemodynamics by means of intravital microscopy. *Thorac Cardiovasc Surgeon*, **38**, 139-145.

Mills Jr., R.M., Billett J.M. and Nichols W.W. (1992) Endothelial dysfunction early after heart transplantation. *Circulation*, **86**, 1171-1174.

Miwa H., Matsuda T., Kondo K., Tani N., Fukaya Y., Morimoto M. and Iida F. (1992) Improved patency of an elastomeric vascular graft by hybridization. *ASAIO Trans*, **38, No. 3**, M512-M515.

Nabel E.G., Plautz G. and Nabel G.J. (1992) Transduction of a foreign histocompatibility gene into the arterial wall induces vasculitis. *Proc.Natl.Acad.Sci.USA*, **89**, 5157-5161.

Niu S., Matsuda T. and Oka T. (1990) In vitro model of endothelialization at anastomotic sites. *Trans Am Soc Artif Intern Organs*, **36**, M757-M760.

Niu S. and Matsuda T. (1992) Endothelial cell senescence inhibits unidirectional endothelialization in vitro. *Cell Transplantation*, **1**, 355-364.

Noishiki Y., Yamane Y., Tomizawa Y., Okoshi T., Satoh S., Wildevuur C.R.H. and Suzuki K. (1992) Rapid endothelialization of vascular prostheses by seeding autologous venous tissue fragments. *J Thorac Cardiovasc Surg*, **104, No. 3**, 770-778.

Papadimitriou E., Unsworth B.R., Maragoudakis M.E. and Lelkes P.I. (1993) Time-course and quantitation of extracellular matrix maturation in the chick chorioallantoic membrane and in cultured endothelial cells. *Endothelium*,

Park P.K., Jarrell B.E., Williams S.K., Carter T.L., Rose D.G., Martinez-Hernandez A. and Carabasi R.A.,III (1990) Thrombus-free, human endothelial surface in the midregion of a Dacron vascular graft in the splanchnic venous circuit--observations after nine months of implantation. *J Vasc Surg*, **11**, 468-475.

Pette D. (1991) Changes in phenotype expression of stimulated skeletal muscle. In *Cardiomyoplasty* (eds. A. Carpentier, J.C. Chachques and P.A. Grandjean), Bakken Reserach Center Series, volume 3, Futura Publishing Company, Inc., pp. 19-31. Mount Kisco, NY

Pochettino A., Anderson D.R., Hammond R.L., Spanta A.D., Hohenhause E., Niinami H., Huiping L., Ruggiero R., Hooper T.L., Baars M., Devireddy C. and Stephenson L.W. (1991) Skeletal muscle ventricles: a promising treatment option for heart failure. *J Card Surg*, **6(1)**, 145-153.

Rosengart T.K., Kupferschmid J.P., Ferrans V.J., Casscells W., Maciag T. and Clark R.E. (1988) Heparin-binding growth factor-I (endothelial cell growth factor) binds to endothelium in vivo. *J Vasc Surg*, **7**, 311-317.

Rupnick M.A., Hubbard F.A., Pratt K., Jarrell B.E. and Williams S.K. (1991) Endothelialization of vascular prosthetic surfaces after seeding or sodding with human microvascular endothelial cells. *J Vasc Surg*, **9 No 6**, 788-795.

Salmons S. and Jarvis J.C. (1992) Cardiac assistance from skeletal muscle: a critical appraisal of the various approaches. *Br Heart J*, **68**, 333-338.

Schneider P.A., Hanson S.R., Price T.M. and Harker L.A. (1990) Confluent durable endothelialization of endarterectomized baboon aorta by early attachment of cultured endothelial cells. *J Vasc Surg*, **11**, 365-372.

Seeger J.M. and Klingman N. (1988) Improved in vivo endothelialization of prosthetic grafts by surface modification with fibronectin. *J Vasc Surg*, **8**, 476-482.

Thomas G.A., Lelkes P.I., Chick D.M., Isoda S., Lu H., Nakajima H., Hammond R.L., Walters H.L. and Stephenson L.W. (1993a) Skeletal muscle ventricles seeded with autogenous endothelium. *ASAIO J.*, (In Press)

Thomas G.A., Lelkes P.I., Chick D.M., Lu H., Kowal T.A., Hammond R.L., Nakajima H., Spanta A.D. and Stephenson L.W. (1993b) Skeletal muscle ventricles seeded with allogenic endothelial cells via open and percutaneous techniques. *J Surg Res*, (In Press)

Wilson J.M., Birinyi L.K., Salomon R.N., Libby P., Callow A.D. and Mulligan R.C. (1989) Implantation of vascular grafts lined with genetically modified endothelial cells. *Science*, **244**, 1344-1346.

Zilla P., Fasol R., Deutsch M., Fischlein T., Minar E., Hammerle A., Krapicka O. and Kadletz M. (1987) Endothelial cell seeding of dacron and polytetrafluoroethylene grafts: the cellular events of healing. *Surgery*, **6**, 535-541.

SOME PROBLEMS OF TRIAL DESIGN FOR ANTI-ANGIOGENIC AGENTS IN CANCER THERAPY

Frank Arnold

Dept. of Dermatology
Churchill Hospital Headington
Oxford, OX3 7LJ, United Kingdom

The propensity of tumours to generate resistant clones is a major barrier to cytotoxic and hormone therapy. The vasculature of malignancies is a more appealing target, since tumour endothelial cells are not known to be tranformed or subject to somatic mutation or random phenotypic variation. Thus an effective anti-angiogenic therapy need not be vitiated by the acquisition of resistance. Although several anti-angiogenic agents have been described and used in experimental models, there remain severe problems about optimising their use and about setting up informative clinical trials. These difficulties arise predominantly because of the "invisible" nature of tumour angiogenesis.

When a tumour regresses under influence of effective chemotherapy of hormone therapy (or spontaneously) the vessels regress in parallel. We are not left with a fibrovascular "ghost" (Arnold, 1985). This is a matter of frequent clinical experience, and has been described in detail in both experimenal (Gullino et al., 1962) and human tumours. Because of parallel regression, it is difficult to distinguish between anti-angiogenic and anti-tumour cell effects. Furthermore, many existing treatments thought to act by anti-angiogenic machanisms (protamine, heparin/cortisone, tetracycline derivatives, fumagillin, cromolyn) reduce the growth rate of tumours, rather than causing regression. These two factors make the conduct of conventional phase II trials difficult, and mandate new approaches to the monitoring of the cellular effect and clinical outcome of therapy.

Agents which can cause prolonged growth restraint (rather than temporary regression limited by resistance) may actually confer greater survival benefits, even where conventional indices of effect (eg percentage "responders") are non-existant (figure 1). Thus real benefits of anti-angiogenic therapy might not be recognised by conventional phase II trial designs. At least three alternatives are possible:

a) To use survival time (rather than proportion of regressions) as the primary endpoint. This is scientifically and clinically valid (Fraser et al., 1990) but a slow and expensive way to conduct early trials.

b) To chose patients with measureable lesions (eg lung meatstases from colorectal and renal cancer) who, unless symptomatic, are not conventional candidates for therapy. It is ethical to obtain serial growth curves, before and after the commencement of therapy in such patients. An inflection point in the logarithmic growth curve is taken as evidence of effective therapy.

c) To conduct trials using patients such as those with refractory lung-metastatic breast cancer, the growth of whose metastases is predictable from histogical data. To compare the rate of progression of the disease at a defined and prognostic time point with those expected from the existing data.

We are currently conducting two early phase II clinical trials using methods (b) and (c) to examine the benefits of the mast cell stabiliser sodium cromoglycate in patients with lung metastases. The drung is wide used for asthma prophylaxis, and has previously been shown to reduce tumour growth rates in rats and mice (Arnold, 1991). These designs are the first to have been established to deal with problems specific to anti-angiogenic therapy, and may be of general usefulness.

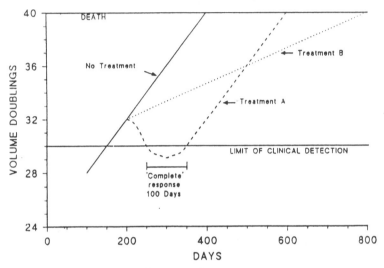

Figure 1. Theoretical response of metastatic tumour treatments.
Treatments: None (survival, 100 days)
(A) Effective chemotherapy (survival 400 days)
(B) Hypothetical agent causing persistent reduction in growth rate (survival, 600 days)

REFERENCES

Arnold F., 1985, Tumour Angiogenesis. Ann. Roy Coll. Surgeones (Eng) 67:295-298.
Gullino P.M. and Grantham F.M., 1962, Studies on the exchange of fluids between host and tumour. J.N.C.I. 28: 211-221.
Fraser S.C.A. et al., 1990, The design of advanced breast cancer trials, Acta Oncologica 29: 397-400.
Arnold F., 1991, Exploiting angiogenesis, Lancet 337: 856-856.

REGULATION OF BLOOD VESSEL GROWTH IN TUMORS AND WOUNDS

H. Brem[1], G. Tamavokopulis[1], D. Tsakayannis[1], I. Gresser[2], A. Budson[1], and J. Folkman[1]

[1]The Ohio State University Hospital, Dept. of Surgery, Children's Hospital and Harvard Med. School, Columbus, Ohio, USA
[2]Groupe de Laboratoires de l' Institut de Recherches Scientififiques sur le Cancer, Villejuif, France

We previously reported the identification of a novel fungal-derived angiogenesis inhibitor, AGM-1470, a member of a family of: "angioinhibins" (Nature 348:555, 1990). We now show that AGM-1470 inhibits tumor growth of more than 15 animal tumors as well as human tumors in athymic mice, suppresses metastasis, does not induce drug resistance and has very low toxicity with long-term administration. Alpha/beta interferon (murine), also an angiogenesis inhibitor, potentiates the anti-tumor effects of AGM-1470. These experimental findings and its implications in wound healing and tumors will be discussed. Examples are as follows:

Inhibition of primary tumor: Mean tumor volume of mice with Lewis lung tumors was, treated/control (T/C)=0.35, with untreated tumors having a mean weight of 5.7±0.5 grams. Lewis lung tumors were treated with AGM-1470 and/or murine alpha/beta interferon 200,000 units. This combination resulted in an additive antitumor (TC/0.2) and antimetastatic activity.

Resistance studies: *In vivo*: 10 consecutive tumor passages in mice that were treated with AGM-1470 resulted in 57-72% reduction in tumor volume. Pretreatment of mice for 100 days before tumor implantation did not effect the subsequent T/C. *In vitro*: The dose to obtain half maximal inhibition (50 pg/ml) and the time to confluence (8 days), did not change after 8 consecutive serial passages of capillary endothelial cells with AGM-1470.

Wound healing: Wound healing is angiogenesis dependent. After linear wounds were made (n=160), AGM-1470 given on post wound days 0-10 (every other day) resulted in tensile strengths of: 2.47±0.25 and 7.73±1.4 lbs compared to saline controls of 4.72±0.66 and 13.3±0.77 lbs assayed on days 7 (P=0.014) and 12 post incision, respectively (P=0.017). In contrast, treatment with AGM-1470 for 20 days prior to wounding did not significantly effect wound healing.

Time-dependent expression of bFGF during wound healing: Basic fibroblast growht factor (bFGF) is a ubiquitous, mitogenic and potent angiogenic molecule. An increase in endogenous bFGF occurred during the healing of full thickness murine wounds (n=204). The peak level of bFGF in the wound as measured by ELISA or heparin affinity chromatography occurs between days 10-14 post wounding. AGM-1470 decreased the concentration of bFGF in the day 10 wounds by 55±12%.

<u>Conclusions</u>: (1) The biological properties of AGM-1470 provide a paradigm of the potential therapeutic efficacy of antiangiogenic therapy for solid tumors and their metastases; (2) Such therapy is likely to be long-term, of low toxicity, equally effective in males or females, independent of immune status, and potentiated when two angiogenesis inhibitors are administered together; (3) Tumor resistance did not develop *in vivo* after prolonged antiangiogenic therapy with AGM-1470; (4) The angiogenesis inhibitor, AGM-1470, suppresses wound healing in a time dependent manner; (5) AGM-1470 does not effect wound healing if given before, or five days after the wound is made; (6) Endogenous bFGF concentrations peak in the wound during days 10-14. This elevation appears to follow the influx of endothelial cells into the wound. These results suggest that invading vascular endothelium may contribute additional bFGF to the wound over and above the level of bFGF that is present in the wound before neovascularization; (7) AGM-1470 decreased the peak bFGF concentration in the wound suggesting that neovascularization significantly contributes to the amount of endogenous bFGF in the wound.

THE CHARACTERIZATION OF VASCULAR ENDOTHELIAL GROWTH FACTOR (VEGF) PRODUCED BY THE BACULOVIRUS EXPRESSION SYSTEM

Tzafra Cohen[1], Hela Gitay-Goren[2], Gera Neufeld[2] and Ben Zion Levi[1]

[1]Department of Food Engineering and Biotechnology

[2]Department of Biology Technion,

Israel Institute of Technology,

Haifa 32000, Israel

Vascular endothelial growth factor (VEGF) is a recently discovered mitogen for endothelial cells *in vitro*, and a potent angiogenesis promoting factor *in vivo*. VEGF is secreted from producing cells as a homodimer, binds to specific receptors on the cell surface of endothelial cells, and is produced in four forms as a result of alternative splicing. We have expressed the cDNAs encoding the 165, 121 and 189 amino-acid long isoform of VEGF in insect cells using the baculovirus based expression vector. We show that infected insect cells produce large amounts of the VEGFs. Antibodies directed against a synthetic peptide prepared from human VEGF identify the secreted factor. The baculovirus derived $VEGF_{165}$ expressed in insect cells binds directly to the VEGF receptors. $VEGF_{165}$ competes with pure mammalian cells derived $[^{125}I]$-VEGF for binding to the VEGF receptors that are present on the cell surface of endothelial cells. Furthermore, $VEGF_{165}$ is biologically active and induces the proliferation of human umbilical vein derived endothelial cells. The purification and the biological activities of the other forms of VEGF produced in insect cells will be discussed.

SCANNING LASER-DOPPLER IMAGING: A NEW, NON-INVASIVE MEASURE OF ANGIOGENESIS

George Cherry and Frank Arnold

Department of Dermatology
Churchill Hospital
Oxford, OX3 7LJ United Kingdom

Single point laser-doppler measurement has been used to demonstrate changes in blood flow resulting from a variety of stimuli. However, flow on the surface of skin and other organs is heterogenous; estimation of flow indices at a single point are of limited value. The scanning laser-doppler harvests information about backscattered, doppler-shifted light from a series of (up to 4096) points and displays the data as a two dimensional image in which flow is represented by colour.

We have used this technique:

a)   To examine postural vasoregulation of skin blood flow in patients with venous leg ulcers annd normal controls.

b)   To demonstrate and, analyse the growth of a functional vasculature in experimantal full thickness wounds under conditions of angiogenic stimulation.

c)   To visualise the sequence of events during reperfusion of ischaemic skin island flaps.

We are also applying it to examine the effects of conventional therapy on the vascularity of skin-recurrent tumour nodules. This method provides a non-invasive index of vascularity, which can be coupled to ultrasonic tumour volume measurement to determine the sequence of events during the parallel regression of umours and vessels. It provides a powerful tool for dissecting the "invisible" relationship, and may therefore help in the optimisation on anti-angiogenic agents.

TWO MOLECULES RELATED TO THE VEFG-RECEPTOR ARE EXPRESSED IN EARLY ENDOTHELIAL CELLS DURING AVIAN EMBRYONIC DEVELOPMENT

Anne Eichman, Christophe Marcelle, Christiane Bréant and Nicole M. LeDouarin

Institut d' Embryologie du Centre National de la Recherche Scientifique et du Collège de France,
94736 Nogent-sur-Marne, Cedex, France

We present the partial cloning and the expression patterns of two putative growth factor receptor molecules named Quek1 and Quek2 (for quail endothelial kinase) in chick and quail embryos for gastrulation to embryonic day 9 (E9). Quek1 and Quek2 show high homology to three interrelated murine and human genes, flk-1, KDR and flt. Flt was recently shown to be the receptor for the endothelial cell mitogen vascular endothelial growth factor (VEGF). In situ hybridization of Quek1 and Quek2 to sections of avian embryos showed that they are both expressed essentially by endothelial cells, that we identified with a monoclonal antibody (Mab) QH1 specific for endothelial and white blood cells of the quail. Quek1 is expressed in the mesoderm from the onset of gastrulation, whereas Quek2 message is first detected on QH1-expressing endothelial cells. The expression pattern of Quek1 suggests that it could identify the putative precursor of both endothelial and hematopoietic lineages, the hemangioblast. Quek1 and Quek2 are not expressed in all endothelial cells throughout life. At E9, after the initial phase of vasculogenesis, these genes are switched off in various compartments of the vascular network.

ANALYSIS OF BRAIN CAPILLARY AND BLOOD BRAIN BARRIER FUNCTION *IN VITRO*

Nam D. Tran[1], Vicky L.Y. Wong[1], James Bready[2], Judith Berliner[3] and Mark Fisher[1]

[1]University of Southern California School of Medicine, Los Angeles
[2]Amgen Corporation
[3]University of California at Los Angeles School of Medicine
Los Angeles, U.S.A.

Monolayer endothelial cell (EC) cultures are less than ideal for studying the functions of brain capillary endothelium. EC in the central nervous system share the basement membrane with astrocytes and this cell-cell interaction is thought to play a role in regulating microvessel formation and differentiation. We are utilizing an *in vitro* blood-brain barrier (BBB) system in which capillaries develop tight junctions (Laboratory Investigation 65:32, 1991). Bovine brain capillary EC grown on gelatin-covered slide chambers and covered with a layer of type I collagen can form capillary-like structures (CS). The cells of the CS express von Willebrand factor, an endothelial marker. Our EC model can be used for co-culture with astrocytes and thus provide a means for studying astrocyte regulation of EC. Two days after the formation of CS, astrocytes are added to the culture. Staining for glial fibrillary acidic protein reveals astrocytic processes with CS within five hours of co-culture. After 48 hours, astrocytic processes envelop the CS. The juxtaposition of the astrocytes to the CS appears to correlate with expression of the enzyme gamma glutamyl transpeptidase, a putative marker for the BBB. Quantification of the CS is performed with photographs of four representatitve fields from each slide chamber, and digitized by an image analyzer to calculate the two dimensional surface area of the CS and monolayer. EC activity from different samples can thus be adjusted for the surface area of the CS. We propose to use this BBB model for quantitative delineation of astrocyte-endothelial interactions.

EFFECTS OF X-RAY IRRADIATION OF ANGIOGENESIS AND TUBE FORMATION BY ENDOTHELIAL CELLS *IN VITRO*

O. Hadjiconti[1], S. Papaioannou[2], G.C. Haralabopoulos[1], I. Demopoulos[3] and M. E. Maragoudakis[1]

[1]Department of Pharmacology, School of Medicine
[2]Department of Molecular Pharmacology, School of Pharmacy
[3]Department of Radiology, School of Medicine
 University of Patras, Patras 261 10, Greece

Angiogenesis was assessed *in vivo* in the chick embryo chorioallantoic membrane (CAM) by measurements of vascular density and collagenous protein biosynthesis. CAM membranes were irradiated with 10 or 15 Gy on the 9th or the 14th day of the chick embryo development, when the angiogenesis of the non-irradiated control CAM is maximal or minimal, respectively, and evaluated 6, 24 and 48 hours post-irradiation.

Irradiation on the 9th day of the embryo development resulted in significant reduction of CAM vascular density 6, 24 and 48 hours post-irradiation. The inhibition of collagenous protein-biosynthesis 6 hours post-irradiadion was also significant, whereas at 48 hours post-irradiation a significant induction of collagenous protein biosynthesis was observed.

On the 14th day of chick embryo development there were no significant changes in vascular density or collagenous protein biosynthesis after 6 or 24 hours for either the irradiated or the control CAM.

Histological studies corraborate and provide additional details to the above studies. Preliminary irradiation experiments of HUVECs of Matrigel with 10 Gy showed no evidence of tube formation inhibition. Studies are in progress for comparison of angiogenesis in CAM with HUVECS tube formation *in vitro* toward a better understanding of X-ray effects on angiogenesis.

These results suggest that the destruction of CAM vessels by X-ray irradiation is correlated with an inhibition of basement membrane biosynthesis for the first 6 hours post-irradiation. At later post-irradiation times the induction of collagenous protein biosynthesis may be due to accelerated nonvascular collagenous protein biosynthesis during the active repair process of CAM. Collagenous protein biosynthesis is inhibited by X-ray irradiation during periods of maximal angiogenesis and not during quiscent angiogenesis periods.

COMPUTER ASSISTED IMAGE ANALYSIS OF ANGIOGENESIS IN THE CAM SYSTEM *IN VIVO*

George C. Haralabopoulos, Nikos E. Tsopanoglou, Eva Pipili-Synetos and Michael E. Maragoudakis

Department of Pharmacology
University of Patras
Medical School
261 10 Rio Patras, Greece

A technique was developed for the asssessment of vascular area in the chorioallantoic membrane system (CAM) *in vivo* by computer assisted image analysis and density slicing which allows for evaluation of the area of small and medium sized vessels excluding larger vessels. This process involves the fixation and the mounting of the CAM containing the area where the pellets were placed on a slide, and measurement of the area between 0.08 and 0.53 relative optical densities on the 4.12 version of the MCID image analysis software (Imaging Research Inc., Brock University).

This technique was compared to the method developed in our laboratory which uses the rate of basement membrane synthesis as a biochemical index of angiogenesis, as well as the Harris-Hooker morphological evaluation which involves the counting of all vessels intersecting three concentric rings of 4mm, 5mm, and 6mm in diameter [1,2]. In order to evaluate promotion or inhibition of angiogenesis, the promoters of angiogenesis 4-β-PMA, α-thrombin, or the inhibitors of angiogenesis Ro 318220, D609, and GPA 1734 were used [3,4,5].

Evaluation by image analysis indicated a 3-fold increase in vascular area from day 7 to day 11 whereupon a plateau was reached, which correlated to results obtained from measurement of blood vessels intersecting three concentric rings. α-thrombin 1I.U. and PMA 60ng caused a 44% and 50% increase in vascular area respectively, compared to a 32% and 28% increase by counting vessels and an increase of 88% and 35% in collagenous proteins. Addition of Ro 318220 10µg, D609 25µg, and GPA 1734 20µg resulted in a 42%, 22%, and 33% inhibition of vascular area respectively as measured by image analysis, compared to a 30%, 25%, and 33% reduction by counting vessels and a 40%, 65%, and 43% reduction in collagenous proteins.

These results indicate that angiogenesis in the CAM can be quantified by using image analysis and density slicing techniques in order to evaluate the relative area of small and medium sized blood vessels.

REFERENCES

Harris-Hooker, S.A. et. al., 1983, Neovascular responses induced by cultured aortic endothelial cells, J. Cell. Physiol. 114:302-310.

Maragoudakis, M.E. et. al., 1988b, Inhibition of basement membrane biosynthesis prevents angiogenesis, J. Pharm. Exp. Ther. 244:729-733.

Maragoudakis, M.E. et. al., 1992, Evaluation of promotors and inhibitors of angiogenesis using basement membrane biosynthesis as an index. Nato ASI 227:275-286. Plenum Press.

Tsopanoglou, N.E. et. al., 1993, Thrombin promotes angiogensis by a method independent of fibrin formation, Am. J. Physiol. (in press).

Tsopanoglou, N.E., et. al., 1993, Protein Kinase C involvement in the regulation of angiogenesis, J. Vasc. Res. (in press).

IN SITU ANALYSIS OF NATURAL ANGIOGENIC PROCESSES: FACTORS MEDIATING RESPONSE TO HYPOXIA AND BALANCING OF PROTEOLYSIS

E. Keshet, D. Shweiki, E. Bacharach, A. Itin and S. Banai

Dept. of Molecular Biology
Hebrew University-Hadassah Medical School
Jerusalem 91010, Israel

A number of factors are capable of eliciting an angiogenic response in a model system. Yet, little is known about angiogenic factors operating in the context of natural neovascularization. We have, therefore, evaluate the *in vivo* relevance of putative mediators of angiogenesis through retrieving specimens that represent sequential stages in natural neovascularization, and elucidating spatio-temporal patterns of expression of candidate genes during the given process. This procedure also provides a molecular access to downstream responses to the angiogenic stimulus, identifying patterns of expression that distinguish angiogenic- from a quiescent endothelium.

Here we report the following observations: 1. Vascular endothelial growth factor (VEGF), a secreted endothelial cell-specific mitogen, is expressed during gour independent angiogenic processes taking place in the female reproductive system (neovascularization of ovarian follicles, neovascularization of the corpus luteum, repair of endometrial vessels, and angiogenesis in embryonic implantation sites), specifically in cells surrounding the expanding vasculature. VEGF is predominantly produced in tissues that acquire new capillary networks. VEGF-binding activity, on the other hand, was foudn exclusively on endothelial cells of both quiescent and proliferating blood vessels. These findings are consistent with a role for VEGF in the targeting of angiogenic responses to specific areas. VEGF is expressed in ten different steroidogenic and/or steroid-responsive cell types, and in certain cells up-regulation of VEGF expression is concurrent with the acquisition of steroidogenic activity. These findings suggest that expression of VEGF is hormonally-regulated, and further suggest that VEGF may mediate hormonally-triggered cycles of neovascularization in the reproductive system (1).

2. Inefficient vascular supply and the resultant reduction in tissue oxygen tension, often lead to neovascularization in order to satisfy the needs of the tissue. Examples include the compensatory development of collateral blood vessels in ischemic tissues that are, otherwise, angiogenesis-quiescent, and angiogenesis associated with the healing of hypoxic wounds. Yet, the nature of the presumptive hypoxia-induced angiogenic factors that mediate this feedback response has not been elucidated. We show that VEGF is likely to function as an hypoxia-inducible angiogenic factor (2). VEFG mRNA levels are dramatically elevated within a few hours of exposing different cell cultures to hypoxia, and mRNA levels revert to background levels upon

resumption of normal oxygen supply. *In situ* analysis of tumor specimens undergoing neovascularization showed that the production of VEGF is specifically induced in a subset of tumor cells distinguished by their immediate proximity to necrotic foci (presumably hypoxic regions), and the clustering of capillaries alongside VEGF-producing cells. VEGF production is also upregulated by ischemia in cardiac myocytes. Occlusion of a major porcine coronary artery (LAD) results in augumentation of VEGF expression, specifically in the ischemic myocardial territory. Thus, VEGF may potentially mediate the natural process of ischemia-induced coronary collateral formation.

3. In order to evaluate the role of plasminogen activators (PA) in physiological angiogenesis, we have investigated the in vivo patterns of expression of urokinase type-PA (uPA) and the PA-inhibitor type I (PAI-1) during ongoing neovascularization, using in situ hybridization analysis. uPA mRNA expression was detected in endothelial cell cords engaged in organization of vascular networks, but could not be detected in endothelial cells upon completion of neovascularization, thus suggesting that uPA expression is a part of the angiogenic response. Expression of PAI-1, on the other hand, was preferentially activated in cells in the vicinity of uPA-expressing capillary-like structures (3). These findings suggest a functional interplay between uPA- and PAI-1-expressing cells, and is consistent with the notion that natural PA-inhibitors function during angiogenesis to protect neovascularized tissues from excessive proteolysis. In co-culture experiments, PAI-I expression was specifically induced in fibroblasts juxtapositioned next to uPA-producing endothelial cells, supporting the thesis that regulation of PAI-1 is apposition-dependent.

REFERENCES

1. Shweiki, D., Itin, A., Gitay-Goren, H., Neufeld, G. and E. Keshet, *J. Clin. Invest.* (in press) (1993).
2. Shweiki, D., Itin, A., Soffer, D., and Keshet, E., *Nature* 395:843-845 (1992).
3. Bacharach, E., Itin, A., and Keshet, E., *Proc. Natl. Acad. Sci., USA* 89:10686-10690 (1992).

MORPHOLOGICAL ASPECTS OF ANGIOGENESIS: NETWORK FORMATION IN DIFFERENT TUMOR ENTITIES AND WOUND HEALING

M.A. Konerding[1], C. van Ackern[1], B. Klapthor[1], F. Steinberg[2], M. Lehmann[2]

[1]Institüt für Anatomie
[2]Institut für Med. Strahlenbiologie
Universitätsklinikum Essen
D-4300 Essen 1, Germany

Vascular systems of individual organs develop and differentiate in parallel with or with only a slight delay to the tissues to be supplied and drained since the tissues blood flow is critically dependent on the structure and 3-D architecture of the vascular network. Despite the tumor biological and therapeutic implications, studies on primary tumors and comparisons of different tumor models, which confirm their reliability and validity, are still very scarce. This is also true for comparisons of network formation in tumor angiogenesis and wound healing.

Against this background, the aim of this study is to compare the vascularity of different tumor entities and with that of normal tissues during wound healing. In total, 16 different types of carcinomas and 17 different types of sarcomas were considered. Of those, two types of carcinomas and one type of sarcoma were human primary tumors, 9 types of carcinomas and 11 sarcomas were human tumors xenografted onto nude mice. The remainder were chemically induced or spontaneous murine tumors. Network formaiton in reparative angiogenesis was studied on peripheral nerves after epineural suture. For this, the sciatic nerves of 24 chicken were dissected and sutured after sharp cutting. The nerves were examined 3 to 22 days after surgery. The 3-D architecture of the vascular networks both of the tumors and sciatic nerves was studied by scanning electron microscopy of microvascular corrosion casts.

All tumors share the following common features irrspective of their origin: plexuses of flat vessels with abrupt changes in diameter and blind endings, lack of differentiation, and vessel outpouchings. The plexuses predominantly border avascular tumour areas. They consist of dilated venules and capillaries of varying diameters. Venules and capillaries are intensively interconnected. Quantitative analysis of these casts reveals similar ranges of parameters such as diameters, intervascular and interbranching distances. Diameters of vessels forming sinusoidal plexuses range from 6 μm to 55 μm in the human primary tumors, and form 4.5 μm to 80 μm in xenografted tumors. Intervascular distances in the human primary tumors range from 1.7 μm to 52 μm, and from 11 μm to 105 μm in the xenografts. Interbranching distances range from 34 μm to 258 μm in the former, and from 11 μm to 160 μm in the latter. It has to be pointed out, however, that diameters, intervascular and interbranching distances mainly lie within the lower dimensional ranges. In summary, vascular plexuses are found in all tumors studied, i.e. carcinomas and sarcomas. They show only negligible qualitative and quantitative differences, making a differentiation between CAs and SAs based on this feature impossible. However, architectonic characteristics could be established in the peripheral regions of individual tumors: after rendering the specimens anonymous, the individual entities and cell lines could be correctly assigned in over 90% of cases based solely on the peripheral vascular architecture.

On comparing the vascular networks in tumors to those in the sciatic nerve after epineural suture, it becomes evident that all structural characteristics of tumor angiogenesis such as glomeruloidal arrangement and tortuous course of peripheral vessels, formation of sisnuses of differing diameter etc. can be seen in wound healing as well. However, the range of vessel diameter and the amount of changes in diameter is lower and the time of vascular reconstruction is much shorter than in tumor angiogenesis. The most significant architectural difference is the earlier onset of vascular reconstruction in terms of replacement of unsufficient vascular segments and the rapid remodeling.

bFGF AND TNFα COOPERATE OF TUBULAR STRUCTURES OF HUMAN ENDOTHELIAL CELLS IN AN *IN VITRO* SYSTEM

P. Koolwijk, W.J.A. de Vree, C. Zurcher, and V.W.M. van Hinsbergh

Gaubius Laboratory, IVVO-TNO, Zernickedreef 9, 233 CK Leiden, The Netherlands

Angiogenesis plays an important role in several human diseases, e.g. tumour development and rheumatoid arthritis. It is generally assumed that local proteolysis and matrix degradation by plasmin and several metalloproteinases play an important role in the regulation of angiogenesis, and that u-PA bound to its cellular receptor (u-PAR) is involved in this process. We have investigated the effect of bFGF and TNFα, two mediators known to induce angiogenesis *in vivo* , on i) the induction of u-PA and the u-PA receptor by/on human endothelial cells (EC), and ii) on the induction of EC tube formation in an *in vitro* angiogenesis system.

TNFα markedly increased the amount of u-PA mRNA and the production of u-PA antigen by human foreskin microvascular EC (MVEC) and human umbilical vein EC (HUVEC), whereas it did not affect or even slightly decresed the binding of ^{125}I-DIP-u-PA to the cells. In contrast bFGF increased specific binding of ^{125}I-DIP-u-PA up to 2-fold, whereas the accumulation of u-PA antigen in the medium did not change or slightly decreased (both with and without additional stimulation by TNFα)

To investigate the ability of bFGF and TNFα to induce EC tube formation *in vitro* MVEC and HUVEC were cultured on three-dimensional fibrin gels and stimulated by bFGF, TNFα, or a combination of bFGF and TNFα for two to ten days. Tube formation was analyzed by phase contrast microscopy and, after fixation of the gels, by semi-thin cross-section analysis. bFGF induced EC to form capillary-like tubes in the fibrin gel. TNFa showed an additive effect on the bFGF-induced tube formation, whereas it had no significant effect itself on the invasion of EC in the fibrin gel. When the amount of u-PA, t-PA, and PAl-1 was determined in the supernatants of the EC cultured on the fibrin gels, we found a correlation between the amount of tube formation and u-PA but not t-PA and PAI-1.

The data presented here suggest that - in addition to its mitogenic effect on EC -, bFGF can influence the migration of EC and formation of EC tubes. Basic FGF is capable to increase the number of u-PAR, and thereby the amount of u-PA, available on the EC surface, which may result in an increased local proteolysis. The separate regulation of u-PAR expression and u-PA production by bFGF and TNFα provides the EC with a mechanism to regulate local proteolytic activity. We speculate that in addition to the u-PA inducing inflammatory mediators (e.g. TNFα, IL-1), other (growth) factors (such as, aFGF, bFGF, and VEGF) control local u-PA activity on human endothelial cells, via the enhancement and direction of the u-PAR.

THE ROLE OF ANGIOGENESIS IN CEREBRAL ISCHEMIA (STROKE)

Jerzy Krupinski and Jozef Kaluza

Department of Neuropathology
Institute of Neurology
University of Cracow Medical School
Poland

Focal cerebral ischemia is one of the commonest causes of morbidity in the Western world   and results from a reduction in cerebral blood flow. Stroke occurs as a result of occlusion of the middle cerebral artery which is followed by severe disturbances in blood distribution. Some areas become ischemic, while others become hyperemic. Variability in capillary changes has prompted us to undertake quantitative morphometric studies of infarcted areas of brain to correlate these changes with patient survival after stroke.

In the initial study, brains were obtained at autopsy from 10 patients (ages 46-85). Samples were collected from infarcted hemisphere and controls from the contralateral hemisphere.   Formalin fixed and paraffin embedded sections were stained after Pickworth and with H&E. Altogether 6,520 microvessels, representing 10,801 microscopic fields were counted. The Wilcoxon range test was used for statistical analysis. The results showed significant differences both in morphology and blood vessel density between infarcted and control tissues (Table 1 & 2). In all 10 patients in the infarcted brains, there was a marked increase in the capillary density ($p<0.01$) when compared with normal brain tissues. In addition a positive correlation was also found between the time of survival and both total density and density of non-perfused blood vessels.

More recently, a new murine MAb E-9 has been raised, using "activated-proliferating" human umbilical vein endothelial cells (HUVEC). The antigen recognized by MAb E-9 is present on vascular endothelial cells of all foetal organs, tumours, regenerating an inflammed tissue. In contrast, this antibody stains only a few normal tissues (Wang et al 1993, In J. Cancer- in press). The E-9 antigen is upregulated following irradiation and phorbol ester treatment of endothelial cells. We have now initiated a further study to examine localisation of E-9 antigen in normal and infarcted brain and brain tumours. The preliminary results are promising. MAb E-9 strongly stains blood vessels of infarcted brains and brain tumours.

Table 1. Details of morphometric study of blood vessel counts in normal and infarcted brain and its correlation with the patient survival

Case No	BLOOD VESSEL COUNTS IN		No of Areas Studied	Survival after stroke (Days)
	Infarcted hemisphere	Control hemisphere		
1	181	144	288	15
2	861	1000	2000	5
3	1408	1200	2598	10
4	533	460	920	16
5	1572	1000	2000	90
6	961	750	1500	21
7	161	68	135	8
8	590	500	1000	5
9	199	150	300	5
10	54	30	60	92

Table 2. Details of morphological differences in blood vessels in normal and infarcted brains of 10 patients

SOURCE OF TISSUE	TYPE OF BLOOD VESSELS +- SD	
	Empty Capillaries	Normal Capillaries
Infarcted hemisphere	0.111 +- 0.054	0.170 +- 0.108
Normal hemisphere	0.058 +- 0.007	0.350 +- 0.057

350

THE ANTITUMOR AGENT TITANOCENE DICHLORIDE SUPPRESSES ANGIOGENESIS

E Missirli[1], M. Bastaki[1], G. Karakiulakis[2] and M.E. Maragoudakis[1]

[1]Dept. Pharmacol., Univ. of Patras Med. Schl., Patras 26110
[2]Dept Pharmacol., Schl. Pharm. Aristotle Univ., Thessaloniki 54006,
Greece

Titanocene dichloride (TD), cyclopentadienyl titanium dichloride ($(C_5H_5)_2TiCl_2$), has been shown to possess antitumor activity against solid Ehrlich ascites tumor, Lewis lung carcinoma and sarcoma 180 (Köpf-Maier etal, 1986), while it exhibits only marginal activity against the experimental leukemia cell lines L1210 and P388 (Köpf-Maier et al., 1981). The agent is of low toxicity profile (Köpf-Maier et al., 1986) and its antitumor mechanism of action remains unknown. Since this agent is an active antitumor agent against solid but not blood-born tumors and in view of the fact that solid tumor growth and metastasis are angiogenesis dependent (Folkman, 1985), we considered angiogenesis as a possible target. We investigated the effect of TD on the formation of new blood vessels in the chorioallantoic membrane of the chick embryo (CAM) (Folkman, 1985), modified as previously reported (Maragoudakis et al., 1990).

We found that TD produced a dose related inhibition of angiogenesis in CAM *in vivo* (as observed morphologically). In the same system, the agent also produced a dose related inhibition in the biosynthesis of collagenous proteins, while total protein biosynthesis was not significantly affected. Parallel experiments using cis-platinum, which has similar antitumor potency as TD, revealed that this agent had no effect on the number of blood vessels in CAM *in vivo* and did not inhibit biosynthesis of collagenous proteins. TD had no effect on biosynthesis of collagenous proteins in the CAM system *in vitro*.

Using the animal model of i.p. carcinomatosis for testing agents that may prevent tumor growth and metastasis (Missirlis et al., 1990) we observed that the agent induced a pattern of tumor growth similar to that observed in case of deprivation of neovascularization (Folkman, 1985). Bolus i.p. injection of TD (50 mg/kg), on day 2 of implantation (intra-abdominal transplantation of solid Walker 256 carcinosarcoma) caused a significant delay in tumor growth. With the higher dose (40 mg/kg each on days 2 and 4 post-implantation) the delay in tumor growth was 94% and there was inhibition of macroscopically observed metastasis seeded along the spleen hilus and the whole length of the mesenteric attachment to the intestinal track, as compared to controls. Nevertheless in the presence of the agent, the mesentery was occasionally covered in part with a whitish tissue, the origin of which is under current histopathological examination.

The above described action of TD on growth of Walker 256 carcinosarcoma is not due to toxic effects, because at corresponding no-toxic dose regimens which suppressed angiogenesis and inhibited biosynthesis of collagenous proteins, this agent had no repercussion on proliferation, attachment or viability of Walker 256 carcinosarcoma, human A 549 lung adenocarcinoma cells or human umbilical vein endothelial cells in culture.

It appears that the antitumor activity of TD may be attributed, at least in part, to its ability to inhibit biosynthesis of collagenous proteins and suppress angiogenesis. This compound, in preliminary experiments, has also been shown to inhibit the phenotype expression of tube formation in the matrigel system for *in vitro* angiogenesis. Compounds such as TD, protamine (Taylor et al., 1982), heparin plus cortisone (Heuser et al., 1984; Maragoudakis et al., 1989), 8,9-dixydroxy-7-methyl-benzo(b)quinolizinium bromide (Missirlis et al., 1990), tricyclodecan-9-yl-xanthate (Maragoudakis et al., 1990), minocycline (Tamargo et al., 1991) and others which prevent tumor growth and metastasis by inhibiting angiogenesis may be used as experimental tools and be of clinical usefulness for antiangiogenic therapy, providing they lack toxic or undesirable side effects.

REFERENCES

Folkman, J., 1985, *Perspect. Biol. Med.* 29:10.
Folkman, J., et al., 1979, *Proc.Natl.Acad.Sci., USA* 76:5217.
Gross, T.L., et al., 1983, *Proc.Natl.Acad.Sci.,USA* 80:2623.
Heuser, L.S. et al., 1984, *J.Surg.Res.* 36:244.
Köpf-Maier, P., et al., 1981, *Eur.J.Cancer* 17:665.
Köpf-Maier, P., et al., 1986, *Drugs of the Future* 11:297.
Maragoudakis, M.E., et al., 1990, *J.Pharmacol. Exp.Ther.* 252:753.
Maragoudakis, M.E., et al., 1989, *J.Pharamacol.Exp.Ther.* 251:679.
Missirlis, E., et al., 1990, *Invest.New Drugs* 8:145.
Taylor, S., et al., 1982, *Nature* 297:307.
Tamargo, R.J., et al., 1991, *Cancer Res.* 51:672.

THE ROLE OF ESTROGENS IN ANGIOGENIC ACTIVITY

D.E. Morales[1], D.S. Grant[1], S. Maheshwari[1], D. Bhartiya[2], M.C. Cid[1],
H.K. Kleinman[1] and W.H. Schnaper[1]

Laboratory of Developmental Biology
Developmental Endocrinology Branch
NICMD. NIH
Bethesda, MD., USA

Angiogenesis is a critical event in wound healing, tumor growth, and many inflammatory diseases such as the vasculitities. Since many of the vasculitic diseases have a higher incidence in women, we examined the effects of exogenous female sex steroids, particularly estradiol, on endothelial cell attachment, proliferation, migration and angiogenesis *in vitro* and *in vivo*.

Cells were grown in estrogen-free medium and then exposed to physiologic doses for three hours before and during each assay, with the controls receiving no estrogen. Exposure to estradiol increased cell attachment to various compounds including laminin, collagen IV, and fibronectin as well as to tissue culture plastic. Cell proliferation on plastic was also increased significantly in the presence of estrogen. To test for differences in migration between treated and untreated cells, confluent monolayers of cells were "wounded" by scraping an area of cells on the plate, and then quantitating repair of the wound over time. Estradiol-treated cells migrated into the wound faster than untreated cells.

Plating of endothelial cells on a reconstituted basement membrane, Matrigel, allows the cells to differentiate and form capillary-like networks. Estradiol-treated cells form two fold more capillary-like structures than untreated cells. Using an *in vivo* angiogenesis assay in an ovariectomized mouse model, we have investigated the role of estrogens. After ovariectomy, half the mice are estradiol-restored using slow-release estrogen pellets. Mice are injected with Matrigel either lacking or containing a known angiogenic factor, b-FGF. Histological analysis demonstrated a several-fold increase in the neovascularization of Matrigel plugs in mice with restored estrogen levels.

These studies show that cells *in vitro* treated with estradiol show enhanced attachment, proliferation, migration, and morphological differentiation. This increased activity is related to new vessel formation *in vivo* and may suggest an important physiologic influence of estrogens on angiogenesis.

EXPRESSION OF SPARC DURING EMBRYONIC DEVELOPMENT OF THE CHICK CHORIOALLANTOIC MEMBRANE.

E.Papadimitriou[*][#], B.R.Unsworth[*], M.E.Maragoudakis[#] and P.I.Lelkes[a]

[*]Marquette University,Dept of Biology,Milwaukee,WI;[#]Dept of Pharmacology,University of Patras,Patras,Greece;[a]University of WI Medical School, Milwaukee Clinical Campus, Milwaukee, WI.

The formation of new capillaries from pre-existing ones (angiogenesis) is a fundamental process during embryonic development and is associated with migration and proliferation of endothelial cells, as well as remodelling of the subendothelial extracellular matrix (ECM).SPARC (secreted protein, acidic and rich in cysteine) is a 43kDa calcium-binding noncollagenous glycoprotein, which is associated with morphogenesis, remodelling and cellular migration and proliferation. *In vitro,* in cultures of endothelial cells, SPARC expression has been equated with the formation of tubes by these cells. Therefore, we hypothesized that SPARC expression might be a correlate of *in vivo* angiogenesis. The amounts of SPARC present in the chick chorioallantoic membrane (CAM) in different days of development were quantitated by Western blotting techniques, using quantitative image analysis.During embryonic chick development, the levels of SPARC were maximal at day five and decreased about 50% from day five until day 11, after which they remained stable. This pattern of expression correlates with a suggested role for SPARC during cell proliferation and migration. It has been shown that in the chick CAM, endothelial cell proliferation occurs until day 10, after which the cells become quiescent and differentiate. Immunocytochemical studies showed that SPARC is distributed around epithelial surfaces and throughout fibrous tissue during all stages of gestation and after day 13 of embryo development, it was localized along the blood vessel wall. Molecular biological studies are under way to assess the temporal sequence of the gene expression for SPARC.

VASCULARISATION OF EMBRYONIC LONG BONE CARTILAGE

P. Rooney, J. Smith and S. Kumar

Department of Pathological Sciences
University of Machester and Christie Hospital
Manchester, England

Cartilaginous long bone rudiments are composed of three types of chondrocyte-rounded, flattened and hypertrophic. The transition from cartilage to bone during development requires the vascularisation of the hypertrophic cell region, flattened and rounded cell regions do not become vascularised. The regulatory steps involved in the initiation of this spatially regulated vascularisation are not known. In the chick embryo, cartilage vascularisation begins at approximately 9-10 days of incubation, however an avascular cartilage rudiment consisting of the three cellular types is present from at least day 7 of incubation, thus, cell hypertrophy *per se* is not the triggering factor.

In an attempt to investigate whether this localised vascularisation could be recapitulated under experimental conditions, we have grown intact cartilage rudiments, and isolated cellular regions, on the chick embryo chorioallantoic membrane (CAM). Vascularisation was determined by an angiogenic response and histologically by the presence of blood vessels within the explanted tissue.

Intact cartilaginous rudiments rapidly became vascularised on the CAM but, only within the hypertrophic cell region where the chondrocytes were resorbed and replaced with a bone marrow and surrounded by a thin sheath of sub-periosteal bone. Isolated hypertrophic regions were also vascularised and with increasing time of culture, the cartilage calcified and traces of bone could be seen. In contrast, isolated rounded cell regions did not become vascularised at either the cut end or anywhere around the periphery of the explant, however, discrete cartilage canals formed within the centre of the tissue.

These results suggest that the pattern of vascularisation is a property intrinsic to the cartilage rudiment. Since cell hypertrophia is not the triggering factor, this implies that some change within the extracellular matrix initiates vascularisation. A loss of metachromatic staining can be observed immediately prior to vascular invasion and we would suggest that this loss is due to an alteration in the properties of hyaluronan possibly related to the synthesis of hypertrophic cell unique type X collagen.

PULSED ELECTROMAGNETIC FIELDS INCREASE PROLIFERATION OF ENDOTHELIAL CELLS: A HYPOTHESIS FOR MOLECULAR MECHANISM

Cliff Stevens, Suzanne Harley, Rajdip Marok, Tulin Sahinoglu, Stewart Abbot, David Blake.

The Bone & Joint Research Unit
The London Hospital Medical College
London E1, United Kingdom

The effects of Pulsed Electromagnetic Fields (PEMF) on the treatment of a variety of bone and vascular disorders has frequently been reported. It has been shown, for example, to be beneficial in treating delayed union fractures and osteoarthritis in a surgically non-invasive manner. Apart from these clinically-based studies, many attempts have been made to elucidate the biological influences of such fields at the cellular or molecular level *in vitro* . These have been largely based on the promotion of angiogenesis by applied PEMF on cultured endothelial cells. Studies have shown in models of wound healing that even relatively short exposure to PEMF (5 hours) causes significant increases in growth rate and vascular reorganisation. We have reproduced the effect of endothelial cell reorganisation into tube or vessel-like structures and, in addition, have reproducibly shown that significant augmentation of endothelial cell proliferation can be achieved by as littele as 4 hours exposure to PEMF (Electro Biology Inc. Bone Healing System) To begin to look at the possible molecular mechanisms involved, we have investigated the PEMF effect on the activation of the transcription factor NFkB which leads to transcription of genes which encode a variety of cytokines of relevance to angiogenesis.

Human umbilical vein endothelial cells (HUVECs) were isolated and cultured using standard techniques and seeded at a concentration of 4000 per coverslip (Thermanox) and allowed to adhere for > 1hr. BrdU was then added to the cells to give a final concentration of 10 mM/ well. The stimulated and control plates were generally left for 8 hrs at 37^0 C in and out of the PEMF respectively. Incorporated BrdU was detected using a mouse anti-BrdU antibody incubated at $+4^0$ C overnight. The staining was completed using avidin/biotin peroxidase complex and visualised DAB/H_2O_2. The percentage of proliferating cells was then assessed for each coverslip by light microscopy. Confluent monolayers of HUVECs in flasks were also exposed to PEMF and H_2O_2, subsequent lysates of these and control cells where assessed by electrophoretic shift assay (EMSA) for the activation of NFkB.

RESULTS of three separate experiments mean (SD) (n=6)

Proliferation	Experimental 1	Experimental 2	Experimental 3
Control (out of field)	8.265%(0,854)	13.010% (0.939)	18.052% (2.066)
In fiel	11.552 (0.835)	16.084% (1.558)	21.552% (1.596)
	39.9% increase	25.59% increase	19.39% increase
	p= 0.002	p=0.003	p=0.011

Exposure of HUVECs to either H_2O_2 or PEMF induced the DNA binding activity of NFkB. This activity is enhanced when exposed to a combined stimulus of H_2O_2 and PEMF.

We hypothesise that PEMF may, under appropriate circumstances have the effect of altering the reactivity of reactive oxygen species (ROS) and that might account for the results of our preliminary studies. The effects of ROS in biological systems are diverse and have been implicated in many disease states as well as providing vital bacteriocidal defence and signal transduction in a variety of molecular mechanisms. Cellular proliferation in one such mechanism where ROS are involved in the activation of growth competent and genes and nuclear transcription factors. Low levels of ROS have been shown to increase the mRNA for protooncogenes c-*fos* and c-*jun* in conjunction with the trsnsactivation of the proliferation related transcription factor AP-1. Related oncogenes such as c-*ets*, important in angiogenesis, are likely to share common signal transduction pathways involving ROS. It is therefore likely that PEMF, which undoubtedly have an influence on ROS activity, could effect the finely controlled functioning of proliferative mechanisms. Our future studies are designed to investigate this possibility.

IDENTIFICATION OF THE KDR TYROSINE KINASE AS A RECEPTOR FOR VASCULAR ENDOTHELIAL CELL GROWTH FACTOR

Maureen Dougher-Vermazen[1], Denis Gospodarowicz[2] and Bruce I. Terman[1]

[1]Medical Research Division, Lederle Labs, Pearl River, NY 10965
[2]Cancer Research Institute, School of Medicine, Univ. of California, San Francisco, CA 94143, USA

Vascular endothelial cell growth factor (VEGF), also known as vascular permeability factor, is an endothelial cell mitogen which stimulates angiogenesis. Here we report that a previously identified receptor tyrosine kinase gene, KDR (Terman et al., 1991, Oncogene 6, 1677), encodes a receptor for VEGF. cDNA for the coding portion of the KDR gene was cloned from a HUVE cell library. DNA sequencing data predicts that the coding portion of KDR contains 1365 amino acids. The amino acid sequence contains many of the features associated with type III receptor tyrosine kinases, including seven extracellular immunoglobulin like domains, and an intracellular kinase insert domain. Expression of KDR in either CMT-3 or NIH3T3 cells allows for saturable ^{125}I-VEGF binding with high affinity (KD=75 pM and 15 pM respectively for the two cell types). Affinity cross-linking of ^{125}I-VEGF to KDR-transfected CMT-3 cells results in specific labeling of two proteins of M_r=195 and 235 kDa. Cross-linking studies using KDR-transfected NIH3T3 fibroblasts results in specific labeling of two proteins of M_r=170 and 235 kDa. The KDR receptor tyrosine kinase shares structural similarities with a recently reported receptor for VEGF, flt, in a manner reminiscent of the similarities between the α and β forms of the PDGF receptors.

Participants of the NATO Advanced Studies Institute "Angiogenesis: Molecular Biology, Clinical Aspects" held at Paradise Hotel, Rhodos, Greece, during 16-27 June, 1993.

1. Morales, David
2. Tsopanoglou, Nikos
3. Konerding, Moritz
4. Ziche, Marina
5. Kleinman, Hynda
6. Haralabopoulos, G.
7. Keshet, Eli
8. Peristeris, Platon
9. Grant, Derrick
10. Karakiulakis, George
11. Tsilibary, Effie
12. Andriopoulou, Par.
13. Dimopoulos, John

14. Sponza, Delia-Teresa
15. Maglione, Domenico
16. Charonis, Aris
17. Maragoudaki, Jane
18. Persico, Maria
19. Van Hinsbergh, V.
20. Ozben, Tomris
21. Vehar, Gordon
22. Poole, Thomas
23. Hla, Timothy
24. Haaksma, Monika
25. Schnürch, Harald
26. Eichman, Anne

27. Krupinski, Jerzy
28. Mironov, Vladimir
29. Haudenschild-Chen
30. Haudenschild, Chr.
31. Terman, Bruce
32. Maragoudakis, Man.
33. Plendl, Johanna
34. Fenton II, John
35. Denekamp, Juliana
36. Stevens, Cliff
37. Maragoudakis, Mich.
38. Lelkes, Peter
39. Perricone, Michael

40. Narayan, Keneth
41. Weiss, Jacqueline
42. Cherry, George
43. Littbrand, Bo
44. Sutherland, D.
45. Kefalides, Nick
46. Kumar, Shant
47. Hellergvist, Carl
48. Thompson, D.W.
49. Papaioannou, Stam.
50. Polverini, Peter
51. Baydanoff, St.
52. Sgoutas, Dimitrios

53. McLaughlin, Barry
54. Auerbach, Robert
55. Gullino, Peter
56. Augustin-Voss, H.
57. Marmara, Anna
58. Tsanaclis, Ana-Maria
59. Argyropoulou, Lydia
60. Antoniades, Harry
61. Fisher, Mark
62. Licholai, Gregory
63. Solomon, Scott

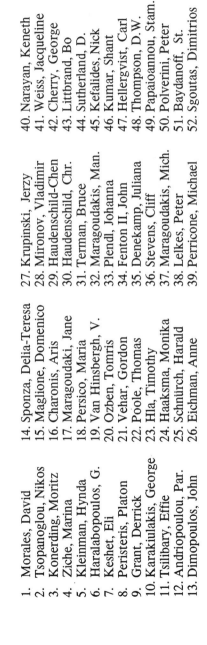
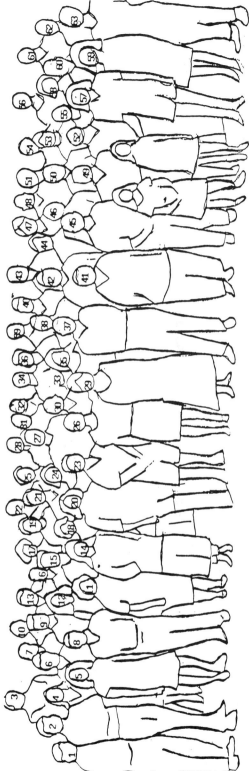

361

PARTICIPANTS

AL-HYMAYYD, M.S.

Social Insurance Hospital, P.O. Box 42142, Riyadh 11541, Kindom of Saudi Arabia

ANDREOPOULOU, P.

Department of Pharmacology University of Patras Medical School, Rio, Patras 261 10 , GREECE

ANTONIADES, H.

Harvard School of Public Health Cancer Biology, 655 Huntington Avenue, Boston, Mass. 02115, U.S.A.

ARNOLD, F.W.

Dept. of Dermatology, Churchill Hospital Headington, Oxford OX3 7LJ, UNITED KINGDOM

AUERBACH, R.

Laboratory of Development, Biology, Zoology, Research Building, University of Wisconsin, 1117 West Johnson Str. Madison WI53706, U.S.A.

AUGUSTIN-VOSS, H.

Cell Biology Laboratory, Dept. of Gynecology & Obstetrics, University of Göttingen Med. School, 3400 Göttingen, GERMANY

BASTAKI, M.

The Gray Laboratory, PO Box 100, Mount Vernon Hospital, Northwood, Middlesex, HA6 2JR, UNITED KINGDOM

BAYDANOFF, ST.

Dept. of Biology/Immunology, University School of Medicine, 1, St. Kliment Ohridsky str., 5800 Pleven, BULGARY

BENEZRA, D.

Department of Ophthalmology, Hadassah Medical Organization, Kiryat, Hadassah, P.O.B. 12000, Il-91120 Jerusalem, ISRAEL

BERGER, S.

Dept. of Gynecology & Obstetrics University of Mainz, Langenbeckstrabe 1, 6500 Mainz, GERMANY

BREM, H.	Department of Surgery Ohio State University Hospital, 410 West 10th Avenue Columbus, Ohio 43204, U.S.A.
BREM, S.	Director Neurosurgical Oncology, North Western Medical Faculty Found. Inc. Streetville Center, 233 East Evie, Suite 500, Chicago, Ill. 60611, U.S.A.
CHATZICONTI, O.	Department of Pharmacology University of Patras Medical School, Rio, Patras 261 10 , GREECE
CHARONIS, A.	University of Minnesota, Med. School, Dept. of Lab. Medicine & Pathology, Box 609 Mayo Memorial Bldg., 420 Delaware Str. SE, Minneapolis, Minnesota 55455-0315, U.S.A.
CHERRY, G.W.	Dept. of Dermatology, Churchill Hospital Headington, Oxford OX3 7LJ, UNITED KINGDOM
COHEN, J.	Technion Israel Institute of Technology, Dept. of Food Engineering and Biotechnology, Technion City, Haifa 320000, ISRAEL
DENEKAMP, J.	The Gray Laboratory, PO Box 100, Mount Vernon Hospital, Nortwood, Middlesex, HA6 2JR, UNITED KINGDOM
DIETERLEN-LIEVRE, F.	Institut D´Embryologie Cellulaire et Moleculaire UMR 009, 49 bis Avenue de la Gambriello, 94736 Nogent-Sur-Marne Cedex, Paris, FRANCE
DEMOPOULOS, J.	Department of Radiology, University of Patras Medical School, Rio, Patras 261 10 , GREECE
EICHMANN, A.	Institut D´Embryologie , C.N.R.S. et College de France, 49 bis, Avenue de la Belle Gabrielle, 94130 Nogent-Sur-Marne Cedex, FRANCE
FENTON, II, J.W.	State of New York, The Governor N.A. Rockfeller, Empire State Plaza, P.O. Box 509, Albany NY 12201-0509, U.S.A.
FISHER, M.	University of Southern California, School of Medicine, Dept. of Neurology, 1333 San Pablo Str., MCH 246, Los Angeles, U.S.A.
FRISCHMANN-BERGER, R.	Dept. of Gynecology & Obstetrics University of Mainz, Langenbeckstrabe 1, 6500 Mainz, GERMANY

GRANDI, M.	Farmitalia Carlo Erba srl, Research Center R & D, Oncology Department, Via Giovanni XXIII 23, 20014 Nerviano (MI), ITALY
GRANT, D.	Section of Dev. Biol. & Anomalies, National Inst. of Dental Res., Bldg. 30, Room 414, Bethesda MD 20892, U.S.A.
GREGORIOU, V.	Duke University, Dept. of Chemistry, North Carolina, U.S.A.
GULLINO, P.M.	Dipartimento di Scienze Biomediche e Oncologia Umana, Sezione di Anatomia e Istologia Patologia, Via Santena 7, 10126 Torino, Italy
HAAKSMA-HERCZEGH, M.	Dr. Karl Thomae GmbH, Pharmacol. Research, Pharmacology E, Birkendorfer Strabe 65, Postfach 1755, D-7950 Biberach an der Riss 1, GERMANY
HARALABOPOULOS, G.C.	Department of Pharmacology University of Patras Medical School, Rio, Patras 261 10 , GREECE
HAUDENCHILD, C.C.	American Red Cross, The Jerome Holland Laboratory, 15601 Crabbs Branch Way, Rockville, MD 20855, U.S.A.
HAUDENCHILD, CH.	American Red Cross, The Jerome Holland Laboratory, 15601 Crabbs Branch Way, Rockville, MD 20855, U.S.A.
HELLERGVIST, C.G.	Vanderbilt University, School of Medicine, Dept. of Biochemistry, Nashville, Tennessee 37232-0146, U.S.A.
HLA, T.	Dept. of Molecular Biology, Biomedical Research and Development, The Jerome H. Holland Laboratory, 15601 Crabbs Branch Way, Rockville, MD 20855, U.S.A.
HOCKEL, M.	Frauenklinik der Universität Mainz Langenbeckerstr. 1, 6500 Mainz, GERMANY
HUNT, T.	University of California, Dept. of Surgery, Parnassus Avenue, Room HSE 839, San Francisco, USA
KARAKIULAKIS, G.D.	Aristotle University Thessaloniki, Medical School, Laboratory of Pharmacology, Thessaloniki 540 06, GREECE
KEFALIDES, N.	University of Pennsylvania, Dept. of

	Medicine, University City Science Center, 3624 Market Street, Philadelphia, Pennsylvania 19104, U.S.A.
KESHET, E.	Department of Virology, The Hebrew University Hadassah Medical School, P.O. Box 1172, Jerusalem 91010, ISRAEL
KLEINMAN, H.	Section of Dev. Biol. & Anomalies National Institute of Dental Research, Bldg. 30, Room 414, Bethesda, MD 20892, U.S.A.
KONERDING, M.A.	Institute of Anatomy, University of EssenHufelandstr. 55, 4300 Essen I GERMANY
KOOLWIJK, P.	TNO Institute of Ageing & Vascular Res., Zernikedreef 9, P.O. Box 430, 2300 AK Leiden, THE NETHERLANDS
KRUPINSKI, J.	Institut of Neurology, Dept. of Neuropathology, University of Krakow, ul. Botaniczna 3, 31-503 Krakow, POLAND
KUMAR, P.	Clinical Research Laboratories, Christie Hospital, Manchester M209BX, U.K
KUMAR, S.	Clinical Research Laboratories, Christie Hospital, Manchester M209BX, U.K
LAKKA, L.	Patras Endocrinology Clinic, Pantanassis 45, Patras, GREECE
LELKES, P.I.	Lab of Cell Biology, Univ. of Wisconsin, Dept. of Medicine, Sinai Samaritan Med. Ctr., Mount Sinai Campus, 950 North Twelfth St., P.O. Box 342, Milwaukee WI53201-0342, USA
LICHOLAI, G.P.	Dept. of Health & Human Services, National Institutes of Health, Bldg. 10, Room 5D-37, Bethesda MD 20892, U.S.A.
LITTBRAND, B.	Department of Oncology, University of Umea, S-901 85 Umea, SWEDEN
MAGLIONE, D.	Instituto Internazionale di Genetica e Biofisica, Consiglio Nationale delle Ricerche, Via Guglielmo Marconi, 10, C.P. 3061, 80100 Napoli, ITALY
MARAGOUDAKIS, M.E.	Department of Pharmacology, University of Patras Medical School, Rio, Patras 261 10 , GREECE

MARIANI, M.

Farmitalia Carlo Erba srl, Research Center R & D, Oncology Department, Via Giovanni XXIII 23, 20014 Nerviano (MI), ITALY

MCLAUGHLIN, B.L.

Dept. of Rheumatology, Wolfson Angiogenesis Unit, Clinical Science Bldg., Hope Hospital, Salford M6 8HD, UNITED KINGDOM

MIRONOV, B.

Lehrstuhl für Anatomie II, Klinikum der Rheinisch Westfalische Technische Hochshule (RWTH), Pauwelsstrabe, 5100 Aachen, GERMANY

MISSIRLIS, E.

Department of Pharmacology, University of Patras Medical School, Rio, Patras 261 10 , GREECE

MONGELLI, N.

Farmitalia Carlo Erba, srl, Research Center R & D, Via Carlo Imbonati, 24, 20159 Milano, ITALY

MORALES, D.E.

Lab. of Dev. Biol. & Anomalies, National Inst. of Dental Res., Bldg. 30, Room 407, Bethesda MD 20892, U.S.A.

NARAYAN, K.

Peter MacCallum Cancer Institute, 481 Little Lonsdale Street, Melbourne, Bictoria 3000, AUSTRALIA

OZBEN, T.

Akdeniz Universitesi, Tip Fakultesi, 07058 Kepez Antalya, TURKEY

PAPAIOANNOU, ST.

Department of Pharmacology, University of Patras, School of Pharmacy, Rio, Patras 261 10 , GREECE

PARDANAUD, L.

Institut D´ Embryologie Cellulaire et Moleculaire UMR 009, 49 bis, Avenue de la Belle Gabrielle, 94736 Nogent-Sur-Marne Cedex, FRANCE

PEPPER, M.

University of Geneve, Department of Morphology, CMU-1, Rue Michel-Servet, CH-1211, Geneva 4, SWITZERLAND

PERISTERIS, P.

Department of Pharmacology University of Patras Medical School, Rio, Patras 261 10 , GREECE

PERRICONE, M.

The DuPont Merck Pharmac. Co., Glenolden Laboratory, 500S, Ridgeway Avenue, Glenolden, PA 19036, U.S.A.

PERSICO, M.G.

Instituto Internazionale di Genetica e Biofisica, Consiglio Nationale delle Ricerche, Via Guglielmo Marconi, 10, C.P. 3061, 80100 Napoli, ITALY

PIPILI-SYNETOS, E.

Department of Pharmacology University of Patras Medical School, Rio, Patras 261 10 , GREECE

PIRIANOV, C.

National Centre of Oncology, Medical Academy, 6 Plovdivsco Pole Street, Sofia 1156, BULGARY

PLENDL, J.

Institut für Tieranatomie, Veterinärstr. 13, D-800 München 22, GERMANY

POLVERINI, P.

Lab. of Molecular Pathology, Univ. of Michigan, Sch. of Dentistry, 1011 North Univ., R. 5217, Ann Arbor, Michigan 48109-1078, U.S.A.

POOLE, T.

Dept. of Anatomy & Cell Biology, State University of New York, Health Science Center, 766 Irving Ave., Syracuse, NY 13210, U.S.A.

PRESTA, M.

University degli Studi di Brescia, Facolta di Medic. e Chirurgia, Dipartimento di Scienze, Biomediche e Biotechnologie, Via Valsabbina 19, I-25123 Brescia, ITALY

RAVAZOULA, P.

University of Patras Medical School, Pathology Lab., 261 10 Rio, Patras, GREECE

ROONEY, P.

Clinical Research Laboratories, Christie Hospital, NHS Trust, Wilmslow Road, Manchester M20 9BX, UNITED KINGDOM

ROUSSOS, CH.

Hospital "Evagelismos", Ipsilantou 45-47, 106 76 Athens, GREECE

SAKKOULA, E.

Department of Pharmacology University of Patras Medical School, Rio, Patras 261 10 , GREECE

SCHLENGER, K.

Dept. of Gynecology & Obstetrics, University of Mainz, Langenbeckstrabe 1, 6500 Mainz, GERMANY

SCHNURCH, H.

Max-Planck Institute für Psychiatrie Abteilung Neurochemie, 8033 Planegg-Martinsried, Am Klopferspitz 18A, GERMANY

SGOUTAS, D.

Clinical Chemistry Laboratory, Emory

	University Hospital, Room F-153C, Atlanta, Georgia 30322, U.S.A.
SKOTSIMARA, P.	Department of Toxicology and Pharmacokinetics, University of Patras Medical School, General Hospital of Patras "AGIOS ANDREAS", Patras, GREECE
SOLA F.	Farmitalia Carlo Erba, srl, Research Center R & D, Oncology Department, Via Giovanni XXIII 23, 20014 Nerviano (MI), ITALY
SOLOMON, S.R.	National Institute of Health, Clinical Neuroendocrinology Branch, Bldg. 10, Bethesda MD 20892, U.S.A.
SOUTHERLAND, D.	Toronto-Bayview Regional Cancer Ctr., 2075 Bayview Ave., Toronto, Ontario M4N 3M5, CANADA
SPONZA, D.T.	Dokuz Eylül University, Müh Fak. Genze Müh. Böl, Bornova, Izmir, TURKEY
STEVENS C.	London Hospital Medical College, University of London, Bone & Joint Res. Unit, Turner Str., London E1 2AD, UNITED KINGDOM
TERMAN, B.I.	American Cyanamid Company, Lederle Laboratories, Bldg. 205/227, Pearl River, N.Y. 10965, U.S.A.
THOMPSON, D.W.	University of Aberdeen, Department of Pathology, University Medical Buildings, Foresterhill, Aberdeen AB9 2ZD, SCOTLAND
TSANACLIS, A.M.	University of Sao Paolo, School of Medicine Av. Dr. Arnaldo 455, Dept. of Pathology, 01246 Sao Paulo, BRAZIL
TSILIBARY, E.	University of Minnesota, Med. School, Dept. of Lab. Medicine & Pathology, Box 609 Mayo Memorial Bldg., 420 Delaware Str. SE, Minneapolis, Minnesota 55455-0315, U.S.A.
TSOPANOGLOU, N.	Department of Pharmacology University of Patras Medical School, Rio, Patras 261 10 , GREECE
VAN HINSBERGH, V.W.M.	TNO Gaubius Institute, P.O. Box 430, 2300 AK Leiden, NETHERLANDS
VEHAR, G.A.	Genentech Inc., 460 Point San Bruno Boulevard, South San Francisco, CA 94080, U.S.A.

WEISS, J.B.

Dept. of Rheumatology, Wolfson
Angiogenesis Unit, Clinical Science Bldg.,
Hope Hospital, Salford M6 8HD, UNITED
KINGDOM

WOJTUKIEWICZ, M. Z.

Dept. of Hematology, Medical Academy, ul.
M. Sklodowskiej-Curie 24, 15-276 Bialystok,
POLAND

ZICHE, M.

Universita degli Studi di Firenze,
Dipartimento di Farmacologia, Preclinica e
Clinica "Mario Aiazzi-Mancini", Vale
Morgagni 65, 50134 Firenze, ITALY

INDEX